T0320611

Mathematics: Theory & Applications

Series Editor
Nolan Wallach

Manfredo Perdigão do Carmo

Riemannian Geometry

Translated by Francis Flaherty

Birkhäuser

Boston · Basel · Berlin

Manfredo Perdigão do Carmo
Instituto de Matematica Pura
e Aplicada
Edificio Lelio Gama
Rio de Janeiro
Brazil

Translated by
Francis Flaherty
Department of Mathematics
Oregon State University
Corvallis, OR 97331
USA

Library of Congress Cataloging-in-Publication Data
Carmo, Manfredo Perdigão do.
[Geometria riemanniana. English]
Riemannian geometry / Manfredo do Carmo : translated by Francis Flaherty.
p. cm. — (Mathematics. Theory and applications)
Translation of: Geometria riemanniana.
Includes bibliographical references and index.
ISBN 0-8176-3490-8 (acid-free)
1. Geometry, Riemannian. I. Title. II. Series: Mathematics (Boston, Mass.)

QA649.C2913 1992
516.3'73-dc20
91-37377
CIP

ISBN 978-0-8176-3490-2 ISBN 978-1-4757-2201-7 (eBook)
DOI 10.1007/978-1-4757-2201-7

Originally published in Portuguese as *Geometria Riemanniana* by Instituto de Matematica Pura e Aplicada, 1979 first edition; 1988 second edition.

Printed in the United States of America.

19 18 17 16 15 14 (Corrected at 14th printing 2013)

Birkhäuser Boston is part of Springer Science+Business Media (www.birkhauser-science.com)

CONTENTS

PREFACE TO THE 1st EDITION

The object of this book is to familiarize the reader with the basic language of and some fundamental theorems in Riemannian Geometry. To avoid referring to previous knowledge of differentiable manifolds, we include Chapter 0, which contains those concepts and results on differentiable manifolds which are used in an essential way in the rest of the book.

The first four chapters of the book present the basic concepts of Riemannian Geometry (Riemannian metrics, Riemannian connections, geodesics and curvature). A good part of the study of Riemannian Geometry consists of understanding the relationship between geodesics and curvature. Jacobi fields, an essential tool for this understanding, are introduced in Chapter 5. In Chapter 6 we introduce the second fundamental form associated with an isometric immersion, and prove a generalization of the Theorem Egregium of Gauss. This allows us to relate the notion of curvature in Riemannian manifolds to the classical concept of Gaussian curvature for surfaces.

Starting in Chapter 7, we begin the study of global questions. We emphasize techniques of the Calculus of Variations which we present without assuming a previous knowledge of the subject. Among other things, we prove the Theorems of Hadamard (Chap. 7), Myers (Chap. 9) and Synge (Chap. 9), the Rauch Comparison Theorem (Chap. 10), and the Morse Index Theorem (Chap. 11). One of the most remarkable applications of these techniques of the Calculus of Variations, the Sphere Theorem, is presented in Chapter 13. In addition, we include a "uniformization" theorem for manifolds of constant curvature (Chap. 8) and a study of the fundamental group of compact manifolds of negative curvature (Chap. 12).

Many important topics are absent. Because of limitations of time and space, a choice was necessary; we hope that the references mentioned in each chapter stimulate the reader to complete his knowledge in the direction of his own taste.

Our debt to existing sources (written and oral) is enormous and impossible to catalog. We mention only Chern [Ch 1], Klingenberg-Gromoll-Meyer [KGM] and Milnor [Mi] as main influences.

This book had its origin in notes of a course given in Berkeley in 1968. Later, with the help of students at IMPA (Instituto de Matemática Pura e Aplicada), the notes were translated into Portuguese and published in the Monograph collection of IMPA in 1971. Finally, in a form very close to the present, it was given as a course in the School of Differential Geometry at Fortaleza in July 1978. Throughout all these years, various colleagues and students contributed criticisms and suggestions to improve the text. I want to express, in a most special way, my gratitude to Professor Lucio Rodriguez who, in my absence assumed the unpleasant task of correcting the proofs and organizing the alphabetical index. To all, my sincerest thanks.

Manfredo Perdigão do Carmo

Rio de Janeiro, June 1979

PREFACE TO THE 2nd EDITION

Besides the numerous corrections and modifications throughout the text, the second edition differs from the first in the following aspects:

Chapter 13 has been entirely rewritten. For the benefit of readers less familiar with Morse Theory, the proof of the sphere theorem in even dimension (which does not depend on Morse Theory) can be dealt with in an independent manner.

In Chapter 4 a concise exposition of tensors on a Riemannian manifold was added. The goal is to show that, on a Riemannian manifold, tensors can be differentiated covariantly. Among other applications, this allows us to introduce, in Chapter 6, the fundamental equations of an isometric immersion.

In Chapter 8 a section on the isometries of hyperbolic space and their relationship with conformal transformations of Euclidean space was added.

The number of exercises has grown considerably. Some topics which are not encountered in the text appear in the form of exercises: Riemannian submersions, the complex projective space, Einstein manifolds, the 2nd Bianchi identity, etc.

In spite of the initial plan, it was not possible to include a chapter on Partial Differential Equations and Geometry; this will have to wait for another occasion.

It remains for us to thank an enormous list of persons who, through corrections, criticisms and suggestions, contributed to the improvement of this book; special thanks are due Jonas Gomes, J. Gilvan de Oliveira and Gudlaugur Thorbergsson. Thanks are also due to Professor Lucio Rodriguez, who with dedication looked after the system of TEX used here, and to Wilson Goes who was in charge of the final presentation of the text.

Manfredo Perdigão do Carmo

Rio de Janeiro, 4 July 1988

PREFACE TO THE ENGLISH EDITION

This is a translation of the second edition of a book published originally in Portuguese. Except for minor corrections and the substitution of some references, no changes were made.

I am indebted to several persons whose cooperation was essential to bring the present edition into existence. First, to my friends at the University of Pennsylvania who used the Portuguese edition for a number of years and convinced me that a translation was worthwhile. Second, to Frank Flaherty, of Oregon State University, who volunteered to, and worked hard at, the arduous task of the translation. Third, to the staff of Birkhäuser, for their patience and interest. Finally, to Bill Firey, Jerry Kazdan, Juha Pohjanpelto, Walcy Santos and Beth Stahelin, for a critical reading of the English manuscript.

I would like to use this opportunity to express my deep appreciation to my colleagues and students at IMPA, who made this book possible.

Manfredo Perdigão do Carmo

Rio de Janeiro, February 1991

HOW TO USE THIS BOOK

The prerequisites for the reader of this book are:

1) A good knowledge of Calculus, including the geometric formulation of the notion of the differential and the inverse function theorem.
2) A certain familiarity with the elements of the Differential Geometry of surfaces. For example, Chapter 2 (2.1 to 2.4), 3 (3.2 and 3.3) and 4 (4.1 to 4.6) of M. do Carmo [dC 2] are sufficient.

If the reader is familiar with the basic definitions of differentiable manifolds, he can omit Chapter 0 entirely. Otherwise, this chapter should be considered as part of the course.

Starting with Chapter 6, properties of covering spaces and of the fundamental group are used. For the elements of covering spaces, we use §5.6 of Chapter 5 of M. do Carmo [dC 2], and for the fundamental group and its relationship to covering spaces, we use Chapters 2 and 5 of Massey [Ma].

A few exercises (never, however, in the text) assume some knowledge of differential forms. Chapters 1, 2 and 3 of M. do Carmo [dC 3] are sufficient. Those exercises are indicated by †.

Chapters 1 to 7 are indispensable to the rest of the book. From there, a course which aims at the sphere theorem could omit Chapters 8 and 12. As an alternative, Chapters 8 and 11 could be omitted and the course could finish with Chapter 12. A minimal course would contain Chapters 0 to 7 of this book, Section 5.6 of M. do Carmo [dC 2], Sections 1, 2 and 3 of Chapter 8 and Chapter 9 up to (including) the Theorem of Bonnet-Myers.

TRANSLATOR'S NOTE

It is intended that this translation follow the original Portuguese closely.

Frank Flaherty

Corvallis, Oregon, 30 June 1990

To S. S. Chern

LIST OF FREQUENTLY USED SYMBOLS

M^n	An n-dimensional differentiable or Riemannian manifold
X, Y	Vector fields on M
$\mathbb{R}^n$	Space of n-tuples of real numbers $(x_1, \ldots, x_n)$
T_pM	Tangent space at a point $p \in M$
$\langle\, , \rangle_p$	A positive definite inner product in T_pM
TM	The tangent bundle over M
$\{X_i\} = \left\{\frac{\partial}{\partial x_i}\right\}$	Basis of $T_p(M)$ corresponding to coordinates $(x_1, \ldots, x_n)$
$\mathfrak{X}(M)$	The space of vector fields on M
$\mathcal{D}(M)$	The space of differentiable functions on M
$\nabla_X Y$	Covariant derivative of the vector field Y along the vector field X
$P_{c(t)}(y)$	Parallel transport of the vector y along the curve $c(t)$
γ	A geodesic on M
J	A Jacobi field
R	The curvature tensor
$K(p, \sigma)$	Sectional curvature of M at $p \in M$ along a plane $\sigma \subset T_pM$
$\mathrm{Ric}_p(x)$	Ricci curvature at $p \in M$ along a unit vector $x \in T_pM$
$U \subset M$	An open set in M
$\{E_i\}$	A local moving frame
$T\colon \mathfrak{X}(M) \times \cdots \times \mathfrak{X}(M) \to D(M)$	A tensor on M
$f\colon M^n \to \overline{M}^{n+m}$	Immersion of a manifold M into another manifold $\overline{M}$ with codimension m
$(TM)^\perp$	The normal bundle
$\mathfrak{X}(M)^\perp$	The space of normal vector fields
$B\colon \mathfrak{X}(M) \times \mathfrak{X}(M) \to \mathfrak{X}(M)^\perp$	Second fundamental form of an immersion
$R^\perp$	Normal curvature tensor
$k_1, \ldots, k_n$	Principal curvature of a hypersurface
S_1^n	Unit sphere in $\mathbb{R}^{n+1}$
$g\colon M^n \to S_1^n$	Gauss map of a hypersurface in $\mathbb{R}^{n+1}$

CHAPTER 0

DIFFERENTIABLE MANIFOLDS

1. Introduction

The notion of a differentiable manifold is necessary for extending the methods of differential calculus to spaces more general than $\mathbf{R}^n$. The first example of a manifold, accessible to our experience, is a regular surface in $\mathbf{R}^3$. Recall that a subset $S \subset \mathbf{R}^3$ is a regular surface if, for every point $p \in S$, there exist a neighborhood V of p in $\mathbf{R}^3$ and a mapping $\mathbf{x}: U \subset \mathbf{R}^2 \to V \cap S$ of an open set $U \subset \mathbf{R}^2$ onto $V \cap S$, such that:

(a) $\mathbf{x}$ is a differentiable homeomorphism;

(b) The differential $(d\mathbf{x})_q: \mathbf{R}^2 \to \mathbf{R}^3$ is injective for all $q \in U$ (See M. do Carmo, [dC 2], Chap. 2).

The mapping $\mathbf{x}$ is called a parametrization of S at p. The most important consequence of the definition of regular surface is the fact that the transition from one parametrization to another is a diffeomorphism (M. do Carmo, [dC 2], §2.3. Cf. also Example 4.2 below). More precisely, if $\mathbf{x}_\alpha: U_\alpha \to S$ and $\mathbf{x}_\beta: U_\beta \to S$ are two parametrizations such that $\mathbf{x}_\alpha(U_\alpha) \cap \mathbf{x}_\beta(U_\beta) = W \neq \phi$, then the mappings $\mathbf{x}_\beta^{-1} \circ \mathbf{x}_\alpha: \mathbf{x}_\alpha^{-1}(W) \to \mathbf{R}^2$ and $\mathbf{x}_\alpha^{-1} \circ \mathbf{x}_\beta: \mathbf{x}_\beta^{-1}(W) \to \mathbf{R}^2$ are differentiable.

Thus, a regular surface is intuitively a union of open sets of $\mathbf{R}^2$, organized in such a way that when two such open sets intersect the change from one to the other can be made in a differentiable manner. As a consequence, it makes sense to speak of differentiable functions on a regular surface and, in that situation, apply the methods of differential calculus.

The major defect of the definition of regular surface is its dependence on $\mathbf{R}^3$. Indeed, the natural idea of a surface is of a set which is two-dimensional (in a certain sense) and to which the differential calculus of $\mathbf{R}^2$ can be applied; the unnecessary presence of $\mathbf{R}^3$ is simply an imposition of our physical nature.

Although the necessity of an abstract idea of surface (that is, without involving the ambient space) is clear since Gauss ([Ga], p. 21), it was nearly a century before such an idea attained the definitive form that we present here. One of the reasons for this delay is that the fundamental role of the change of parameters was not well understood, even for surfaces in $\mathbf{R}^3$ (cf. Rem. 2.2 of the next section).

The explicit definition of a differentiable manifold will be presented in the next section. Since there is no advantage in restricting ourselves to two dimensions, the definition will be given for an arbitrary dimension n. Differentiable always signifies of class C^∞.

2. Differentiable manifolds; tangent space

2.1 DEFINITION. A *differentiable manifold* of dimension n is a set M and a family of injective mappings $\mathbf{x}_\alpha: U_\alpha \subset \mathbf{R}^n \to M$ of open sets U_α of $\mathbf{R}^n$ into M such that:

(1) $\bigcup_\alpha \mathbf{x}_\alpha(U_\alpha) = M$.

(2) for any pair α, β, with $\mathbf{x}_\alpha(U_\alpha) \cap \mathbf{x}_\beta(U_\beta) = W \neq \phi$, the sets $\mathbf{x}_\alpha^{-1}(W)$ and $\mathbf{x}_\beta^{-1}(W)$ are open sets in $\mathbf{R}^n$ and the mappings $\mathbf{x}_\beta^{-1} \circ \mathbf{x}_\alpha$ are differentiable (Fig. 1).

(3) The family $\{(U_\alpha, \mathbf{x}_\alpha)\}$ is maximal relative to the conditions (1) and (2).

The pair $(U_\alpha, \mathbf{x}_\alpha)$ (or the mapping $\mathbf{x}_\alpha$) with $p \in \mathbf{x}_\alpha(U_\alpha)$ is called a *parametrization* (or *system of coordinates*) of M at p; $\mathbf{x}_\alpha(U_\alpha)$ is then called a *coordinate neighborhood* at p. A family $\{(U_\alpha, \mathbf{x}_\alpha)\}$ satisfying (1) and (2) is called a *differentiable structure* on M.

The condition (3) is included for purely technical reasons. Indeed, given a differentiable structure on M, we can easily complete it to a maximal one, by taking the union of all the parametrizations that, together with any of the parametrizations of the given structure, satisfy condition (2). Therefore, with a certain abuse of language, we can say that a differentiable manifold is a set provided with a differentiable structure. In general, the extension to the maximal structure will be done without further comment.

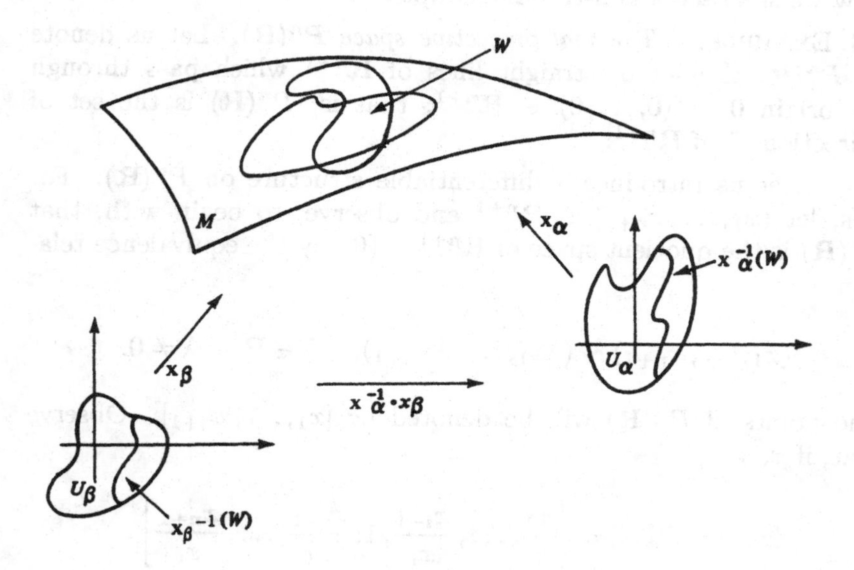

Figure 1

2.2 REMARK. A comparison between the definition 2.1 and the definition of a regular surface in $\mathbf{R}^3$ shows that the essential point (except for the change of dimension from 2 to n) was to distinguish the fundamental property of the change of parameters (which is a theorem for surfaces in $\mathbf{R}^3$) and incorporate it as an axiom. This is precisely condition 2 of Definition 2.1. As we shall soon see, this is the condition that allows us to carry over all of the ideas of differential calculus in $\mathbf{R}^n$ to differentiable manifolds.

2.3 REMARK. A differentiable structure on a set M induces a natural topology on M. It suffices to define $A \subset M$ to be an *open set* in M if and only if $\mathbf{x}_\alpha^{-1}(A \cap \mathbf{x}_\alpha(U_\alpha))$ is an open set in $\mathbf{R}^n$ for all α. It is easy to verify that M and the empty set are open sets, that a union of open sets is again an open set and that the finite intersection of open sets remains an open set. Observe that the topology is defined in such a way that the sets $\mathbf{x}_\alpha(U_\alpha)$ are open and that the mappings $\mathbf{x}_\alpha$ are continuous.

The Euclidean space $\mathbf{R}^n$, with the differentiable structure given by the identity, is a trivial example of a differentiable manifold. Now we shall see a non-trivial example.

2.4 EXAMPLE. *The real projective space* $P^n(\mathbf{R})$. Let us denote by $P^n(\mathbf{R})$ the set of straight lines of $\mathbf{R}^{n+1}$ which pass through the origin $0 = (0,\dots,0) \in \mathbf{R}^{n+1}$; that is, $P^n(\mathbf{R})$ is the set of "directions" of $\mathbf{R}^{n+1}$.

Let us introduce a differentiable structure on $P^n(\mathbf{R})$. For this, let $(x_1,\dots,x_{n+1}) \in \mathbf{R}^{n+1}$ and observe, to begin with, that $P^n(\mathbf{R})$ is the quotient space of $\mathbf{R}^{n+1} - \{0\}$ by the equivalence relation:

$$(x_1,\dots,x_{n+1}) \sim (\lambda x_1,\dots,\lambda x_{n+1}), \quad \lambda \in \mathbf{R}, \quad \lambda \neq 0.$$

The points of $P^n(\mathbf{R})$ will be denoted by $[x_1,\dots,x_{n+1}]$. Observe that, if $x_i \neq 0$,

$$[x_1,\dots,x_{n+1}] = \left[\frac{x_1}{x_i},\dots,\frac{x_{i-1}}{x_i},1,\frac{x_{i+1}}{x_i},\dots,\frac{x_{n+1}}{x_i}\right].$$

Define subsets $V_1,\dots,V_{n+1}$, of $P^n(\mathbf{R})$, by:

$$V_i = \{[x_1,\dots,x_{n+1}]; x_i \neq 0\}, \quad i = 1,\dots,n+1.$$

Geometrically, V_i is the set of straight lines $\mathbf{R}^{n+1}$ which pass through the origin and do not belong to the hyperplane $x_i = 0$. We are now going to show that we can take the V_i's as coordinate neighborhoods, where the coordinates on V_i are

$$y_1 = \frac{x_1}{x_i},\dots,y_{i-1} = \frac{x_{i-1}}{x_i}, \quad y_i = \frac{x_{i+1}}{x_i},\dots,y_n = \frac{x_{n+1}}{x_i}.$$

For this, we will define mappings $\mathbf{x}_i\colon \mathbf{R}^n \to V_i$ by

$$\mathbf{x}_i(y_1,\dots,y_n) = [y_1,\dots,y_{i-1},1,y_i,\dots,y_n], \quad (y_1,\dots,y_n) \in \mathbf{R}^n,$$

and will show that the family $\{(\mathbf{R}^n,\mathbf{x}_i)\}$ is a differentiable structure on $P^n(\mathbf{R})$.

Indeed, any mapping $\mathbf{x}_i$ is clearly bijective while $\bigcup \mathbf{x}_i(\mathbf{R}^n) = P^n(\mathbf{R})$. It remains to show that $\mathbf{x}_i^{-1}(V_i \cap V_j)$ is an open set in

$\mathbf{R}^n$ and that $\mathbf{x}_j^{-1} \circ \mathbf{x}_i$, $j = 1, \dots, n+1$, is differentiable there. Now, if $i > j$, the points in $\mathbf{x}_i^{-1}(V_i \cap V_j)$ are of the form:

$$\{(y_1, \dots, y_n) \in \mathbf{R}^n; \quad y_j \neq 0\}.$$

Therefore $\mathbf{x}_i^{-1}(V_i \cap V_j)$ is an open set in $\mathbf{R}^n$, and supposing that $i > j$ (the case $i < j$ is similar),

$$\begin{aligned}
\mathbf{x}_j^{-1} \circ \mathbf{x}_i(y_1, \dots, y_n) &= \mathbf{x}_j^{-1}[y_1, \dots, y_{i-1}, 1, y_i, \dots, y_n] \\
&= \mathbf{x}_j^{-1}\left[\frac{y_1}{y_j}, \dots, \frac{y_{j-1}}{y_j}, 1, \frac{y_{j+1}}{y_j}, \dots, \frac{y_{i-1}}{y_j}, \frac{1}{y_j}, \frac{y_i}{y_j}, \dots, \frac{y_n}{y_j}\right] \\
&= \left(\frac{y_1}{y_j}, \dots, \frac{y_{j-1}}{y_j}, \frac{y_{j+1}}{y_j}, \dots, \frac{y_{i-1}}{y_j}, \frac{1}{y_j}, \frac{y_i}{y_j}, \dots, \frac{y_n}{y_j}\right),
\end{aligned}$$

which is clearly differentiable.

In summary, the space of directions of $\mathbf{R}^{n+1}$ (real projective space $P^n(\mathbf{R})$) can be covered by $n+1$ coordinate neighborhoods V_i, where the V_i are made up of those directions of $\mathbf{R}^{n+1}$ that are not in the hyperplane $x_i = 0$; in addition, in each V_i we have coordinates

$$\left(\frac{x_1}{x_i}, \dots, \frac{x_{i-1}}{x_i}, \frac{x_{i+1}}{x_i}, \dots, \frac{x_{n+1}}{x_i}\right),$$

where $(x_1, \dots, x_{n+1})$ are the coordinates of $\mathbf{R}^{n+1}$. It is customary, in the classical terminology, to call the coordinates of V_i "inhomogeneous coordinates" corresponding to the "homogeneous coordinates" $(x_1, \dots, x_{n+1}) \in \mathbf{R}^{n+1}$.

Before presenting further examples of differentiable manifolds we should present a few more consequences of Definition 2.1. From now on, when we denote a differentiable manifold by M^n, the upper index n indicates the dimension of M.

First, let us extend the idea of differentiability to mappings between manifolds.

2.5 DEFINITION. Let M_1^n and M_2^m be differentiable manifolds. A mapping $\varphi\colon M_1 \to M_2$ is *differentiable at* $p \in M_1$ if given a parametrization $\mathbf{y}\colon V \subset \mathbf{R}^m \to M_2$ at $\varphi(p)$ there exists a parametrization $\mathbf{x}\colon U \subset \mathbf{R}^n \to M_1$ at p such that $\varphi(\mathbf{x}(U)) \subset \mathbf{y}(V)$ and the mapping

$$\mathbf{y}^{-1} \circ \varphi \circ \mathbf{x}\colon U \subset \mathbf{R}^n \to \mathbf{R}^m \tag{1}$$

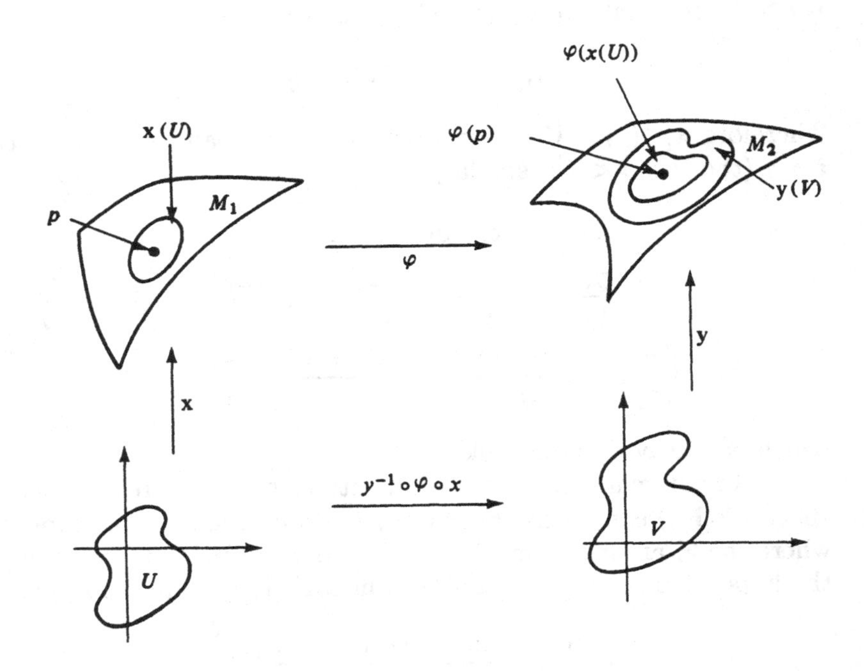

Figure 2

is differentiable at $\mathbf{x}^{-1}(p)$ (Fig. 2). φ is differentiable on an open set of M_1 if it is differentiable at all of the points of this open set.

It follows from condition (2) of Definition 2.1 that the given definition is independent of the choice of the parametrizations. The mapping (1) is called the *expression* of φ in the parametrizations $\mathbf{x}$ and $\mathbf{y}$.

Next, we would like to extend the idea of tangent vector to differentiable manifolds. It is convenient, as usual, to use our experience with regular surfaces in $\mathbf{R}^3$. For surfaces in $\mathbf{R}^3$, a tangent vector at a point p of the surface is defined as the "velocity" in $\mathbf{R}^3$ of a curve in the surface passing through p. Since we do not have at our disposal the support of the ambient space, we have to find a characteristic property of the tangent vector which will substitute for the idea of velocity.

The next considerations will motivate the definition that we are going to present below. Let $\alpha: (-\varepsilon, \varepsilon) \to \mathbf{R}^n$ be a differentiable curve in $\mathbf{R}^n$, with $\alpha(0) = p$. Write

$$\alpha(t) = (x_1(t), \dots, x_n(t)),\ t \in (-\varepsilon, \varepsilon), \quad (x_1, \dots, x_n) \in \mathbf{R}^n.$$

Then $\alpha'(0) = (x_1'(0), \dots, x_n'(0)) = v \in \mathbf{R}^n$. Now let f be a differentiable function defined in a neighborhood of p. We can restrict f to the curve α and express the directional derivative with respect to the vector $v \in \mathbf{R}^n$ as

$$\left.\frac{d(f \circ \alpha)}{dt}\right|_{t=0} = \sum_{i=1}^{n} \left.\frac{\partial f}{\partial x_i}\right|_{t=0} \left.\frac{dx_i}{dt}\right|_{t=0} = (\sum_i x_i'(0) \frac{\partial}{\partial x_i}) f.$$

Therefore, the directional derivative with respect to v is an operator on differentiable functions that depends uniquely on v. This is the characteristic property that we are going to use to define tangent vectors on a manifold.

2.6 DEFINITION. Let M be a differentiable manifold. A differentiable function $\alpha: (-\varepsilon, \varepsilon) \to M$ is called a (differentiable) *curve* in M. Suppose that $\alpha(0) = p \in M$, and let $\mathcal{D}$ be the set of functions on M that are differentiable at p. The *tangent vector to the curve* α at $t = 0$ is a function $\alpha'(0): \mathcal{D} \to \mathbf{R}$ given by

$$\alpha'(0) f = \left.\frac{d(f \circ \alpha)}{dt}\right|_{t=0}, \qquad f \in \mathcal{D}.$$

A *tangent vector at* p is the tangent vector at $t = 0$ of some curve $\alpha: (-\varepsilon, \varepsilon) \to M$ with $\alpha(0) = p$. The set of all tangent vectors to M at p will be indicated by T_pM.

If we choose a parametrization $\mathbf{x}: U \to M^n$ at $p = \mathbf{x}(0)$, we can express the function f and the curve α in this parametrization by

$$f \circ \mathbf{x}(q) = f(x_1, \dots, x_n), \quad q = (x_1, \dots, x_n) \in U,$$

and

$$\mathbf{x}^{-1} \circ \alpha(t) = (x_1(t), \dots, x_n(t)),$$

respectively. Therefore, restricting f to α, we obtain

$$\begin{aligned} \alpha'(0) f &= \left.\frac{d}{dt}(f \circ \alpha)\right|_{t=0} = \left.\frac{d}{dt} f(x_1(t), \dots, x_n(t))\right|_{t=0} \\ &= \sum_{i=1}^{n} x_i'(0) \left(\frac{\partial f}{\partial x_i}\right) = \left(\sum_i x_i'(0) \left(\frac{\partial}{\partial x_i}\right)_0\right) f. \end{aligned}$$

In other words, the vector $\alpha'(0)$ can be expressed in the parametrization $\mathbf{x}$ by

$$\alpha'(0) = \sum_i x_i'(0) \left(\frac{\partial}{\partial x_i}\right)_0. \tag{2}$$

Observe that $\left(\frac{\partial}{\partial x_i}\right)_0$ is the tangent vector at p of the "coordinate curve" (Fig. 3):

$$x_i \to \mathbf{x}(0, \dots, 0, x_i, 0, \dots, 0).$$

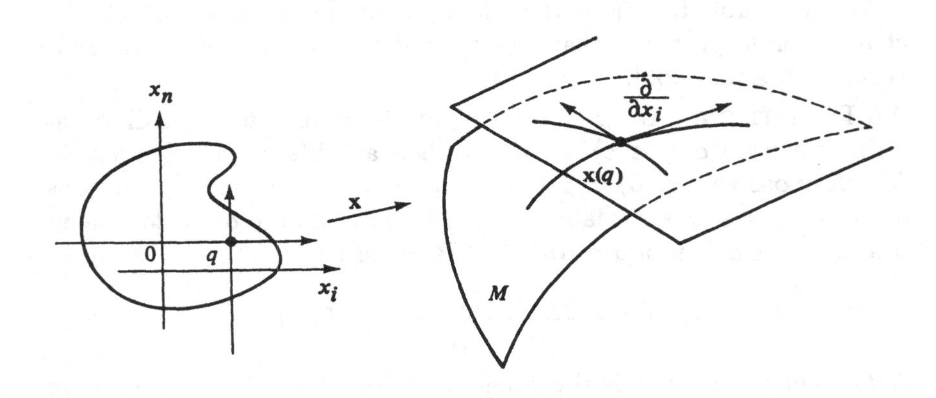

Figure 3

The expression (2) shows that the tangent vector to the curve α at p depends only the derivative of α in a coordinate system. It follows also from (2) that the set T_pM, with the usual operations of functions, forms a vector space of dimension n, and that the choice of a parametrization $\mathbf{x}: U \to M$ determines an *associated basis* $\left\{\left(\frac{\partial}{\partial x_1}\right)_0, \dots, \left(\frac{\partial}{\partial x_n}\right)_0\right\}$ in T_pM (Fig. 3). It is immediate that the linear structure in T_pM defined above does not depend on the parametrization $\mathbf{x}$. The vector space T_pM is called the *tangent space* of M at p.

With the idea of tangent space we can extend to differentiable manifolds the notion of the differential of a differentiable mapping.

2.7 PROPOSITION. *Let M_1^n and M_2^m be differentiable manifolds and let $\varphi: M_1 \to M_2$ be a differentiable mapping. For every $p \in M_1$ and for each $v \in T_pM_1$, choose a differentiable curve $\alpha: (-\varepsilon, \varepsilon) \to M_1$ with $\alpha(0) = p$, $\alpha'(0) = v$. Take $\beta = \varphi \circ \alpha$. The mapping $d\varphi_p: T_pM_1 \to T_{\varphi(p)}M_2$ given by $d\varphi_p(v) = \beta'(0)$ is a linear mapping that does not depend on the choice of α (Fig. 4).*

Proof. Let $\mathbf{x}: U \to M_1$ and $\mathbf{y}: V \to M_2$ be parametrizations at p and $\varphi(p)$, respectively. Expressing φ in these parametrizations, we can write

$$\mathbf{y}^{-1} \circ \varphi \circ \mathbf{x}(q) = (y_1(x_1, \ldots, x_n), \ldots, y_m(x_1, \ldots, x_n))$$

$$q = (x_1, \ldots, x_n) \in U, \quad (y_1, \ldots, y_m) \in V.$$

On the other hand, expressing α in the parametrization $\mathbf{x}$, we obtain

$$\mathbf{x}^{-1} \circ \alpha(t) = (x_1(t), \ldots, x_n(t)).$$

Therefore,

$$\mathbf{y}^{-1} \circ \beta(t) = (y_1(x_1(t), \ldots, x_n(t)), \ldots, y_m(x_1(t), \ldots, x_n(t))).$$

It follows that the expression for $\beta'(0)$ with respect to the basis $\left\{\left(\frac{\partial}{\partial y_i}\right)_0\right\}$ of $T_{\varphi(p)}M_2$, associated to the parametrization $\mathbf{y}$, is given by

$$\beta'(0) = \left(\sum_{i=1}^{n} \frac{\partial y_1}{\partial x_i} x_i'(0), \ldots, \sum_{i=1}^{n} \frac{\partial y_m}{\partial x_i} x_i'(0)\right). \tag{3}$$

The relation (3) shows immediately that $\beta'(0)$ does not depend on the choice of α. In addition, (3) can be written as

$$\beta'(0) = d\varphi_p(v) = \left(\frac{\partial y_i}{\partial x_j}\right)(x_j'(0)),$$

$$i = 1, \ldots, m; \qquad j = 1, \ldots, n,$$

where $\left(\frac{\partial y_i}{\partial x_j}\right)$ denotes an $m \times n$ matrix and $x_j'(0)$ denotes a column matrix with n elements. Therefore, $d\varphi_p$ is a linear mapping of T_pM_1

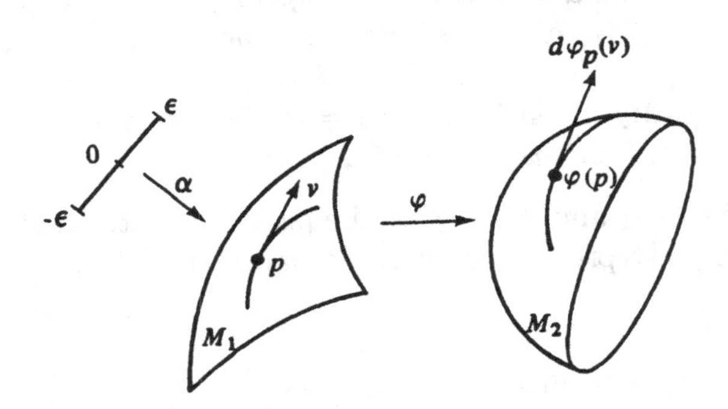

Figure 4

into $T_{\varphi(p)}M_2$ whose matrix in the associated bases obtained from the parametrizations $\mathbf{x}$ and $\mathbf{y}$ is precisely the matrix $(\frac{\partial y_i}{\partial x_j})$. □

2.8 DEFINITION. The linear mapping $d\varphi_p$ defined by Proposition 2.7 is called the *differential* of φ at p.

2.9 DEFINITION. Let M_1 and M_2 be differentiable manifolds. A mapping $\varphi\colon M_1 \to M_2$ is a *diffeomorphism* if it is differentiable, bijective, and its inverse φ^{-1} is differentiable. φ is said to be a *local diffeomorphism* at $p \in M$ if there exist neighborhoods U of p and V of $\varphi(p)$ such that $\varphi\colon U \to V$ is a diffeomorphism.

The notion of diffeomorphism is the natural idea of equivalence between differentiable manifolds. It is an immediate consequence of the chain rule that if $\varphi\colon M_1 \to M_2$ is a diffeomorphism, then $d\varphi_p\colon T_pM_1 \to T_{\varphi(p)}M_2$ is an isomorphism for all $p \in M_1$; in particular, the dimensions of M_1 and M_2 are equal. A local converse to this fact is the following theorem.

2.10 Theorem. *Let $\varphi\colon M_1^n \to M_2^n$ be a differentiable mapping and let $p \in M_1$ be such that $d\varphi_p\colon T_pM_1 \to T_{\varphi(p)}M_2$ is an isomorphism. Then φ is a local diffeomorphism at p.*

The proof follows from an immediate application of the inverse function theorem in $\mathbf{R}^n$.

3. Immersions and embeddings; examples

3.1 DEFINITION. Let M^m and N^n be differentiable manifolds. A differentiable mapping $\varphi: M \to N$ is said to be an *immersion* if $d\varphi_p: T_pM \to T_{\varphi(p)}N$ is injective for all $p \in M$. If, in addition, φ is a homeomorphism onto $\varphi(M) \subset N$, where $\varphi(M)$ has the subspace topology induced from N, we say that φ is an *embedding*. If $M \subset N$ and the inclusion $i: M \subset N$ is an embedding, we say that M is a *submanifold* of N.

It can be seen that if $\varphi: M^m \to N^n$ is an immersion, then $m \leq n$; the difference $n - m$ is called the *codimension* of the immersion φ.

3.2 EXAMPLE. The curve $\alpha: \mathbf{R} \to \mathbf{R}^2$ given by $\alpha(t) = (t, |t|)$ is not differentiable at $t = 0$ (Fig. 5).

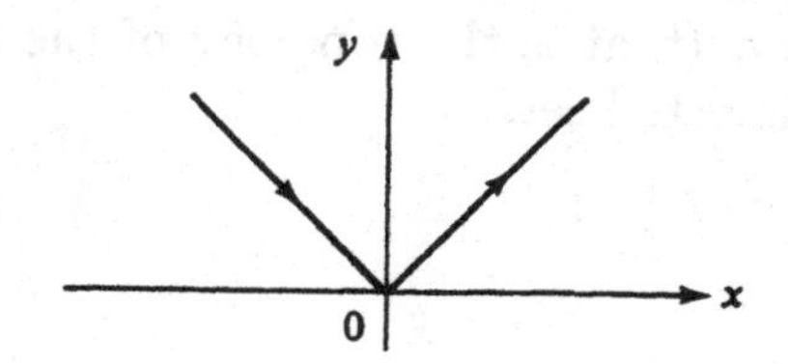

Figure 5

3.3 EXAMPLE. The curve $\alpha: \mathbf{R} \to \mathbf{R}^2$ given by $\alpha(t) = (t^3, t^2)$ is a differentiable mapping but is not an immersion. Indeed, the condition for the map to be an immersion in this case is equivalent to the fact that $\alpha'(t) \neq 0$, which does not occur for $t = 0$ (Fig. 6).

3.4 EXAMPLE. The curve $\alpha(t) = (t^3 - 4t, t^2 - 4)$ (Fig. 7) is an immersion $\alpha: \mathbf{R} \to \mathbf{R}^2$ which has a self-intersection for $t = 2, t = -2$. Therefore, α is not an embedding.

3.5 EXAMPLE. The curve (Fig. 8)

$$\alpha(t) \begin{cases} = (0, -(t+2)), & t \in (-3, -1), \\ = \text{regular curve (see Fig. 8)}, & t \in (-1, -\frac{1}{\pi}) \\ = (-t, -\sin \frac{1}{t}), & t \in (-\frac{1}{\pi}, 0) \end{cases}$$

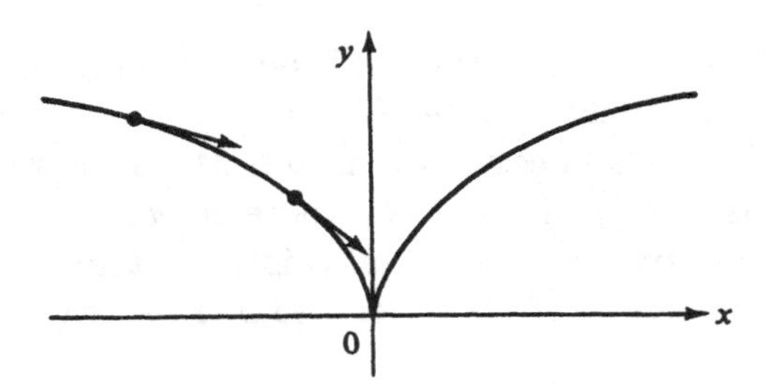

Figure 6

is an immersion $\alpha: (-3, 0) \to \mathbf{R}^2$ without self-intersections. Nevertheless, α is not an embedding. Indeed, a neighborhood of a point p, in the vertical part of the curve (Fig. 8) consists of an infinite number of connected components in the topology induced from $\mathbf{R}^2$. On the other hand, a neighborhood of such a point in the topology "induced" from α (that is the topology of the line) is an open interval, hence a connected set.

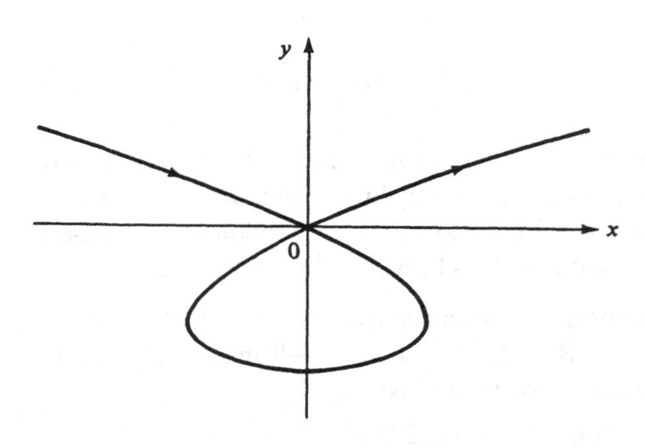

Figure 7

3.6 EXAMPLE. It is clear that a regular surface $S \subset \mathbf{R}^3$ has a differentiable structure given by its parametrizations $\mathbf{x}_\alpha: U_\alpha \to S$. With

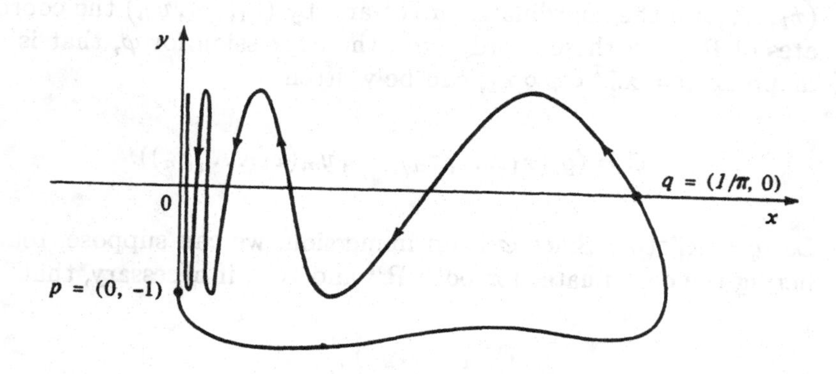

Figure 8

such a structure, the mappings $\mathbf{x}_\alpha$ are differentiable and, indeed, are embeddings of U_α into S; that is an immediate consequence of conditions (a) and (b) of the definition of regular surface given in the introduction. We are going to show that the inclusion $i: S \subset \mathbf{R}^3$ is an embedding, that is, S is a submanifold of $\mathbf{R}^3$.

In fact, i is differentiable, because for all $p \in S$ there exists a parametrization $\mathbf{x}: U \subset \mathbf{R}^2 \to S$ of S at p and a parametrization $j: V \subset \mathbf{R}^3 \to V$ of $\mathbf{R}^3$ at $i(p)$ (V is a neighborhood of p in $\mathbf{R}^3$ and j is the identity mapping), such that $j^{-1} \circ i \circ \mathbf{x} = \mathbf{x}$ is differentiable. In addition, from condition (b), i is an immersion and, from condition (a), i is a homeomorphism onto its image, which proves the claim.

For most local questions of geometry, it is the same to work with either immersions or embeddings. This comes from the following proposition which shows that every immersion is locally (in a certain sense) an embedding.

3.7 PROPOSITION. *Let $\varphi: M_1^n \to M_2^m$, $n \leq m$, be an immersion of the differentiable manifold M_1 into the differentiable manifold M_2. For every point $p \in M_1$, there exists a neighborhood $V \subset M_1$ of p such that the restriction $\varphi \mid V \to M_2$ is an embedding.*

Proof. This fact is a consequence of the inverse function theorem. Let $\mathbf{x}_1: U_1 \subset \mathbf{R}^n \to M_1$ and $\mathbf{x}_2: U_2 \subset \mathbf{R}^m \to M_2$ be a system of coordinates at p and at $\varphi(p)$, respectively, and let us denote by $(x_1, \ldots, x_n)$ the coordinates of $\mathbf{R}^n$ and by $(y_1, \ldots, y_m)$ the coordinates of $\mathbf{R}^m$. In these coordinates, the expression for φ, that is, the mapping $\tilde{\varphi} = \mathbf{x}_2^{-1} \circ \varphi \circ \mathbf{x}_1$, can be written

$$\tilde{\varphi} = (y_1(x_1, \ldots, x_n), \ldots, y_m(x_1, \ldots, x_n)).$$

Let $q = \mathbf{x}_1^{-1}(p)$. Since φ is an immersion, we can suppose, renumbering the coordinates for both $\mathbf{R}^n$ and $\mathbf{R}^m$, if necessary, that

$$\frac{\partial(y_1, \ldots, y_n)}{\partial(x_1, \ldots, x_n)}(q) \neq 0.$$

To apply the inverse function theorem, we introduce the mapping $\phi = U_1 \times R^{m-n=k} \to R^m$ given by

$$\phi(x_1, \ldots, x_n, t_1, \ldots, t_k) =$$

$$= (y_1(x_1, \ldots, x_n), \ldots, y_n(x_1, \ldots, x_n), y_{n+1}(x_1, \ldots, x_n) + t_1, \ldots,$$

$$, \ldots, y_{n+k}(x_1, \ldots, x_n) + t_k),$$

where $(t_1, \ldots, t_k) \in \mathbf{R}^{m-n=k}$. It is easy to verify that ϕ restricted to U_1 coincides with $\tilde{\varphi}$ and that

$$\det(d\phi_q) = \frac{\partial(y_1, \ldots, y_n)}{\partial(x_1, \ldots, x_n)}(q) \neq 0.$$

It follows from the inverse function theorem, that there exist neighborhoods $W_1 \subset U_1 \times \mathbf{R}^k$ of q and $W_2 \subset \mathbf{R}^m$ of $\phi(q)$ such that the restriction $\phi \mid W_1$ is a diffeomorphism onto W_2. Let $\tilde{V} = W_1 \cap U_1$. Since $\phi \mid \tilde{V} = \tilde{\varphi} \mid \tilde{V}$ and $\mathbf{x}_i$ is a diffeomorphism, for $i = 1, 2$, we conclude that the restriction to $V = \mathbf{x}_1(\tilde{V})$ of the mapping $\phi = \mathbf{x}_2 \circ \tilde{\varphi} \circ \mathbf{x}_1^{-1} : V \to \varphi(V) \subset M_2$ is a diffeomorphism, hence an embedding. □

4. Other examples of manifolds. Orientation

4.1 EXAMPLE. (The tangent bundle). Let M^n be a differentiable manifold and let $TM = \{(p,v); p \in M, v \in T_pM\}$. We are going to provide the set TM with a differentiable structure (of dimension $2n$); with such a structure TM will be called the *tangent bundle* of M. This is the natural space to work with when treating questions that involve positions and velocities, as in the case of mechanics.

Let $\{(U_\alpha, \mathbf{x}_\alpha)\}$ be a maximal differentiable structure on M. Denote by $(x_1^\alpha, \ldots, x_n^\alpha)$ the coordinates of U_α and by $\{\frac{\partial}{\partial x_1^\alpha}, \ldots, \frac{\partial}{\partial x_n^\alpha}\}$ the associated bases to the tangent spaces of $\mathbf{x}_\alpha(U_\alpha)$. For every α, define

$$\mathbf{y}_\alpha : U_\alpha \times \mathbf{R}^n \to TM,$$

by

$$\mathbf{y}_\alpha(x_1^\alpha, \ldots, x_n^\alpha, u_1, \ldots, u_n) =$$

$$= (\mathbf{x}_\alpha(x_1^\alpha, \ldots, x_n^\alpha), \sum_{i=1}^n u_i \frac{\partial}{\partial x_i^\alpha}), \quad (u_1, \ldots, u_n) \in \mathbf{R}^n.$$

Geometrically, this means that we are taking as coordinates of a point $(p,v) \in TM$ the coordinates $x_1^\alpha, \ldots, x_n^\alpha$ of p together with the coordinates of v in the basis $\left\{\frac{\partial}{\partial x_1^\alpha}, \ldots, \frac{\partial}{\partial x_n^\alpha}\right\}$.

We are going to show that $\{(U_\alpha \times \mathbf{R}^n, \mathbf{y}_\alpha)\}$ is a differentiable structure on TM. Since $\bigcup_\alpha \mathbf{x}_\alpha(U_\alpha) = M$ and $(d\mathbf{x}_\alpha)_q(\mathbf{R}^n) = T_{\mathbf{x}_\alpha(q)}M, \quad q \in U_\alpha$, we have that

$$\bigcup_\alpha \mathbf{y}_\alpha(U_\alpha \times \mathbf{R}^n) = TM,$$

which verifies condition (1) of Definition 2.1. Now let

$$(p,v) \in \mathbf{y}_\alpha(U_\alpha \times \mathbf{R}^n) \cap \mathbf{y}_\beta(U_\beta \times \mathbf{R}^n).$$

Then

$$(p,v) = (\mathbf{x}_\alpha(q_\alpha), d\mathbf{x}_\alpha(v_\alpha)) = (\mathbf{x}_\beta(q_\beta), d\mathbf{x}_\beta(v_\beta)),$$

where $q_\alpha \in U_\alpha$, $q_\beta \in U_\beta$, $v_\alpha, v_\beta \in \mathbf{R}^n$. Therefore,

$$\mathbf{y}_\beta^{-1} \circ \mathbf{y}_\alpha(q_\alpha, v_\alpha) = \mathbf{y}_\beta^{-1}(\mathbf{x}_\alpha(q_\alpha), d\mathbf{x}_\alpha(v_\alpha)) =$$

$$= ((\mathbf{x}_\beta^{-1} \circ \mathbf{x}_\alpha)(q_\alpha), d(\mathbf{x}_\beta^{-1} \circ \mathbf{x}_\alpha)(v_\alpha)).$$

Since $\mathbf{x}_\beta^{-1} \circ \mathbf{x}_\alpha$ is differentiable, $d(\mathbf{x}_\beta^{-1} \circ \mathbf{x}_\alpha)$ is as well. It follows that $\mathbf{y}_\beta^{-1} \circ \mathbf{y}_\alpha$ is differentiable, which verifies condition (2) of the definition 2.1 and completes the example.

4.2 EXAMPLE. (*Regular surfaces in* $\mathbf{R}^n$). The natural generalization of the notion of a regular surface in $\mathbf{R}^3$ is the idea of a surface of dimension k in $\mathbf{R}^n$, $k \leq n$. A subset $M^k \subset \mathbf{R}^n$ is a *regular surface of dimension* k if for every $p \in M^k$ there exists a neighborhood V of p in $\mathbf{R}^n$ and a mapping $\mathbf{x}: U \subset \mathbf{R}^k \to M \cap V$ of an open set $U \subset \mathbf{R}^k$ onto $M \cap V$ such that:

(a) $\mathbf{x}$ is a differentiable homeomorphism.

(b) $(d\mathbf{x})_q: \mathbf{R}^k \to \mathbf{R}^n$ is injective for all $q \in U$.

Except for the dimensions involved, the definition is exactly the same as was given in the Introduction for a regular surface in $\mathbf{R}^3$.

In a similar way as was done for surfaces in $\mathbf{R}^3$ (M. do Carmo [dC 2], p. 71), it can be proved that if $\mathbf{x}: U \subset \mathbf{R}^k \to M^k$ and $\mathbf{y}: V \subset \mathbf{R}^k \to M^k$ *are two parametrizations with* $\mathbf{x}(U) \cap \mathbf{y}(V) = W \neq \phi$, *then the mapping* $h = \mathbf{x}^{-1} \circ \mathbf{y}: \mathbf{y}^{-1}(W) \to \mathbf{x}^{-1}(W)$ *is a diffeomorphism.* For completeness, we give a sketch of this proof in what follows.

First, we observe that h is a homeomorphism, being a composition of homeomorphisms. Let $r \in \mathbf{y}^{-1}(W)$ and put $q = h(r)$. Let $(u_1, \ldots, u_k) \in U$ and $(v_1, \ldots, v_n) \in \mathbf{R}^n$, and write $\mathbf{x}$ in these coordinates as

$$\mathbf{x}(u_1, \ldots, u_k) = (v_1(u_1, \ldots, u_k), \ldots, v_n(u_1, \ldots, u_k)).$$

From condition (b), we can suppose that

$$\frac{\partial(v_1, \ldots, v_k)}{\partial(u_1, \ldots, u_k)}(q) \neq 0.$$

Extend $\mathbf{x}$ to a mapping $F: U \times \mathbf{R}^{n-k} \to \mathbf{R}^n$ given by

$$\begin{aligned} &F(u_1, \ldots, u_k, t_{k+1}, \ldots, t_n) \\ &\quad = (v_1(u_1, \ldots, u_k), \ldots, v_k(u_1, \ldots, u_k), \\ &\qquad v_{k+1}(u_1, \ldots, u_k) + t_{k+1}, \ldots, v_n(u_1, \ldots, u_k) + t_n), \end{aligned}$$

where $(t_{k+1}, ..., \ldots t_n) \in \mathbf{R}^{n-k}$. It is clear that F is differentiable and the restriction of F to $U \times \{(0, \ldots, 0)\}$ coincides with $\mathbf{x}$. By a simple calculation, we obtain that

$$\det(dF_q) = \frac{\partial(v_1, \ldots, v_k)}{\partial(u_1, \ldots, u_k)}(q) \neq 0.$$

We are then able to apply the inverse function theorem, which guarantees the existence of a neighborhood Q of $\mathbf{x}(q)$ where F^{-1} exists and is differentiable. By the continuity of $\mathbf{y}$, there exists a neighborhood $R \subset V$ of r such that $\mathbf{y}(R) \subset Q$. Note that the restriction of h to $R, h \mid R = F^{-1} \circ \mathbf{y} \mid R$ is a composition of differentiable mappings. Thus h is differentiable at r, hence in $\mathbf{y}^{-1}(W)$. A similar argument would show that h^{-1} is differentiable as well, proving the assertion. □

From what we have just proved, it follows by an entirely similar argument as in Example 3.6 that M^k is a differentiable manifold of dimension k and that the inclusion $i: M^k \subset \mathbf{R}^n$ is an embedding, that is, M^k is a submanifold of $\mathbf{R}^n$.

4.3 Example. (*Inverse image of a regular value*). Before discussing the next example, we need some definitions.

Let $F: U \subset \mathbf{R}^n \to \mathbf{R}^m$ be a differentiable mapping of an open set U of $\mathbf{R}^n$. A point $p \in U$ is defined to be a *critical point* of F if the differential $dF_p: \mathbf{R}^n \to \mathbf{R}^m$ is not surjective. The image $F(p)$ of a critical point is called a *critical value* of F. A point $a \in \mathbf{R}^m$ that is not a critical value is said to be a *regular value* of F. Note that any point $a \notin F(U)$ is trivially a regular value of F and that if there exists a regular value of F in $\mathbf{R}^m$, then $n \geq m$.

Now let $a \in F(U)$ be a regular value of F. We are going to show that the *inverse image* $F^{-1}(a) \subset \mathbf{R}^n$ is a *regular surface of dimension* $n - m = k$. From what was seen in Example 4.2, $F^{-1}(a)$ is then a submanifold of $\mathbf{R}^n$.

To prove the assertion we use, again, the inverse function theorem. Let $p \in F^{-1}(a)$. Denote by $q = (y_1, \ldots, y_m, x_1, \ldots x_k)$ an arbitrary point of $\mathbf{R}^{n=m+k}$ and by $F(q) = (f_1(q), \ldots, f_m(q))$ its image by the mapping F. Since a is a regular value of F, dF_p is surjective. Therefore, we can suppose that

$$\frac{\partial(f_1, \ldots, f_m)}{\partial(y_1, \ldots, y_m)}(p) \neq 0.$$

Define a mapping $\varphi: U \subset \mathbf{R}^n \to \mathbf{R}^{n=m+k}$ by

$$\varphi(y_1, \ldots, y_m, x_1, \ldots, x_k) = (f_1(q), \ldots, f_m(q), x_1, \ldots, x_k).$$

Then

$$\det(d\varphi)_p = \frac{\partial(f_1, \ldots, f_m)}{\partial(y_1, \ldots, y_m)}(p) \neq 0.$$

By the inverse function theorem, φ is a diffeomorphism of a neighborhood Q of p onto a neighborhood W of $\varphi(p)$. Let $K^{m+k} \subset W \subset \mathbf{R}^{m+k}$ be a cube of center $\varphi(p)$ and put $V = \varphi^{-1}(K^{m+k}) \cap Q$. Then φ maps the neighborhood V diffeomorphically onto $K^{m+k} = K^m \times K^k$. Define a mapping $\mathbf{x}: K^k \to V$ by

$$\mathbf{x}(x_1, \ldots, x_k) = \varphi^{-1}(a_1, \ldots, a_m, x_1, \ldots, x_k),$$

where $(a_1, \ldots, a_m) = a$. It is easy to check that φ satisfies conditions (a) and (b) of the definition of regular surface given in Example 4.2. Since p is arbitrary, $F^{-1}(a)$ is a regular surface in $\mathbf{R}^n$, as asserted.

Before going on to other examples of differentiable manifolds, we should introduce the important global notion of orientation.

4.4 DEFINITION. Let M be a differentiable manifold. We say that M is *orientable* if M admits a differentiable structure $\{(U_\alpha, \mathbf{x}_\alpha)\}$ such that:

(i) for every pair α, β, with $\mathbf{x}_\alpha(U_\alpha) \cap \mathbf{x}_\beta(U_\beta) = W \neq \phi$, the differential of the change of coordinates $\mathbf{x}_\beta^{-1} \circ \mathbf{x}_\alpha$ has positive determinant.

In the opposite case, we say that M is *non-orientable*. If M is orientable, a choice of a differentiable structure satisfying (i) is called an *orientation* of M. M is then said to be *oriented*. Two differentiable structures that satisfy (i) *determine the same orientation* if their union again satisfies (i).

It is not difficult to verify that if M is orientable and connected there exist exactly two distinct orientations on M.

Now let M_1 and M_2 be differentiable manifolds and let $\varphi: M_1 \to M_2$ be a diffeomorphism. It is easy to verify that M_1 is orientable if and only if M_2 is orientable. If, additionally, M_1 and M_2 are connected and are oriented, φ induces an orientation on M_2 which may or may not coincide with the initial orientation of M_2. In the first case, we say that φ *preserves the orientation* and in the second case, that φ *reverses the orientation*.

4.5 EXAMPLE. *If M can be covered by two coordinate neighborhoods V_1 and V_2 in such a way that the intersection $V_1 \cap V_2$ is connected, then M is orientable.* Indeed, since the determinant of the differential of the coordinate change is $\neq 0$, it does not change sign in $V_1 \cap V_2$; if it is negative at a single point, it suffices to change the sign of one of the coordinates to make it positive at that point, hence on $V_1 \cap V_2$.

4.6 EXAMPLE. The simple criterion of the previous example can be used to show that the sphere

$$S^n = \left\{ (x_1, \ldots, x_{n+1}) \in \mathbf{R}^{n+1}; \sum_{i=1}^{n+1} x_i^2 = 1 \right\} \subset \mathbf{R}^{n+1}$$

is orientable. Indeed, let $N = (0, \ldots, 0, 1)$ be the north pole and $S = (0, \ldots, 0, -1)$ the south pole of S^n. Define a mapping $\pi_1 \colon S^n - \{N\} \to \mathbf{R}^n$ (*stereographic projection from the north pole*) that takes $p = (x_1, \ldots x_{n+1})$ in $S^n - \{N\}$ into the intersection of the hyperplane $x_{n+1} = 0$ with the line that passes through p and N. It is easy to verify that (Fig. 9)

$$\pi_1(x_1, \ldots, x_{n+1}) = \left(\frac{x_1}{1 - x_{n+1}}, \ldots, \frac{x_n}{1 - x_{n+1}}\right).$$

The mapping π_1 is differentiable, injective and maps $S^n - \{N\}$ onto the hyperplane $x_{n+1} = 0$. The stereographic projection $\pi_2 \colon S^n - \{S\} \to \mathbf{R}^n$ from the south pole onto the hyperplane $x_{n+1} = 0$ has the same properties.

Therefore, the parametrizations $(\mathbf{R}^n, \pi_1^{-1})$, $(\mathbf{R}^n, \pi_2^{-1})$ cover S^n. In addition, the change of coordinates:

$$y_j = \frac{x_j}{1 - x_{n+1}} \leftrightarrow y'_j = \frac{x_j}{1 + x_{n+1}},$$

$$(y_1, \ldots, y_n) \in \mathbf{R}^n, \quad j = 1, \ldots, n,$$

is given by

$$y'_j = \frac{y_j}{\sum_{i=1}^n y_i^2}$$

(here we use the fact that $\sum_{k=1}^{n+1} x_k^2 = 1$). Therefore, the family $\{(\mathbf{R}^n, \pi_1^{-1}), (\mathbf{R}^n, \pi_2^{-1})\}$ is a differentiable structure on S^n. Observe that the intersection $\pi_1^{-1}(\mathbf{R}^n) \cap \pi_2^{-1}(\mathbf{R}^n) = S^n - \{N \cup S\}$ is connected, thus S^n is orientable and the family given determines an orientation of S^n.

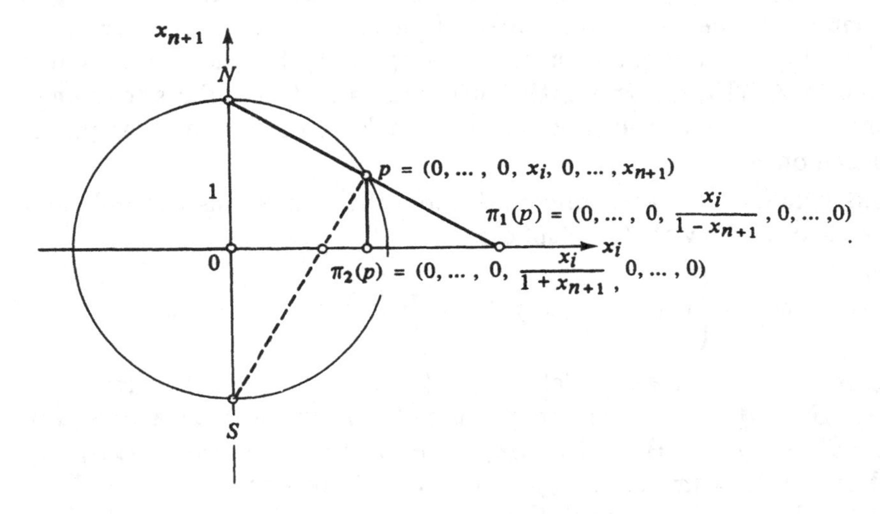

Figure 9

Now let $A: S^n \to S^n$ be the antipodal map given by $A(p) = -p$, $p \in \mathbf{R}^{n+1}$. A is differentiable and $A^2 =$ ident. Therefore, A is a diffeomorphism of S^n. Observe that when n is even, A reverses the orientation of S^n and when n is odd, A preserves the orientation of S^n.

We are now in a position to exhibit some other examples of differentiable manifolds.

4.7 EXAMPLE. (*Another description of projective space*). The set $P^n(\mathbf{R})$ of lines of $\mathbf{R}^{n+1}$ that pass through the origin can be thought of as the quotient space of the unit sphere $S^n = \{p \in \mathbf{R}^{n+1}; |p| = 1\}$ by the equivalence relation that identifies $p \in S^n$ with its antipodal point, $A(p) = -p$. Indeed, each line that passes through the origin determines two antipodal points and the correspondence so obtained is evidently bijective.

Taking into account this fact, we are going to introduce another differentiable structure on $P^n(\mathbf{R})$ (Cf. Example 2.4). For this, we initially introduce on $S^n \subset \mathbf{R}^{n+1}$ the structure of a regular surface, defining parametrizations

$$\mathbf{x}_i^+ : U_i \to S^n, \quad \mathbf{x}_i^- : U_i \to S^n, \qquad i = 1, \ldots, n+1,$$

in the following way:

$$U_i = \{(x_1, \ldots, x_{n+1}) \in \mathbf{R}^{n+1}; x_i = 0,$$
$$x_1^2 + \ldots + x_{i-1}^2 + x_{i+1}^2 + \ldots + x_{n+1}^2 < 1\},$$

$$\mathbf{x}_i^+(x_1, \ldots, x_{i-1}, x_{i+1}, \ldots, x_{n+1})$$
$$= (x_1, \ldots, x_{i-1}, D_i, x_{i+1}, \ldots, x_{n+1}),$$

$$\mathbf{x}_i^-(x_1, \ldots, x_{i-1}, x_{i+1}, \ldots, x_{n+1})$$
$$= (x_1, \ldots, x_{i-1}, -D_i, x_{i+1}, \ldots, x_{n+1}),$$

where $D_i = \sqrt{1 - (x_1^2 + \ldots + x_{i-1}^2 + x_{i+1}^2 + \ldots + x_{n+1}^2)}$. It is easy to verify that conditions (a) and (b) of the definition in Example 4.2 are satisfied. Therefore, the family

$$\{(U_i, \mathbf{x}_i^+), (U_i, \mathbf{x}_i^-)\}, \qquad i = 1, \ldots, n+1$$

is a differentiable structure on S^n. Geometrically, this is equivalent to covering the sphere S^n with coordinate neighborhoods that are hemi-spheres perpendicular to the axes x_i and taking as coordinates on, for example, $\mathbf{x}_i^+(U_i)$, the coordinates of the orthogonal projection of $\mathbf{x}_i^+(U_i)$ on the hyperplane $x_i = 0$ (Fig. 10).

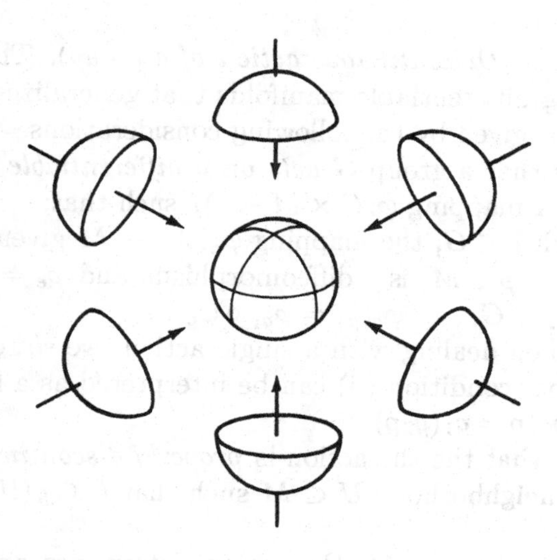

Figure 10

Let $\pi: S^n \to P^n(\mathbf{R})$ be the canonical projection, that is, $\pi(p) = \{p, -p\}$; observe that $\pi(\mathbf{x}_i^+(U_i)) = \pi(\mathbf{x}_i^-(U_i))$. We are going to define a mapping $\mathbf{y}_i: U_i \to P^n(\mathbf{R})$ by

$$\mathbf{y}_i = \pi \circ \mathbf{x}_i^+.$$

Since π restricted to $\mathbf{x}_i^+(U_i)$ is one-to-one, we have that

$$\mathbf{y}_i^{-1} \circ \mathbf{y}_j = (\pi \circ \mathbf{x}_i^+)^{-1} \circ (\pi \circ \mathbf{x}_j^+) = (\mathbf{x}_i^+)^{-1} \circ \mathbf{x}_j^+,$$

which yields the differentiability of $\mathbf{y}_i^{-1} \circ \mathbf{y}_j$, for all $i, j = 1, \dots, n+1$. Thus the family $\{(U_i, \mathbf{y}_i\}$ is a differentiable structure for $P^n(\mathbf{R})$.

In fact, this differentiable structure and that of Example 2.4 give rise to the same maximal structure. Indeed, the coordinate neighborhoods are the same and the change of coordinates are given by:

$$(\frac{x_1}{x_i}, \dots, \frac{x_{i-1}}{x_i}, 1, \frac{x_{i+1}}{x_i}, \dots, \frac{x_{n+1}}{x_i}) \leftrightarrow$$
$$\leftrightarrow (x_1, \dots, x_{i-1}, D_i, x_{i+1}, \dots, x_{n+1})$$

which, since $x_i \neq 0$ and $D_i \neq 0$, is differentiable.

As we shall see in Exercise 9, $P^n(\mathbf{R})$ is orientable if and only if n is odd.

4.8 Example. (*Discontinuous action of a group*). There is a way of constructing differentiable manifolds that generalizes the process above, which is given by the following considerations.

We say that a group G *acts on a differentiable manifold* M if there exists a mapping $\varphi: G \times M \to M$ such that:

(i) For each $g \in G$, the mapping $\varphi_g: M \to M$ given by $\varphi_g(p) = \varphi(g, p)$, $p \in M$, is a diffeomorphism, and $\varphi_e =$ identity.

(ii) If $g_1, g_2 \in G$, $\varphi_{g_1 g_2} = \varphi_{g_1} \circ \varphi_{g_2}$.

Frequently, when dealing with a single action, we set $\varphi(g, p) = gp$; in this notation, condition (ii) can be interpreted as a form of associativity: $(g_1 g_2)p = g_1(g_2 p)$.

We say that the the action is *properly discontinuous* if every $p \in M$ has a neighborhood $U \subset M$ such that $U \cap g(U) = \phi$ for all $g \neq e$.

When G acts on M, the action determines an equivalence relation $\sim$ on M, in which $p_1 \sim p_2$ if and only if $p_2 = g p_1$, for some

$g \in G$. Denote the quotient space of M by this equivalence relation by M/G. The mapping $\pi: M \to M/G$, given by

$$\pi(p) = \text{equiv. class of } p = Gp$$

will be called the *projection* of M onto M/G.

Now let M be a differentiable manifold and let $G \times M \to M$ be a properly discontinuous action of a group G on M. We are going to show that M/G has a differentiable structure with respect to which the projection $\pi: M \to M/G$ is a local diffeomorphism.

For each $p \in M$ choose a parametrization $\mathbf{x}: V \to M$ at p so that $\mathbf{x}(V) \subset U$, where $U \subset M$ is a neighborhood of p such that $U \cap g(U) = \phi$, $g \neq e$. Clearly $\pi \mid U$ is injective, hence $\mathbf{y} = \pi \circ \mathbf{x}: V \to M/G$ is injective. The family $\{(V, \mathbf{y})\}$ clearly covers M/G; for such a family to be a differentiable structure, it suffices to show that given two mappings $\mathbf{y}_1 = \pi \circ \mathbf{x}_1: V_1 \to M/G$ and $\mathbf{y}_2 = \pi \circ \mathbf{x}_2: V_2 \to M/G$ with $\mathbf{y}_1(V_1) \cap \mathbf{y}_2(V_2) \neq \phi$, then $\mathbf{y}_1^{-1} \circ \mathbf{y}_2$ is differentiable.

For this, let π_i be the restriction of π to $x_i(V_i)$, $i = 1, 2$. Let $q \in \mathbf{y}_1(V_1) \cap \mathbf{y}_2(V_2)$ and let $r = \mathbf{x}_2^{-1} \circ \pi_2^{-1}(q)$. Let $W \subset V_2$ be a neighborhood of r such that $(\pi_2 \circ \mathbf{x}_2)(W) \subset \mathbf{y}_1(V_1) \cap \mathbf{y}_2(V_2)$ (Fig. 11). Then, the restriction to W is given by

$$\mathbf{y}_1^{-1} \circ \mathbf{y}_2 \mid W = \mathbf{x}_1^{-1} \circ \pi_1^{-1} \circ \pi_2 \circ \mathbf{x}_2.$$

Therefore, it is enough to show that $\pi_1^{-1} \circ \pi_2$ is differentiable at $p_2 = \pi_2^{-1}(q)$. Let $p_1 = \pi_1^{-1} \circ \pi_2(p_2)$. Then p_1 and p_2 are equivalent in M, hence there is a $g \in G$ such that $gp_2 = p_1$. It follows easily that the restriction $\pi_1^{-1} \circ \pi_2 \mid \mathbf{x}_2(W)$ coincides with the diffeomorphism $\varphi_g \mid \mathbf{x}_2(W)$, which proves that $\pi_1^{-1} \circ \pi_2$ is differentiable at p_2, as stated.

From the very way in which this differentiable structure is constructed, $\pi: M \to M/G$ is a local diffeomorphism. A criterion for the orientability of M/G is given in Exercise 9. Observe that the situation in the previous example reduces to the present one, by taking $M = S^n$ and G the group of diffeomorphisms of S^n formed by the antipodal mapping A and the identity $I = A^2$ of S^n.

4.9 EXAMPLE. (special cases of Example 4.8).

4.9 (a). Consider the group G of "integral" translations of $\mathbf{R}^k$ where the action of G on $\mathbf{R}^k$ is given by

$$\mathbf{y}_1^{-1} \circ \mathbf{y}_2 \mid W = \mathbf{x}_1^{-1} \circ \pi_1^{-1} \circ \pi_2 \circ \mathbf{x}_2.$$

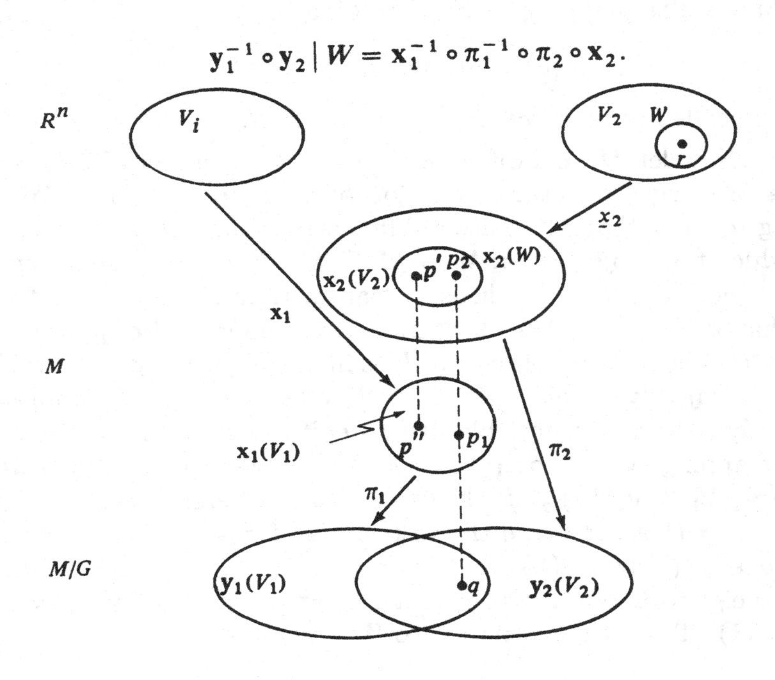

Figure 11

$$G(x_1, \ldots, x_k) = (x_1 + n_1, \ldots, x_k + n_k), \quad n_1, \ldots, n_k \in \mathbb{Z},$$

$$(x_1, \ldots, x_k) \in \mathbf{R}^k.$$

It is easy to check that the mapping above defines an action of G on $\mathbf{R}^k$, which is properly discontinuous. The quotient space $\mathbf{R}^k/G$, with the differentiable structure described in Example 4.8, is called the *k-torus* T^k. When $k = 2$, the 2-torus T^2 is diffeomorphic to the torus of revolution in $\mathbf{R}^3$ obtained as the inverse image of zero of the function $f\colon \mathbf{R}^3 \to \mathbf{R}$

$$f(x, y, z) = z^2 + (\sqrt{x^2 + y^2} - a)^2 - r^2.$$

(Cf. M. do Carmo [dC 2], p. 62).

4.9 (b). Let $S \subset \mathbf{R}^3$ be a regular surface in $\mathbf{R}^3$, symmetric relative to the origin $0 \in \mathbf{R}^3$, that is, if $p \in S$ then $-p = A(p) \in S$. The group of diffeomorphisms of S formed by $\{A, Id.\}$ acts on S in a properly discontinuous manner. Introduce on S/G the differentiable structure given by Example 4.8. When S is the torus of revolution T^2, $S/G = K$ is called *the Klein bottle*; when S is the right circular cylinder given by $C = \{(x,y,z) \in \mathbf{R}^3; x^2+y^2 = 1, \quad -1 < z < 1\}$, S/G is called the *Möbius band.* As we shall see in Exercise 9, the Klein bottle and the Möbius band are non-orientable. In Exercise 6, we shall indicate how the Klein bottle can be embedded in $\mathbf{R}^4$.

5. Vector fields; brackets. Topology of manifolds

5.1 DEFINITION. A *vector field* X on a differentiable manifold M is a correspondence that associates to each point $p \in M$ a vector $X(p) \in T_pM$. In terms of mappings, X is a mapping of M into the tangent bundle TM (see Example 4.1). The field is *differentiable* if the mapping $X: M \to TM$ is differentiable.

Considering a parametrization $\mathbf{x}: U \subset \mathbf{R}^n \to M$ we can write

$$X(p) = \sum_{i=1}^{n} a_i(p) \frac{\partial}{\partial x_i}, \tag{4}$$

where each $a_i: U \to \mathbf{R}$ is a function on U and $\left\{\frac{\partial}{\partial x_i}\right\}$ is the basis associated to $\mathbf{x}$, $\quad i = 1, \dots, n$. It is clear that X is differentiable if and only if the functions a_i are differentiable for some (and, therefore, for any) parametrization.

Occasionally, it is convenient to use the idea suggested by (4) and think of a vector field as a mapping $X: \mathcal{D} \to \mathcal{F}$ from the set $\mathcal{D}$ of differentiable functions on M to the set $\mathcal{F}$ of functions on M, defined in the following way

$$(Xf)(p) = \sum_{i} a_i(p) \frac{\partial f}{\partial x_i}(p), \tag{5}$$

where f denotes, by abuse of notation, the expression of f in the parametrization $\mathbf{x}$. Indeed, this idea of a vector as a directional derivative was precisely what was used to define the notion of tangent vector. It is easy to check that the function Xf obtained in (5) does not depend on the choice of parametrization $\mathbf{x}$. In this context, it is immediate that X is differentiable if and only if $X: \mathcal{D} \to \mathcal{D}$, that is, $Xf \in \mathcal{D}$ for all $f \in \mathcal{D}$.

Observe that if $\varphi: M \to M$ is a diffeomorphism, $v \in T_pM$ and f is a differentiable function in a neighborhood of $\varphi(p)$, we have

$$(d\varphi(v)f)\varphi(p) = v(f \circ \varphi)(p).$$

Indeed, let $\alpha: (-\varepsilon, \varepsilon) \to M$ be a differentiable curve with $\alpha'(0) = v$, $\alpha(0) = p$. Then

$$(d\varphi(v)f)\varphi(p) = \frac{d}{dt}(f \circ \varphi \circ \alpha)\Big|_{t=0} = v(f \circ \varphi)(p).$$

The interpretation of X as an operator on $\mathcal{D}$ permits us to consider the iterates of X. For example, if X and Y are differentiable fields on M and $f: M \to \mathbf{R}$ is a differentiable function, we can consider the functions $X(Yf)$ and $Y(Xf)$. In general, such operations do not lead to vector fields, because they involve derivatives of order higher than one. Nevertheless, we can affirm the following.

5.2 LEMMA. *Let X and Y be differentiable vector fields on a differentiable manifold M. Then there exists a unique vector field Z such that, for all $f \in \mathcal{D}$, $Zf = (XY - YX)f$.*

Proof. First, we prove that if Z exists, then it is unique. Assume, therefore, the existence of such a Z. Let $p \in M$ and let $\mathbf{x}: U \to M$ be a parametrization at p, and let

$$X = \sum_i a_i \frac{\partial}{\partial x_i}, \qquad Y = \sum_j b_j \frac{\partial}{\partial x_j}$$

be the expressions for X and Y in these parametrizations. Then for all $f \in \mathcal{D}$,

$$XYf = X\Big(\sum_j b_j \frac{\partial f}{\partial x_j}\Big) = \sum_{i,j} a_i \frac{\partial b_j}{\partial x_i}\frac{\partial f}{\partial x_j} + \sum_{i,j} a_i b_j \frac{\partial^2 f}{\partial x_i \partial x_j},$$

$$YXf = Y\Big(\sum_i a_i \frac{\partial f}{\partial x_i}\Big) = \sum_{i,j} b_j \frac{\partial a_i}{\partial x_j}\frac{\partial f}{\partial x_i} + \sum_{i,j} a_i b_j \frac{\partial^2 f}{\partial x_i \partial x_j}.$$

Therefore, Z is given, in the parametrization $\mathbf{x}$, by

$$Zf = XYf - YXf = \sum_{i,j}(a_i \frac{\partial b_j}{\partial x_i} - b_i \frac{\partial a_j}{\partial x_i})\frac{\partial f}{\partial x_j}$$

which proves the uniqueness of Z.

To show existence, define Z_α in each coordinate neighborhood $\mathbf{x}_\alpha(U_\alpha)$ of a differentiable structure $\{(U_\alpha, \mathbf{x}_\alpha)\}$ on M by the previous expression. By uniqueness, $Z_\alpha = Z_\beta$ on $x_\alpha(U_\alpha) \cap x_\beta(U_\beta) \neq \phi$, which allows us to define Z over the entire manifold M. □

The vector field Z given by Lemma 5.2 is called the *bracket* $[X,Y] = XY - YX$ of X and Y; Z is obviously differentiable.

The bracket operation has the following properties:

5.3 PROPOSITION. *If X, Y and Z are differentiable vector fields on M, a, b are real numbers, and f, g are differentiable functions, then:*
(a) $[X,Y] = -[Y,X]$ (anticommutativity),
(b) $[aX + bY, Z] = a[X,Z] + b[Y,Z]$ (linearity),
(c) $[[X,Y],Z] + [[Y,Z],X] + [[Z,X],Y] = 0$ (Jacobi identity),
(d) $[fX, gY] = fg[X,Y] + fX(g)Y - gY(f)X$.

Proof. (a) and (b) are immediate. In order to prove (c), it suffices to observe that, on the one hand,

$$[[X,Y],Z] = [XY - YX, Z] = XYZ - YXZ - ZXY + ZYX$$

while, on the other hand,

$$[X,[Y,Z]] + [Y,[Z,X]]$$

$$= XYZ - XZY - YZX + ZYX + YZX - YXZ - ZXY + XZY.$$

Because the second members of the expressions above are equal, (c) follows using (a).

Finally, to prove (d), calculate

$$\begin{aligned}[fX, gY] &= fX(gY) - gY(fX) = fgXY + fX(g)Y \\ &- gfYX - gY(f)X = fg[X,Y] + fX(g)X - gY(f)X. \square\end{aligned}$$

The bracket $[X,Y]$ can also be interpretated as a derivation of Y along the "trajectories" of X. To describe this interpretation, we need some preliminary ideas on differential equations.

Since a differentiable manifold is locally diffeomorphic to $\mathbf{R}^n$, the fundamental theorem on existence, uniqueness, and dependence on initial conditions of ordinary differential equations (which is a local theorem) extends naturally to differentiable manifolds. For later use, it is convenient to state it explicitly here. The reader not familiar with differential equations can assume the statement below, which is all that we need.

Let X be a differentiable vector field on a differentiable manifold M, and let $p \in M$. Then there exist a neighborhood $U \subset M$ of p, an interval $(-\delta, \delta)$, $\delta > 0$, and a differentiable mapping $\varphi\colon (-\delta,\delta) \times U \to M$ such that the curve $t \to \varphi(t,q)$, $t \in (-\delta,\delta)$, $q \in U$, is the unique curve which satisfies $\frac{\partial\varphi}{\partial t} = X(\varphi(t,q))$ and $\varphi(0,q) = q$.

A curve $\alpha\colon (-\delta,\delta) \to M$ which satisfies the conditions $\alpha'(t) = X(\alpha(t))$ and $\alpha(0) = q$ is called a *trajectory* of the field X that passes through q for $t = 0$. The theorem above guarantees that for each point of a certain neighborhood there passes a unique trajectory of X and that the mapping so obtained depends differentiably on t and on the "initial condition" q. It is common to use the notation $\varphi_t(q) = \varphi(t,q)$ and call $\varphi_t\colon U \to M$ the *local flow* of X.

The interpretation of the bracket $[X,Y]$, mentioned above, is contained in the following proposition.

5.4 PROPOSITION. *Let X, Y be differentiable vector fields on a differentiable manifold M, let $p \in M$, and let φ_t be the local flow of X in a neighborhood U of p. Then*

$$[X,Y](p) = \lim_{t\to 0} \frac{1}{t}[Y - d\varphi_t Y](\varphi_t(p)).$$

For the proof, we need the following lemma from calculus.

5.5 LEMMA. *Let $h\colon (-\delta,\delta) \times U \to \mathbf{R}$ be a differentiable mapping with $h(0,q) = 0$ for all $q \in U$. Then there exists a differentiable mapping $g\colon (-\delta,\delta) \times U \to \mathbf{R}$ with $h(t,q) = tg(t,q)$; in particular,*

$$g(0,q) = \left.\frac{\partial h(t,q)}{\partial t}\right|_{t=0}.$$

Proof of lemma. It suffices to define, for fixed t,

$$g(t,q) = \int_0^1 \frac{\partial h(ts,q)}{\partial(ts)}\, ds$$

and, after changing variables, observe that

$$tg(t,q) = \int_0^t \frac{\partial h(ts,q)}{\partial (ts)} d(ts) = h(t,q).$$

Proof of the Proposition. Let f be a differentiable function in a neighborhood of p. Putting

$$h(t,q) = f(\varphi_t(q)) - f(q),$$

and applying the lemma we obtain a differentiable function $g(t,q)$ such that

$$f \circ \varphi_t(q) = f(q) + tg(t,q) \quad \text{and} \quad g(0,q) = Xf(q).$$

Accordingly

$$((d\varphi_t Y)f)(\varphi_t(p)) = (Y(f \circ \varphi_t))(p) = Yf(p) + t(Yg(t,p)).$$

Therefore

$$\begin{aligned}
\lim_{t\to 0} \frac{1}{t}[Y - d\varphi_t Y]f(\varphi_t p) &= \lim_{t\to 0} \frac{(Yf)(\varphi_t p) - Yf(p)}{t} - (Yg(0,p)) \\
&= (X(Yf))(p) - (Y(Xf))(p) \\
&= ((XY - YX)f)(p) = ([X,Y]f)(p). \quad \square
\end{aligned}$$

Up till now we have put no restrictions on the topology of a differentiable manifold. In fact, the natural topology of a manifold can be quite strange. In particular, it can happen that one (or both) of the following axioms not be satisfied:

A) *Hausdorff Axiom:* Given two distinct points of M there exist neighborhoods of these two points that do not intersect.

B) *Countable Basis Axiom:* M can be covered by a countable number of coordinate neighborhoods (we say then that M *has a countable basis*).

Axiom A is essential for the uniqueness of limits of convergent sequences and Axiom B is essential for existence of a differentiable partition of unity, an almost indispensable tool for the study of certain questions on manifolds. (Indeed, if M is connected, Axioms

A and B are equivalent to the existence of a partition of unity; see Theorem 5.6 below.)

For example, a natural question in the theory of differentiable manifolds is to know whether a given manifold can be immersed or embedded into some euclidean space. A fundamental result in this direction is the famous theorem of Whitney which states the following: *Any differentiable manifold (which is Hausdorff and has a countable basis!) of dimension n can be immersed in $\mathbf{R}^{2n}$ and embedded in $\mathbf{R}^{2n+1}$*, (in fact, the theorem can be refined to $\mathbf{R}^{2n-1}$, $n > 1$, and $\mathbf{R}^{2n}$, respectively). A proof of this theorem is not compatible with the intent of this introduction and can be found in M. W. Hirsch [Hi].

For the sake of information, we mention without proof the existence theorem for partitions of unity. This requires some definitions.

Let M be a differentiable manifold. A family of open sets $V_\alpha \subset M$ with $\bigcup_\alpha V_\alpha = M$ is said to be *locally finite* if every point $p \in M$ has a neighborhood W such that $W \cap V_\alpha \neq \phi$ for only a finite number of indices. The *support* of a function $f\colon M \to \mathbf{R}$ is the closure of the set of points where f is different from zero.

We say that a family $\{f_\alpha\}$ of differentiable functions $f_\alpha\colon M \to \mathbf{R}$ is a *differentiable partition of unity* if:

(1) For all α, $f_\alpha \geq 0$ and the support of f_α is contained in a coordinate neighborhood $V_\alpha = \mathbf{x}_\alpha(U_\alpha)$ of a differentiable structure $\{(U_\beta, \mathbf{x}_\beta)\}$ of M.

(2) The family $\{V_\alpha\}$ is locally finite.

(3) $\sum_\alpha f_\alpha(p) = 1$, for all $p \in M$ (this condition makes sense, because for each p, $f_\alpha(p) \neq 0$ only for a finite number of indices).

It is customary to say that the partition of unity $\{f_\alpha\}$ is *subordinate to the covering* $\{V_\alpha\}$.

5.6 THEOREM. *A differentiable manifold M has a differentiable partition of unity if and only if every connected component of M is Hausdorff and has a countable basis.*

For a proof see F. Brickell and R.S. Clark, *Differentiable Manifolds*, Van Nostrand Reinhold Co., London 1970, Chap. 3.

5.7 REMARK. Recall that given $p \in \mathbf{R}^n$ and an open ball $B_r(p) \subset \mathbf{R}^n$ centered at p with radius r, there exists a neighborhood U of p

with $\overline{U} \subset B_r(p)$ and a differentiable function $f\colon \mathbf{R}^n \to \mathbf{R}$ such that $0 \le f(q) \le 1$ for all $q \in \mathbf{R}^n$ and

$$f(q) = \begin{cases} 1, & \text{if} \quad q \in \overline{U}, \\ 0, & \text{if} \quad q \notin B_r(p). \end{cases}$$

Indeed, if we take, for simplicity, $r = 2$, we can choose $U = B_1(p)$ and define f by $f(q) = \beta(-|p-q|)$, $q \in \mathbf{R}^n$, where $\beta\colon \mathbf{R} \to \mathbf{R}$ is given by

$$\beta(t) = \frac{\int_{-\infty}^{t} \alpha(s)ds}{\int_{-2}^{-1} \alpha(s)ds},$$

and $\alpha\colon \mathbf{R} \to \mathbf{R}$ is the smooth function equaling $\exp(-\frac{1}{(t+2)(-1-t)})$ on $[-2,-1]$ and zero off this interval. It is easy to check that f satisfies the required conditions.

Clearly, the same thing happens in a neighborhood contained in a coordinate neighborhood on a differentiable manifold M. In other words, if $p \in M$ and $V \subset M$ is a neighborhood of p contained in a coordinate neighborhood of p which is diffeomorphic to an open ball, then there exists a neighborhood U of p with $\overline{U} \subset V$ and a differentiable function $f\colon M \to \mathbf{R}$ with $0 \le f(q) \le 1$ if $q \in M$, $f(q) = 1$ if $q \in \overline{U}$, and $f(q) = 0$ if $q \notin V$. This fact allows us to show that certain globally defined objects on M are, in reality, local, that is, their behavior at p only depends on how M behaves in a neighborhood of p (cf. the definition of the bracket of two vector fields in this chapter and the definition of an affine connection in Chapter 2).

EXERCISES

1. (*Product manifold*). Let M and N be differentiable manifolds and let $\{(U_\alpha, \mathbf{x}_\alpha)\}$, $\{V_\beta, \mathbf{y}_\beta\}$ differentiable structures on M and N, respectively. Consider the cartesian product $M \times N$ and the mappings $\mathbf{z}_{\alpha\beta}(p,q) = (\mathbf{x}_\alpha(p), \mathbf{y}_\beta(q))$, $p \in U_\alpha$, $q \in V_\beta$.
   (a) Prove that $\{(U_\alpha \times V_\beta, \mathbf{z}_{\alpha\beta})\}$ is a differentiable structure on $M \times N$ in which the projections $\pi_1\colon M \times N \to M$ and

$\pi_2\colon M\times N \to N$ are differentiable. With this differentiable structure $M\times N$ is called the *product manifold* of M with N.

(b) Show that the product manifold $S^1 \times \ldots \times S^1$ of n circles S^1, where $S^1 \subset \mathbf{R}^2$ has the usual differentiable structure, is diffeomorphic to the n-torus T^n of example 4.9 (a).

2. Prove that the tangent bundle of a differentiable manifold M is orientable (even though M may not be).

3. Prove that:
   (a) a regular surface $S \subset \mathbf{R}^3$ is an orientable manifold if and only if there exists a differentiable mapping of $N\colon S \to \mathbf{R}^3$ with $N(p)\perp T_p(S)$ and $|N(p)| = 1$, for all $p \in S$.
   (b) the Möbius band (Example 4.9 (b)) is non-orientable.

4. Show that the projective plane $P^2(\mathbf{R})$ is non-orientable. *Hint*: Prove that if the manifold M is orientable, then any open subset of M is an orientable submanifold. Observe that $P^2(\mathbf{R})$ contains an open subset diffeomorphic to a Möbius band, which is non-orientable.

5. (*Embedding of* $P^2(\mathbf{R})$ *in* $\mathbf{R}^4$). Let $F\colon \mathbf{R}^3 \to \mathbf{R}^4$ be given by

$$F(x,y,z) = (x^2 - y^2, xy, xz, yz), \quad (x,y,z) = p \in \mathbf{R}^3.$$

Let $S^2 \subset \mathbf{R}^3$ be the unit sphere with the origin $0 \in \mathbf{R}^3$. Observe that the restriction $\varphi = F \mid S^2$ is such that $\varphi(p) = \varphi(-p)$, and consider the mapping $\tilde{\varphi}\colon P^2(\mathbf{R}) \to \mathbf{R}^4$ given by

$$\tilde{\varphi}([p]) = \varphi(p),\ [p] = \text{equiv. class of } p = \{p, -p\}\,.$$

Prove that:
   (a) $\tilde{\varphi}$ is an immersion.
   (b) $\tilde{\varphi}$ is injective; together with (a) and the compactness of $P^2(\mathbf{R})$, this implies that $\tilde{\varphi}$ is an embedding.

6. (*Embedding of the Klein bottle in* $\mathbf{R}^4$). Show that the mapping $G\colon \mathbf{R}^2 \to \mathbf{R}^4$ given by

$$G(x,y) = ((r\cos y + a)\cos x, (r\cos y + a)\sin x, \\ r\sin y \cos\frac{x}{2}, r\sin y \sin\frac{x}{2}), \quad (x,y) \in \mathbf{R}^2$$

induces an embedding of the Klein bottle (Example 4.9 (b)) into $\mathbf{R}^4$.

7. (*Infinite Möbius band*).
Let $C = \{(x, y, z) \in \mathbf{R}^3; x^2 + y^2 = 1\}$ be a right circular cylinder, and let $A: C \to C$ be the symmetry with respect to the origin $0 \in \mathbf{R}^3$, that is, $A(x, y, z) = (-x, -y, -z)$. Let M be the quotient space of C with respect to the equivalence relation $p \sim A(p)$, and let $\pi: C \to M$ be the projection $\pi(p) = \{p, A(p)\}$.
   (a) Show that it is possible to give M a differentiable structure such that π is a local diffeomorphism.
   (b) Prove that M is non-orientable.

8. Let M_1 and M_2 be differentiable manifolds. Let $\varphi: M_1 \to M_2$ be a local diffeomorphism. Prove that if M_2 is orientable, then M_1 is orientable.

9. Let $G \times M \to M$ be a properly discontinuous action of a group G on a differentiable manifold M.
   (a) Prove that the manifold M/G (E xample 4.8) is orientable if and only if there exists an orientation of M that is preserved by all the diffeomorphisms of G.
   (b) Use (a) to show that the projective plane $P^2(\mathbf{R})$, the Klein bottle and the Möbius band are non-orientable.
   (c) Prove that $P^n(\mathbf{R})$ is orientable if and only if n is odd.

10. Show that the topology of the differentiable manifold M/G of Example 4.8 is Hausdorff if and only if the following condition holds: given two non-equivalent points $p_1, p_2 \in M$, there exist neighborhoods U_1, U_2 of p_1 and p_2, respectively, such that $U_1 \cap gU_2 = \phi$ for all $g \in G$.

11. Let us consider the two following differentiable structures on the real line $\mathbf{R}$: $(\mathbf{R}, \mathbf{x}_1)$, where $\mathbf{x}_1: \mathbf{R} \to \mathbf{R}$ is given by $\mathbf{x}_1(x) = x$, $x \in \mathbf{R}$; $(\mathbf{R}, \mathbf{x}_2)$, where $\mathbf{x}_2: \mathbf{R} \to \mathbf{R}$ is given by $\mathbf{x}_2(x) = x^3$, $x \in \mathbf{R}$. Show that:
   (a) the identity mapping $i: (\mathbf{R}, \mathbf{x}_1) \to (\mathbf{R}, \mathbf{x}_2)$ is not a diffeomorphism; therefore, the maximal structures determined by $(\mathbf{R}, \mathbf{x}_1)$ and $(\mathbf{R}, \mathbf{x}_2)$ are distinct.
   (b) the mapping $f: (\mathbf{R}, \mathbf{x}_1) \to (\mathbf{R}, \mathbf{x}_2)$ given by $f(x) = x^3$ is a diffeomorphism; that is, even though the differentiable

structure $(\mathbf{R}, \mathbf{x}_1)$ and $(\mathbf{R}, \mathbf{x}_2)$ are distinct, they determine diffeomorphic differentiable manifolds.

12. (*The orientable double covering*). Let M^n be a non-orientable differentiable manifold. For each $p \in M$, consider the set B of bases of T_pM and say that two bases are equivalent if they are related by a matrix with positive determinant. This is an equivalence relation and separates B into two disjoint sets. Let $\mathcal{O}_p$ be the quotient space of B with respect to this equivalence relation. $O_p \in \mathcal{O}_p$ will be called an *orientation* of T_pM. Let $\overline{M}$ be the set

$$\overline{M} = \{(p, O_p); p \in M, O_p \in \mathcal{O}_p\}.$$

Let $\{(U_\alpha, \mathbf{x}_\alpha)\}$ be a maximal differentiable structure on M, and define $\overline{\mathbf{x}}_\alpha \colon U_\alpha \to \overline{M}$ by

$$\overline{\mathbf{x}}_\alpha(u_1^\alpha, \ldots, u_n^\alpha) = (\mathbf{x}_\alpha(u_1^\alpha, \ldots, u_n^\alpha), [\frac{\partial}{\partial u_1^\alpha}, \ldots, \frac{\partial}{\partial u_n^\alpha}]),$$

where $(u_1^\alpha, \ldots, u_n^\alpha) \in U_\alpha$ and $[\frac{\partial}{\partial u_1^\alpha}, \ldots, \frac{\partial}{\partial u_n^\alpha}]$ denotes the element of $\mathcal{O}_p$ determined by the basis $\left\{\frac{\partial}{\partial u_1^\alpha}, \ldots, \frac{\partial}{\partial u_n^\alpha}\right\}$. Prove that:

(a) $\{U_\alpha, \overline{\mathbf{x}}_\alpha)\}$ is a differentiable structure on $\overline{M}$ and that the manifold $\overline{M}$ so obtained is orientable.

(b) The mapping $\pi \colon \overline{M} \to M$ given by $\pi(p, O_p) = p$ is differentiable and surjective. In addition, each $p \in M$ has a neighborhood $U \subset M$ such that $\pi^{-1}(U) = V_1 \cup V_2$, where V_1 and V_2 are disjoint open sets in $\overline{M}$ and π restricted to each V_i, $i = 1, 2$, is a diffeomorphism onto U. For this reason, $\overline{M}$ is called the *orientable double cover of M*.

(c) The sphere S^2 is the orientable double cover of $P^2(\mathbf{R})$ and the torus T^2 is the orientable double cover of the Klein bottle K.

CHAPTER 1

RIEMANNIAN METRICS

1. Introduction

Historically, Riemannian geometry was a natural development of the differential geometry of surfaces in $\mathbf{R}^3$. Given a surface $S \subset \mathbf{R}^3$, we have a natural way of measuring the lengths of vectors tangent to S, namely: the inner product $\langle v, w \rangle$ of two vectors tangent to S at a point p of S is simply the inner product of these vectors in $\mathbf{R}^3$. The way to compute the length of a curve is, by definition, to integrate the length of its velocity vector. The definition of $\langle\ ,\ \rangle$ permits us to measure not only the lengths of curves in S but also the area of domains in S, as well as the angle between two curves, and all the other "metric" ideas used in geometry. More generally, these notions lead us to define on S certain special curves, called geodesics, which possess the following property: given any two points p and q on a geodesic, sufficiently close (in a sense to be made precise later, Cf. Chap. 3), the length of such a curve is less than or equal to the length of any other curve joining p to q. Such curves behave, in many situations, as if they were "the straight lines" of S, and, as we shall see later, play an important role in the development of geometry.

Observe that the definition of the inner product at each point $p \in S$, yields, equivalently, a quadratic form I_p, called the first fundamental form of S at p, defined in the tangent plane T_pS by $I_p(v) = \langle v, v \rangle$, $v \in T_pS$.

The crucial point of this development was an observation made by Gauss in his famous work (see Gauss [Ga]) published in 1827. In this work, Gauss defined a notion of curvature for surfaces, which measures the amount that S deviates, at a point $p \in S$, from its tangent plane at p. In modern notation, Gauss' definition can be expressed in the following terms. Define a mapping $g\colon S \to S^2 \subset \mathbf{R}^3$ of S into the unit sphere S^2 of $\mathbf{R}^3$, associating to every $p \in S$ a unit vector $N(p) \in S^2$ normal to T_pS; if S were orientable then g would be well-defined and differentiable on S. During Gauss' time, the notion of orientation of surfaces was not well-understood (in truth, it wasn't until 1865 that Möbius presented his famous example, well-known today as the Möbius band), and so g was defined on "pieces" of S. In any case, g is differentiable and it is possible then to speak of its differential $dg_p: T_pS \to T_{g(p)}S^2$. Since $N(p)$ is normal to T_pS, we can identify the two vector spaces T_pS and $T_{g(p)}S^2$, and thus it makes sense to speak of the determinant of the linear map dg_p. Gauss defined his curvature as $K(p) = \det(dg_p)$ and showed that it agreed with the product of the principal curvatures introduced in 1760 by Euler.

Perhaps it is worthwhile mentioning that Euler defined the principal curvatures k_1 and k_2 of a surface S by considering the curvature k_n of curves obtained by intersecting S with planes normal to S at p and taking $k_1 = \max k_n$ and $k_2 = \min k_n$. At the time of Gauss it was not at all clear that one function or the other of k_1 and k_2, would be an adequate definition of curvature. Gauss considered that the facts which he had obtained about K justified the choice of $K = k_1 k_2$ as the curvature of S.

The facts that Gauss alluded to were the following. In the first place, the curvature, as defined above, depends only on the manner of measuring in S, that is, only on the first fundamental form I. Secondly, the sum of the interior angles of a triangle formed by geodesics differs from 180° by an expression that depends only on the curvature and the area of the triangle.

Everything indicates that Gauss perceived very clearly the profound implications of his discovery. In fact, one of the fundamental problems during Gauss' time was to decide if the fifth postulate of Euclid ("Given a straight line and a point not on the line then there is a straight line through the point which does not meet the given line") was independent of the other postulates of geometry. Although without immediate applications, the question leads to philosophical implications of primary importance. Earlier, it had been established that Euclid's fifth postulate is equivalent to the fact that the sum of the interior angles of a triangle equals 180°. The discovery of Gauss implied, among other things, that it would be possible to imagine a geometry (at least in dimension two) that depended on a fundamental quadratic form given in an arbitrary manner (without regard to the ambient space). In such a geometry, defining straight lines as geodesics, the sum of the interior angles of a triangle would depend on the curvature and, as Gauss actually verified, its difference from 180° would be equal to the integral of the curvature over the triangle. Gauss, however, did not have the necessary mathematical tools available to develop his ideas (what he lacked was essentially the idea of a differentiable manifold) and he preferred not to discuss this topic openly. The actual appearance of a non-euclidean geometry was due, independently, to Lobatchevski (1829) and Bolyai(1831).

The ideas of Gauss were taken up again by Riemann in 1854 (see Riemann [Ri]), even though he was still without an adequate definition of a manifold. Using intuitive language and without proof, Riemann introduced what we call today a differentiable manifold of dimension n. He further associated to every point of the manifold a fundamental quadratic form and then generalized the idea of Gaussian curvature to this situation (cf. Chap. 4). Furthermore, he stated many relations between the first fundamental quadratic form and the curvature that were only proved decades later. The reading of his work makes it clear that Riemann was motivated by the fundamental question implicit in the development of non-euclidean geometries, namely, the relationship between physics and geometry.

It is curious to observe that the concept of differentiable manifold, necessary for the formalization of the work of Riemann, only appeared explicitly in 1913 in the work of H. Weyl which made precise another of Riemann's audacious concepts, namely, Riemann surfaces. But that is another story.

Due to the lack of adequate tools, Riemannian geometry as such developed very slowly. An important outside source of stimulation was the application of these ideas to the theory of relativity in 1916. Another fundamental step was the introduction of the parallelism of Levi-Civita. We shall return to this topic in the next chapter. Our object here is not to write a complete history of Riemannian geometry but simply to trace its origin and supply motivation for what is to follow.

Our point of departure will be a differentiable manifold on which we introduce at each point a way of measuring the length of tangent vectors. This measurement should change differentiably from point to point. The explicit definition will be given in the next section.

For the remainder of this book, the differentiable manifolds considered will be assumed to be Hausdorff spaces with countable bases. "Differentiable" will signify "of class C^∞", and when $M^n = M$ denotes a differentiable manifold, n denotes the dimension of M.

2. Riemannian Metrics

2.1 DEFINITION. A *Riemannian metric* (or *Riemannian structure*) on a differentiable manifold M is a correspondence which associates to each point p of M an inner product $\langle\ ,\ \rangle_p$ (that is, a symmetric, bilinear, positive-definite form) on the tangent space T_pM, which varies differentiably in the following sense: If $\mathbf{x}: U \subset \mathbf{R}^n \to M$ is a system of coordinates around p, with $\mathbf{x}(x_1, x_2, \ldots, x_n) = q \in \mathbf{x}(U)$ and $\frac{\partial}{\partial x_i}(q) = d\mathbf{x}_q(0, \ldots, 1, \ldots, 0)$, then $\langle \frac{\partial}{\partial x_i}(q), \frac{\partial}{\partial x_j}(q)\rangle_q = g_{ij}(x_1, \ldots, x_n)$ is a differentiable function on U.

It is clear this definition does not depend on the choice of coordinate system.

Another way to express the differentiability of the Riemannian metric is to say that for any pair of vector fields X and Y, which are differentiable in a neighborhood V of M, the function $\langle X, Y\rangle$ is differentiable on V. It is immediate that this definition is equivalent to the other.

It is usual to delete the index p in the function $\langle\ ,\ \rangle_p$ whenever there is no possibility of confusion. The function g_{ij} $(= g_{ji})$ is called *the local representation of the Riemannian metric* (or "the g_{ij} of the metric") *in the coordinate system* $\mathbf{x}: U \subset \mathbf{R}^n \to M$. A differentiable manifold with a given Riemannian metric will be called a *Riemannian manifold*.

After introducing any type of mathematical structure, we must introduce a notion of when two objects are the same.

2.2 DEFINITION. Let M and N be Riemannian manifolds. A diffeomorphism $f: M \to N$ (that is, f is a differentiable bijection with a differentiable inverse) is called an *isometry* if:

(1) $\langle u, v\rangle_p = \langle df_p(u), df_p(v)\rangle_{f(p)}$, for all $p \in M$, $u, v \in T_pM$.

2.3 Definition. Let M and N be Riemannian manifolds. A differentiable mapping $f: M \to N$ is a *local isometry* at $p \in M$ if there is a neighborhood $U \subset M$ of p such that $f: U \to f(U)$ is a diffeomorphism satisfying (1).

It is common to say that a Riemannian manifold M is *locally isometric* to a Riemannian manifold N if for every p in M there exists a neighborhood U of p in M and a local isometry $f: U \to f(U) \subset N$.

What follows are some non-trivial examples of the notion of Riemannian manifold.

2.4 Example. *The almost trivial example.* $M = \mathbf{R}^n$ with $\frac{\partial}{\partial x_i}$ identified with $e_i = (0, \ldots, 1, \ldots, 0)$. The metric is given by $\langle e_i, e_j \rangle = \delta_{ij}$. $\mathbf{R}^n$ is called *Euclidean space of dimension n* and the Riemannian geometry of this space is metric Euclidean geometry.

2.5 Example. *Immersed manifolds.* Let $f: M^n \to N^{n+k}$ be an immersion, that is, f is differentiable and $df_p: T_pM \to T_{f(p)}N$ is injective for all p in M. If N has a Riemannian structure, f induces a Riemannian structure on M by defining $\langle u, v \rangle_p = \langle df_p(u), df_p(v) \rangle_{f(p)}$, $u, v \in T_pM$. Since $df_p$is injective, $\langle\, ,\, \rangle_p$ is positive definite. The other conditions of Definition 2.1 are easily verified. This metric on M is then called the metric *induced* by f, and f is an *isometric immersion*.

A particularly important case occurs when we have a differentiable function $h: M^{n+k} \to N^k$ and $q \in N$ is a regular value of h (that is, $dh_p: T_pM \to T_{h(p)}N$ is surjective for all $p \in h^{-1}(q)$). It is known then that $h^{-1}(q) \subset M$ is a submanifold of M of dimension n; hence, we can put a Riemannian metric on it induced by the inclusion.

For example, let $h: \mathbf{R}^n \to \mathbf{R}$ be given by $h(x_1, \ldots, x_n) = \sum_{i=1}^n x_i^2 - 1$. Then 0 is a regular value of h and $h^{-1}(0) = \{x \in \mathbf{R}^n : x_1^2 + \ldots + x_n^2 = 1\} = S^{n-1}$ is the *unit sphere*of $\mathbf{R}^n$. The metric induced from $\mathbf{R}^n$ on S^{n-1} is called the *canonical metric* of S^{n-1}.

2.6 Example. *Lie groups.* A *Lie group* is a group G with a differentiable structure such that the mapping $G \times G \to G$ given by $(x, y) \to xy^{-1}$, $x, y \in G$, is differentiable. It follows then that *translations from the left* L_x and *translations from the right* R_x given by: $L_x: G \to G$, $L_x(y) = xy$; $R_x: G \to G$, $R_x(y) = yx$ are diffeomorphisms.

We say that a Riemannian metric on G is *left invariant* if $\langle u,v\rangle_y = \langle d(L_x)_y u, d(L_x)_y v\rangle_{L_x(y)}$ for all $x,y \in G$, $u,v \in T_yG$, that is, if L_x is an isometry. Analogously, we can define a *right invariant Riemannian metric*. A Riemannian metric on G which is both right and left invariant is said to be *bi-invariant*.

We say that a differentiable vector field X on a Lie group G is *left invariant* if $dL_x X = X$ for all $x \in G$. The left invariant vector fields are completely determined by their values at a single point of G. This allows us to introduce an additional structure on the tangent space to the neutral element $e \in G$ in the following manner. To each vector$X_e \in T_eG$ we associate the left invariant X defined by $X_a = dL_a X_e$, $a \in G$. Let X, Y be left invariant vector fields on G. Since for each $x \in G$ and for any differentiable function f on G,

$$dL_x[X,Y]f = [X,Y](f\circ L_x) = X(dL_xY)f - Y(dL_xX)f =$$

$$= (XY - YX)f = [X,Y]f,$$

we conclude that the bracket of any two left invariant vector fields is again a left invariant vector field. If $X_e, Y_e \in T_eG$, we put $[X_e, Y_e] = [X,Y]_e$. With this operation, T_eG is called the *Lie algebra* of G, denoted by $\mathcal{G}$. From now on, the elements in the Lie algebra $\mathcal{G}$ will be thought of either as vectors in T_eG or as left invariant vector fields on G.

To introduce a left invariant metric on G, take any arbitrary inner product $\langle\ ,\ \rangle_e$ on $\mathcal{G}$ and define

$$(2)\qquad \langle u,v\rangle_x = \langle (dL_{x^{-1}})_x(u), (dL_{x^{-1}})_x(v)\rangle_e, \quad x \in G,\ u,v \in T_xG.$$

Since L_x depends differentiably on x, this construction actually produces a Riemannian metric, which is clearly left invariant.

In an analogous manner we can construct a right invariant metric on G. If G is compact, we will see in Exercise 7 that G possesses a bi-invariant metric.

If G has a bi-invariant metric, the inner product that the metric determines on $\mathcal{G}$ satisfies the following relation: For any $U, V, X \in \mathcal{G}$,

$$(3)\qquad \langle [U,X], V\rangle = -\langle U, [V,X]\rangle.$$

Before proving the relation above, we need some preliminary facts about Lie groups.

For any $a \in G$, let $R_{a^{-1}}L_a \colon G \to G$ be the inner automorphism of G determined by a. Such a mapping is a diffeomorphism that keeps e fixed. Thus, the differential $d(R_{a^{-1}}L_a) = Ad(a) \colon \mathcal{G} \to \mathcal{G}$ is a linear map (in fact, it is a homomorphism of the Lie algebra, but we do not need this fact). Explicitly,

$$Ad(a)Y = dR_{a^{-1}}dL_aY = dR_{a^{-1}}Y, \quad \text{for all } Y \in \mathcal{G}.$$

Let x_t be the flow of $X \in \mathcal{G}$. Then, from Proposition 5.4 of Chapter 0,

$$[Y, X] = \lim_{t \to 0} \frac{1}{t}(dx_t(Y) - Y).$$

On the other hand, since X is left invariant, $L_y \circ x_t = x_t \circ L_y$, giving

$$x_t(y) = x_t(L_y(e)) = L_y(x_t(e)) = yx_t(e) = R_{x_t(e)}(y).$$

Therefore, $dx_t = dR_{x_t(e)}$, and

$$[Y, X] = \lim_{t \to 0} \frac{1}{t}(dR_{x_t(e)}(Y) - Y) = \lim_{t \to 0} \frac{1}{t}(Ad(x_t^{-1}(e))Y - Y).$$

Let us now return to the proof of (3). Let $\langle\, , \rangle$ be a bi-invariant metric on a Lie group G. Then for any $X, U, V \in \mathcal{G}$,

$$\langle U, V \rangle = \langle dR_{x_t(e)} \circ dL_{x_t^{-1}(e)}U, dR_{x_t(e)} \circ dL_{x_t^{-1}(e)}V \rangle =$$

$$= \langle dR_{x_t(e)}U, dR_{x_t(e)}V \rangle.$$

Differentiating the expression above with respect to t, recalling that $\langle\, , \rangle$ is bilinear, and setting $t = 0$ in the expression obtained, we conclude that

$$0 = \langle [U, X], V \rangle + \langle U, [V, X] \rangle,$$

which is the equation (3).

The important point about the relation above is that it characterizes the bi-invariant metrics of G, in the following sense. If a positive bilinear form $\langle\, , \rangle_e$ defined on $\mathcal{G}$ satisfies the relation (3), then the Riemannian metric defined on G by (2) is bi-invariant. It is not difficult to prove this fact but we will not go into the proof here.

2.7 EXAMPLE. *The product metric.* Let M_1 and M_2 be Riemannian manifolds and consider the cartesian product $M_1 \times M_2$ with the product structure. Let $\pi_1 \colon M_1 \times M_2 \to M_1$ and $\pi_2 \colon M_1 \times M_2 \to M_2$ be the natural projections. Introduce on $M_1 \times M_2$ a Riemannian metric as follows:

$$\langle u, v \rangle_{(p,q)} = \langle d\pi_1 \cdot u, d\pi_1 \cdot v \rangle_p + \langle d\pi_2 \cdot u, d\pi_2 \cdot v \rangle_q,$$

$$\text{for all } (p,q) \in M_1 \times M_2,\ u, v \in T_{(p,q)}(M_1 \times M_2).$$

It is easy to verify that this is really a Riemannian metric on the product. For example, the torus $S^1 \times \cdots \times S^1 = T^n$ has a Riemannian structure obtained by choosing the induced Riemannian metric from $\mathbf{R}^2$ on the circle $S^1 \subset \mathbf{R}^2$ and then taking the product metric. The torus T^n with this metric is called the *flat torus*.

We are now going to show how a Riemannian metric can be used to calculate the lengths of curves.

2.8 DEFINITION. A differentiable mapping $c \colon I \to M$ of an open interval $I \subset \mathbf{R}$ into a differentiable manifold M is called a (parametrized) *curve*.

Observe that a parametrized curve can admit self-intersections as well as "corners" (Fig. 1).

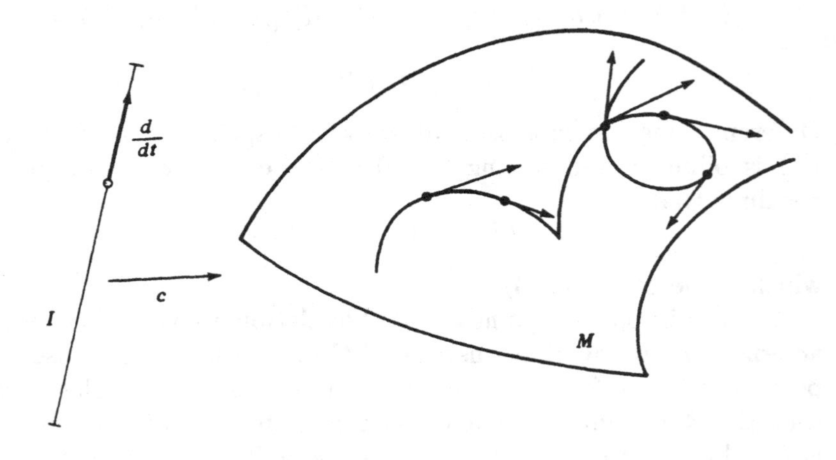

Figure 1

2.9 DEFINITION. *A vector field V along a curve $c: I \to M$* is a differentiable mapping that associates to every $t \in I$ a tangent vector $V(t) \in T_{c(t)}M$. To say that V is *differentiable* means that for any differentiable function f on M, the function $t \to V(t)f$ is a differentiable function on I.

The vector field $dc(\frac{d}{dt})$, denoted by $\frac{dc}{dt}$, is called the *velocity field* (or tangent vector field) of$c$. Observe that a vector field along c cannot necessarily be extended to a vector field on an open set of M.

The restriction of a curve c to a closed interval $[a, b] \subset I$ is called a *segment*. If M is a Riemannian manifold, we define the length of a segment by

$$\ell_a^b(c) = \int_a^b \left\langle \frac{dc}{dt}, \frac{dc}{dt} \right\rangle^{1/2} dt.$$

Let us now prove a theorem on the existence of Riemannian metrics.

2.10 PROPOSITION. *A differentiable manifold M (Hausdorff with countable basis) has a Riemannian metric.*

Proof. Let $\{f_\alpha\}$ be a differentiable partition of unity on M subordinate to a covering $\{V_\alpha\}$ of M by coordinate neighborhoods. This means (See Chap. 0, Sec. 5) that $\{V_\alpha\}$ is a locally finite covering (i.e., any point of M has a neighborhood U such that$U \cap V_\alpha \neq \phi$ at most for a finite number of indices) and $\{f_\alpha\}$ is a family of differentiable functions on M satisfying:

1) $f_\alpha \geq 0$, $f_\alpha = 0$ on the complement of the closed set $\overline{V}_\alpha$.
2) $\sum_\alpha f_\alpha(p) = 1$ for all p on M.

It is clear that we can define a Riemannian metric $\langle\,,\rangle^\alpha$ on each V_α: the metric induced by the system of local coordinates. Let us then set $\langle u, v\rangle_p = \sum_\alpha f_\alpha(p)\langle u, v\rangle_p^\alpha$ for all $p \in M$, $u, v \in T_pM$. It is easy to verify that this construction defines a Riemannian metric on M. □

To conclude this chapter, we are going to show how a Riemannian metric permits us to define a notion of volume on a given oriented manifold M^n.

As usual we need some preliminary facts. Let $p \in M$ and let $\mathbf{x}: U \subset \mathbf{R}^n \to M$ be a parametrization about p which belongs to a family of parametrizations consistent with the orientation of M (we say that such parametrizations are positive). Consider a positive orthonormal basis $\{e_1, \ldots, e_n\}$ of T_pM and write $X_i(p) = \frac{\partial}{\partial x_i}(p)$ in the basis $\{e_i\}$: $X_i(p) = \sum_{ij} a_{ij} e_j$. Then

$$g_{ik}(p) = \langle X_i, X_k \rangle (p) = \sum_{j\ell} a_{ij} a_{k\ell} \langle e_j, e_\ell \rangle = \sum_j a_{ij} a_{kj}.$$

Since the volume $\text{vol}(X_1(p), \ldots, X_n(p))$ of the parallelepiped formed by the vectors $X_1(p), \ldots, X_n(p)$ in T_pM is equal to $\text{vol}(e_1, \ldots, e_n) = 1$ multiplied by the determinant of the matrix (a_{ij}), we obtain

$$\text{vol}(X_1(p), \ldots, X_n(p)) = \det(a_{ij}) = \sqrt{\det(g_{ij})}(p).$$

If $\mathbf{y}: V \subset \mathbf{R}^n \to M$ is another positive parametrization about p, with $Y_i(p) = \frac{\partial}{\partial y_i}(p)$ and $h_{ij}(p) = \langle Y_i, Y_j \rangle (p)$,we obtain

$$\begin{aligned}(4) \qquad \sqrt{\det(g_{ij})}(p) &= \text{vol}(X_1(p), \ldots, X_n(p)) \\ &= J \,\text{vol}(Y_1(p), \ldots, Y_n(p)) = J \sqrt{\det(h_{ij})}(p),\end{aligned}$$

where $J = \det(\frac{\partial x_i}{\partial y_j}) = \det(d\mathbf{y}^{-1} \circ d\mathbf{x})(p) > 0$ is the determinant of the derivative of the change of coordinates.

Now let $R \subset M$ be a region (an open connected subset), whose closure is compact. We suppose that R is contained in a coordinate neighborhood $\mathbf{x}(U)$ with a positive parametrization $\mathbf{x}: U \to M$, and that the boundary of $\mathbf{x}^{-1}(R) \subset U$ has measure zero in $\mathbf{R}^n$ (observe that the notion of measure zero in $\mathbf{R}^n$ is invariant by diffeomorphism). Let us define the *volume* $\text{vol}(R)$ of R by the integral in $\mathbf{R}^n$

$$(5) \qquad \text{vol}(R) = \int_{\mathbf{x}^{-1}(R)} \sqrt{\det(g_{ij})}\, dx_1 \ldots dx_n.$$

The expression above is well-defined. Indeed, if R is contained in another coordinate neighborhood $y(V)$ with a positive parametrization $\mathbf{y}: V \subset \mathbf{R}^n \to M$, we obtain from the change of variable theorem for multiple integrals, (using the same notation as in (4),

$$\int_{\mathbf{x}^{-1}(R)} \sqrt{\det(g_{ij})}dx_1 \dots dx_n$$

$$= \int_{\mathbf{y}^{-1}(R)} \sqrt{\det h_{ij}}dy_1 \dots dy_n = \text{vol}(R),$$

which proves that the definition given by (5) does not depend on the choice of the coordinate system (here the hypothesis of the orientability of M enters by guaranteeing that vol(R) does not change sign).

2.11 Remark. The reader familiar with differential forms will note that equation (4) implies that the integrand in the formula for the volume in expression (5) is a positive differential form of degree n, which is usually called a *volume form* (*or volume element*) ν on M. In order to define the volume of a compact region R, which is not contained in a coordinate neighborhood it is necessary to consider a partition of unity $\{\varphi_i\}$ subordinate to a (finite) covering of R consisting of coordinate neighborhoods $\mathbf{x}(U_i)$ and to take

$$\text{vol}(R) = \sum_i \int_{x_i^{-1}(R)} \varphi_i \nu.$$

It follows immediately that the expression above does not depend on the choice of the partition of unity.

2.12 Remark. It is clear that the existence of a globally defined positive differential form of degree n (volume element) leads to a notion of volume on a differentiable manifold. A Riemannian metric is only one of the ways through which a volume element can be obtained.

EXERCISES

1. Prove that the antipodal mapping $A: S^n \to S^n$ given by $A(p) = -p$ is an isometry of S^n. Use this fact to introduce

a Riemannian metric on the real projective space $P^n(\mathbf{R})$ such that the natural projection $\pi: S^n \to P^n(\mathbf{R})$ is a local isometry.

2. Introduce a Riemannian metric on the torus T^n in such a way that the natural natural projection $\pi: \mathbf{R}^n \to T^n$ given by

$$\pi(x_1, \ldots, x_n) = (e^{ix_1}, \ldots, e^{ix_n}), \quad (x_1, \ldots, x_n) \in \mathbf{R}^n,$$

is a local isometry. Show that with this metric T^n is isometric to the flat torus.

3. Obtain an isometric immersion of the flat torus T^n into $\mathbf{R}^{2n}$.

4. A function $g: \mathbf{R} \to \mathbf{R}$ given by $g(t) = yt + x, \quad t, x, y \in \mathbf{R}, \quad y > 0$, is called a *proper affine function.* The subset of all such functions with respect to the usual composition law forms a Lie group G. As a differentiable manifold G is simply the upper half-plane $\{(x, y) \in \mathbf{R}^2; y > 0\}$ with the differentiable structure induced from$\mathbf{R}^2$. Prove that:
   (a) The left-invariant Riemannian metric of G which at the neutral element $e = (0, 1)$ coincides with the Euclidean metric ($g_{11} = g_{22} = 1$, $g_{12} = 0$) is given by $g_{11} = g_{22} = \frac{1}{y^2}$, $g_{12} = 0$, (this is the metric of the non-euclidean geometry of Lobatchevski).
   (b) Putting $(x, y) = z = x + iy$, $i = \sqrt{-1}$, the transformation $z \to z' = \frac{az+b}{cz+d}$, $a, b, c, d \in \mathbf{R}$, $ad - bc = 1$ is an isometry of G.

   Hint: Observe that the first fundamental form can be written as:

$$ds^2 = \frac{dx^2 + dy^2}{y^2} = -\frac{4dzd\bar{z}}{(z - \bar{z})^2}.$$

5. Prove that the isometries of $S^n \subset \mathbf{R}^{n+1}$, with the induced metric, are the restrictions to S^n of the linear orthogonal maps of $\mathbf{R}^{n+1}$.

6. Show that the relation "M is locally isometric to N" is not a symmetric relation.

†7. Let G be a compact connected Lie group ($\dim G = n$). The object of this exercise is to prove that G has a bi-invariant Riemannian metric. To do this, take the following approach:

(a) Let ω be a differential n-form on G invariant on the left, that is, $L_x^*\omega = \omega$, for all $x \in G$. Prove that ω is right invariant.

Hint: For any $a \in G$, $\mathbf{R}_a^*\omega$ is left invariant. It follows that $\mathbf{R}_a^*\omega = f(a)\omega$. Verify that $f(ab) = f(a)f(b)$, that is, $f: G \to R - \{0\}$ is a (continuous) homomorphism of G into the multiplicative group of real numbers. Since $f(G)$ is a compact connected subgroup, the conclusion $f(G) = 1$ holds. Therefore $\mathbf{R}_a^*\omega = \omega$.

(b) Show that there exists a left invariant differential n-form ω on G.

(c) Let $\langle\, , \rangle$ be a left invariant metric on G. Let ω be a positive differential n-form on G which is invariant on the left, and define a new Riemannian metric $\langle\langle\, , \rangle\rangle$ on G by

$$\langle\langle u, v\rangle\rangle_y = \int_G \langle (dR_x)_y u, (dR_x)_y v\rangle_{yx}\, \omega, \qquad x, y \in G, \quad u, v \in T_y(G).$$

Prove that this new Riemannian metric $\langle\langle\, , \rangle\rangle$ is bi-invariant.

CHAPTER 2

AFFINE CONNECTIONS; RIEMANNIAN CONNECTIONS

1. Introduction

A fundamental event in the development of differential geometry was the introduction, in 1917, of the Levi-Civita parallelism (see Levi-Civita [LC]). For the case of surfaces in $\mathbf{R}^3$, an equivalent idea can be described in the following manner. Let $S \subset \mathbf{R}^3$ be a surface and let $c: I \to S$ be a parametrized curve in S, with $V: I \to \mathbf{R}^3$ a vector field along c tangent to S. The vector $\frac{dV}{dt}(t)$, $t \in I$, does not in general belong to the tangent plane of S, $T_{c(t)}S$. The concept of differentiating a vector field is not therefore an "intrinsic" geometric notion on S. To remedy this state of affairs we consider, instead of the usual derivative $\frac{dV}{dt}(t)$, the orthogonal projection of $\frac{dV}{dt}(t)$ on $T_{c(t)}S$. This orthogonally projected vector we call the covariant derivative and denote it by $\frac{DV}{dt}(t)$. The covariant derivative of V is the derivative of V as seen from the "viewpoint of S".

A basic point is that the covariant derivative depends only on the first fundamental form of S and is therefore a concept which can be considered within Riemannian geometry. In particular, the notion of covariant derivative permits us to take the derivative of the velocity vector of c, which gives the acceleration of the curve c in S. It is possible to show that curves with zero acceleration are precisely the geodesics of S and that the Gaussian curvature of S can be expressed in terms of the notion of the covariant derivative.

We say that a vector field V along c is parallel if $\frac{DV}{dt} \equiv 0$. Conversely, starting from the notion of parallelism it is possible to recover the notion of covariant derivative (Cf. Exercise 2 of this chapter). These notions are then equivalent to each other.

Although nowadays it is preferable to start from the notion of covariant derivative, historically the idea of parallelism came first. For surfaces in $\mathbf{R}^3$, parallelism can be introduced in the following manner. Consider a family of planes tangent to S along the curve c. This family determines a surface E, enveloping these tangent planes, which possesses the property that it will be tangent to S along the curve c and whose Gaussian curvature $K \equiv 0$. (Cf. M. do Carmo [dC 2] pp. 195–197). It is not difficult to show that the parallelism along c, defined through the vanishing of the covariant derivative is the same whether considered relative to S or relative to E. On the other hand, surfaces of zero curvature can be shown to be locally isometric to a plane. Since parallelism is invariant by isometry, we can perform it "euclideanly" in the isometric image of E and then bring it back to S. This was the construction used classically to define parallelism. (M. do Carmo [dC 2] p. 244). It will turn out that it is preferable, technically, to work with the covariant derivative.

The notion of covariant derivative has many important consequences. It makes it clear that the two basic ideas of geodesic and curvature can be defined in more general situations than that of Riemannian manifolds. To this end it suffices that one be able to define a notion of derivation of vector fields with certain properties (which nowadays we call an affine connection, Cf. Definition 2.1 of this chapter). This has stimulated the creation of many different "geometric structures" (on differentiable manifolds) more general than Riemannian geometry. In the same way as metric Euclidean geometry is a particular case of affine geometry and more generally of projective geometry, Riemannian geometry is a particular case of more general geometric structures.

We are not going to enter into the details of these developments. Our interest in affine connections rests in the fact (Cf. Theorem 3.6 of this chapter) that a choice of a Riemannian metric on a manifold M uniquely determines a certain affine connection on M. We are then able, in this fashion, to differentiate vector fields on M.

2. Affine Connections

Let us indicate by $\mathcal{X}(M)$ the set of all vector fields of class C^∞ on M and by $\mathcal{D}(M)$ the ring of real-valued functions of class C^∞ defined on M.

2.1 DEFINITION. An *affine connection* ∇ on a differentiable manifold M is a mapping

$$\nabla: \mathcal{X}(M) \times \mathcal{X}(M) \to \mathcal{X}(M)$$

which is denoted by $(X, Y) \xrightarrow{\nabla} \nabla_X Y$ and which satisfies the following properties :

i) $\nabla_{fX+gY} Z = f\nabla_X Z + g\nabla_Y Z.$
ii) $\nabla_X(Y + Z) = \nabla_X Y + \nabla_X Z.$
iii) $\nabla_X(fY) = f\nabla_X Y + X(f)Y,$

in which $X, Y, Z \in \mathcal{X}(M)$ and $f, g \in \mathcal{D}(M)$.

This definition is not as transparent as that of Riemannian structure. The following proposition, nevertheless, should clarify the situation a little.

2.2 PROPOSITION. *Let M be a differentiable manifold with an affine connection ∇. There exists a unique correspondence which associates to a vector field V along the differentiable curve $c: I \to M$ another vector field $\frac{DV}{dt}$ along c, called the covariant derivative of V along c, such that:*

a) $\frac{D}{dt}(V + W) = \frac{DV}{dt} + \frac{DW}{dt}.$
b) $\frac{D}{dt}(fV) = \frac{df}{dt}V + f\frac{DV}{dt}$, *where W is a vector field along c and f is a differentiable function on I.*

c) *If V is induced by a vector field $Y \in \mathcal{X}(M)$, i.e., $V(t) = Y(c(t))$, then $\frac{DV}{dt} = \nabla_{dc/dt} Y$.*

2.3 REMARK. The last line of (c) makes sense, since $\nabla_X Y(p)$ depends on the value of $X(p)$ and the value Y along a curve, tangent to X at p. In effect, part (iii) of Definition 2.1 allows us to show that the notion of affine connection is actually a local notion (cf. Rem. 5.7 of Chap. 0). Choosing a system of coordinates $(x_1, \ldots, x_n)$ about p and writing

$$X = \sum_i x_i X_i, \qquad Y = \sum_j y_j Xj,$$

where $X_i = \frac{\partial}{\partial x_i}$, we have

$$\nabla_X Y = \sum_i x_i \nabla_{X_i}(\sum_j y_j X_j) = \sum_{ij} x_i y_j \nabla_{X_i} X_j + \sum_{ij} x_i X_i(y_j) X_j.$$

Setting $\nabla_{X_i}X_j = \sum_k \Gamma_{ij}^k X_k$, we conclude that the Γ_{ij}^k are differentiable functions and that

$$\nabla_X Y = \sum_k (\sum_{ij} x_i y_j \Gamma_{ij}^k + X(y_k))X_k,$$

which proves that $\nabla_X Y(p)$ depends on $x_i(p)$, $y_k(p)$ and the derivatives $X(y_k)(p)$ of y_k by X.

2.4 REMARK. The proposition above shows that the choice of an affine connection on M leads to a bona fide (i.e. satisfying (a) and (b)) derivative of vector fields along curves. The notion of connection furnishes, therefore, a manner of differentiating vectors along curves; in particular, it will then be possible to speak of the acceleration of a curve in M.

Proof of Proposition 2.2. Let us suppose initially that there exists a correspondence satisfying (a), (b) and (c). Let $\mathbf{x}: U \subset \mathbf{R}^n \to M$ be a system of coordinates with $c(I) \cap \mathbf{x}(U) \neq \phi$ and let $(x_1(t), x_2(t), \ldots, x_n(t))$ be the local expression of $c(t)$, $t \in I$. Let $X_i = \frac{\partial}{\partial x_i}$. Then we can express the field V locally as $V = \sum_j v^j X_j$, $j = 1, \ldots, n$, where $v^j = v^j(t)$ and $X_j = X_j(c(t))$.

By a) and b), we have

$$\frac{DV}{dt} = \sum_j \frac{dv^j}{dt} X_j + \sum_j v^j \frac{DX_j}{dt}.$$

By c) and (i) of Definition 2.1,

$$\frac{DX_j}{dt} = \nabla_{dc/dt} X_j = \nabla_{(\Sigma \frac{dx_i}{dt} X_i)} X_j$$
$$= \sum_i \frac{dx_i}{dt} \nabla_{X_i} X_j, \quad i, j = 1, \ldots, n.$$

Therefore,

$$\frac{DV}{dt} = \sum_j \frac{dv^j}{dt} X_j + \sum_{i,j} \frac{dx_i}{dt} v^j \nabla_{X_i} X_j. \tag{1}$$

The expression (1) shows us that if there is a correspondence satisfying the conditions of Proposition 2.2, then such a correspondence is unique.

To show existence, define $\frac{DV}{dt}$ in $\mathbf{x}(U)$ by (1). It is easy to verify that (1) possesses the desired properties. If $\mathbf{y}(W)$ is another coordinate neighborhood, with $\mathbf{y}(W) \cap \mathbf{x}(U) \neq \phi$ and we define $\frac{DV}{dt}$ in $\mathbf{y}(W)$ by (1), the definitions agree in $\mathbf{y}(W) \cap \mathbf{x}(U)$, by the uniqueness of $\frac{DV}{dt}$ in $\mathbf{x}(U)$. It follows that the definition can be extended over all of M, and this concludes the proof. □

The concept of parallelism now follows in a natural manner.

2.5 DEFINITION. Let M be a differentiable manifold with an affine connection ∇. A vector field V along a curve $c: I \to M$ is called *parallel* when $\frac{DV}{dt} = 0$, for all $t \in I$.

2.6 PROPOSITION. *Let M be a differentiable manifold with an affine connection ∇. Let $c: I \to M$ be a differentiable curve in M and let V_o be a vector tangent to M at $c(t_o)$, $t_o \in I$ (i.e. $V_o \in T_{c(t_o)}M$). Then there exists a unique parallel vector field V along c, such that $V(t_o) = V_o$, (($V(t)$ is called the parallel transport of $V(t_o)$ along c).*

Proof. Suppose that the theorem was proved for the case in which $c(I)$ is contained in a local coordinate neighborhood. By compactness, for any $t_1 \in I$, the segment $c([t_o, t_1]) \subset M$ can be covered by a finite number of coordinate neighborhoods, in each of which V can be defined, by hypothesis. From uniqueness, the definitions coincide when the intersections are not empty, thus allowing the definition of V along all of $[t_o, t_1]$.

We have only, therefore, to prove the theorem when $c(I)$ is contained in a coordinate neighborhood $\mathbf{x}(U)$ of a system of coordinates $\mathbf{x}: U \subset \mathbf{R}^n \to M$. Let $\mathbf{x}^{-1}(c(t)) = (x_1(t), \ldots, x_n(t))$ be the local expression for $c(t)$ and let $V_o = \sum_j v_o^j X_j$, where $X_j = \frac{\partial}{\partial x_j}(c(t_o))$.

Suppose that there exists a vector field V in $\mathbf{x}(U)$ which is parallel along c with $V(t_o) = V_o$. Then $V = \sum v^j X_j$ satisfies

$$0 = \frac{DV}{dt} = \sum_j \frac{dv^j}{dt} X_j + \sum_{i,j} \frac{dx_i}{dt} v^j \nabla_{X_i} X_j.$$

Putting $\nabla_{X_i} X_j = \sum_k \Gamma_{ij}^k X_k$, and replacing j with k in the first sum, we obtain

$$\frac{DV}{dt} = \sum_k \left\{ \frac{dv^k}{dt} + \sum_{i,j} v^j \frac{dx_i}{dt} \Gamma_{ij}^k \right\} X_k = 0.$$

The system of n differential equations in $v^k(t)$,

$$0 = \frac{dv^k}{dt} + \sum_{i,j} \Gamma^k_{ij} v^j \frac{dx_i}{dt}, \quad k = 1, \ldots, n, \tag{2}$$

possesses a unique solution satisfying the initial conditions $v^k(t_o) = v^k_o$. It then follows that, if V exists, it is unique. Moreover, since the system is linear, any solution is defined for all $t \in I$, which then proves the existence (and uniqueness) of V with the desired properties. □

3. Riemannian Connections

3.1 DEFINITION. Let M be a differentiable manifold with an affine connection ∇ and a Riemannian metric $\langle\, ,\rangle$. A connection is said to be *compatible* with the metric $\langle\, ,\rangle$, when for any smooth curve c and any pair of parallel vector fields P and P' along c, we have $\langle P, P'\rangle = \text{constant}$.

Definition 3.1 is justified by the following proposition which shows that if ∇ is compatible with $\langle\, ,\rangle$, then we are able to differentiate the inner product by the usual "product rule".

3.2 PROPOSITION. *Let M be a Riemannian manifold. A connection ∇ on M is compatible with a metric if and only if for any vector fields V and W along the differentiable curve $c\colon I \to M$ we have*

$$\frac{d}{dt}\langle V, W\rangle = \langle \frac{DV}{dt}, W\rangle + \langle V, \frac{DW}{dt}\rangle, \quad t \in I. \tag{3}$$

Proof. It is obvious that equation (3) implies that ∇ is compatible with $\langle\, ,\rangle$. Therefore, let us prove the converse. Choose an orthonormal basis $\{P_1(t_o), \ldots, P_n(t_o)\}$ of $T_{x(t_o)}(M)$, $t_o \in I$. Using Proposition 2.6, we can extend the vectors $P_i(t_o)$, $i = 1, \ldots, n$, along c by parallel translation. Because ∇ is compatible with the metric, $\{P_1(t), \ldots, P_n(t)\}$ is an orthonormal basis of $T_{c(t)}(M)$, for any $t \in I$. Therefore, we can write

$$V = \sum_i v^i P_i, \qquad W = \sum_i w^i P_i, \qquad i = 1, \ldots, n$$

where v^i and w^i are differentiable functions on I. It follows that

$$\frac{DV}{dt} = \sum_i \frac{dv^i}{dt} P_i, \qquad \frac{DW}{dt} = \sum_i \frac{dw^i}{dt} P_i.$$

Therefore,

$$\langle \frac{DV}{dt}, W \rangle + \langle V, \frac{DW}{dt} \rangle = \sum_i \left\{ \frac{dv^i}{dt} w^i + \frac{dw^i}{dt} v^i \right\}$$

$$= \frac{d}{dt} \left\{ \sum_i v^i w^i \right\} = \frac{d}{dt} \langle V, W \rangle. \quad \square$$

3.3 COROLLARY. *A connection ∇ on a Riemannian manifold M is compatible with the metric if and only if*

(4) $\quad X\langle Y, Z\rangle = \langle \nabla_X Y, Z\rangle + \langle Y, \nabla_X Z\rangle, \qquad X, Y, Z \in \mathcal{X}(M).$

Proof. Suppose that ∇ is compatible with the metric. Let $p \in M$ and let $c\colon I \to M$ be a differentiable curve with $c(t_o) = p$, $t_o \in I$, and with $\frac{dc}{dt}\big|_{t=t_o} = X(p)$. Then

$$X(p)\langle Y, Z\rangle = \frac{d}{dt}\langle Y, Z\rangle\bigg|_{t=t_o} = \langle \nabla_{X(p)} Y, Z\rangle_p + \langle Y, \nabla_{X(p)} Z\rangle_p.$$

Since p is arbitrary, (4) follows. The converse is obvious. $\square$

3.4 DEFINITION. An affine connection ∇ on a smooth manifold M is said to be *symmetric* when

(5) $\quad \nabla_X Y - \nabla_Y X = [X, Y] \quad \text{for all} \quad X, Y \in \mathcal{X}(M).$

3.5 REMARK. In a coordinate system $(U, \mathbf{x})$, the fact that ∇ is symmetric implies that for all $i, j = 1, \ldots, n$,

(5') $\quad \nabla_{X_i} X_j - \nabla_{X_j} X_i = [X_i, X_j] = 0, \quad X_i = \frac{\partial}{\partial x_i},$

which justifies the terminology (observe that (5') is equivalent to the fact that $\Gamma_{ij}^k = \Gamma_{ji}^k$).

We are now able to state the fundamental theorem of this chapter.

3.6 Theorem. *(Levi-Civita). Given a Riemannian manifold M, there exists a unique affine connection ∇ on M satisfying the conditions:*

a) ∇ is symmetric.

b) ∇ is compatible with the Riemannian metric.

Proof. Suppose initially the existence of such a ∇. Then

$$X\langle Y,Z\rangle = \langle\nabla_X Y, Z\rangle + \langle Y, \nabla_X Z\rangle, \tag{6}$$

$$Y\langle Z,X\rangle = \langle\nabla_Y Z, X\rangle + \langle Z, \nabla_Y X\rangle, \tag{7}$$

$$Z\langle X,Y\rangle = \langle\nabla_Z X, Y\rangle + \langle X, \nabla_Z Y\rangle. \tag{8}$$

Adding (6) and (7) and subtracting (8), we have, using the symmetry of ∇, that

$$\begin{aligned} &X\langle Y,Z\rangle + Y\langle Z,X\rangle - Z\langle X,Y\rangle \\ &\qquad = \langle [X,Z],Y\rangle + \langle [Y,Z],X\rangle + \langle [X,Y],Z\rangle + 2\langle Z,\nabla_Y X\rangle. \end{aligned}$$

Therefore

$$\begin{aligned} \langle Z,\nabla_Y X\rangle = \frac{1}{2}\{&X\langle Y,Z\rangle + Y\langle Z,X\rangle - Z\langle X,Y\rangle \\ &-\langle [X,Z],Y\rangle - \langle [Y,Z],X\rangle - \langle [X,Y],Z\rangle\}. \end{aligned} \tag{9}$$

The expression (9) shows that ∇ is uniquely determined from the metric $\langle\, ,\rangle$. Hence, if it exists, it will be unique.

To prove existence, define ∇ by (9). It is easy to verify that ∇ is well-defined and that it satisfies the desired conditions. □

3.7 REMARK. The connection given by the theorem will be referred to, from now on, as the *Levi-Civita* (or *Riemannian*) *connection* on M.

Let us conclude this chapter by writing part of what was shown above in a coordinate system $(U, \mathbf{x})$. It is customary to call the functions Γ^k_{ij} defined on U by $\nabla_{X_i} X_j = \sum_k \Gamma^k_{ij} X_k$, the *coefficients of the connection* ∇ on U or the *Christoffel symbols* of the connection. From (9) it follows that

$$\sum_\ell \Gamma^\ell_{ij} g_{\ell k} = \frac{1}{2}\left\{\frac{\partial}{\partial x_i} g_{jk} + \frac{\partial}{\partial x_j} g_{ki} - \frac{\partial}{\partial x_k} g_{ij}\right\},$$

where $g_{ij} = \langle X_i, X_j \rangle$.

Since the matrix (g_{km}) admits an inverse (g^{km}), we obtain that

$$\Gamma^m_{ij} = \frac{1}{2}\sum_k \left\{ \frac{\partial}{\partial x_i} g_{jk} + \frac{\partial}{\partial x_j} g_{ki} - \frac{\partial}{\partial x_k} g_{ij} \right\} g^{km}. \tag{10}$$

The equation (10) is a classical expression for the Christoffel symbols of the Riemannian connection in terms of the g_{ij} (given by the metric).

Observe that for the Euclidean space $\mathbf{R}^n$, we have $\Gamma^k_{ij} = 0$.

In terms of the Christoffel symbols, the covariant derivative has the classical expression

$$\frac{DV}{dt} = \sum_k \left\{ \frac{dv^k}{dt} + \sum_{i,j} \Gamma^k_{ij} v^j \frac{dx_i}{dt} \right\} X_k$$

which follows from (1). Observe that $\frac{DV}{dt}$ differs from the usual derivative in Euclidean space by terms which involve the Christoffel symbols. Therefore, in Euclidean spaces the covariant derivative coincides with the usual derivative.

EXERCISES

1. Let M be a Riemannian manifold. Consider the mapping

$$P = P_{c,t_o,t} \colon T_{c(t_o)}M \to T_{c(t)}M$$

defined by: $P_{c,t_o,t}(v)$, $v \in T_{c(t_o)}M$, is the vector obtained by parallel transporting the vector v along the curve c. Show that P is an isometry and that, if M is oriented, P preserves the orientation.

2. Let X and Y be differentiable vector fields on a Riemannian manifold M. Let $p \in M$ and let $c\colon I \to M$ be an integral curve

of X through p, i.e. $c(t_o) = p$ and $\frac{dc}{dt} = X(c(t))$. Prove that the Riemannian connection of M is

$$(\nabla_X Y)(p) = \frac{d}{dt}(P^{-1}_{c,t_o,t}(Y(c(t)))\Big|_{t=t_o},$$

where $P_{c,t_o,t}\colon T_{c(t_o)}M \to T_{c(t)}M$ is the parallel transport along c, from t_o to t (this shows how the connection can be reobtained from the concept of parallelism).

3. Let $f\colon M^n \to \overline{M}^{n+k}$ be an immersion of a differentiable manifold M into a Riemannian manifold $\overline{M}$. Assume that M has the Riemannian metric induced by f (cf. Example 2.5 of Chap. 1). Let $p \in M$ and let $U \subset M$ be a neighborhood of p such that $f(U) \subset \overline{M}$ is a submanifold of $\overline{M}$. Further, suppose that X, Y are differentiable vector fields on $f(U)$ which extend to differentiable vector fields $\overline{X}, \overline{Y}$ on an open set of $\overline{M}$. Define $(\nabla_X Y)(p) =$ tangential component of $\overline{\nabla}_{\overline{X}}\overline{Y}(p)$, where $\overline{\nabla}$ is the Riemannian connection of $\overline{M}$. Prove that ∇ is the Riemannian connection of M.

4. Let $M^2 \subset \mathbf{R}^3$ be a surface in $\mathbf{R}^3$ with the induced Riemannian metric. Let $c\colon I \to M$ be a differentiable curve on M and let V be vector field tangent to M along c; V can be thought of as a smooth function $V\colon I \to \mathbf{R}^3$, with $V(t) \in T_{c(t)}M$.
   a) Show that V is parallel if and only if $\frac{dV}{dt}$ is perpendicular to $T_{c(t)}M \subset \mathbf{R}^3$ where $\frac{dV}{dt}$ is the usual derivative of $V\colon I \to \mathbf{R}^3$.
   b) If $S^2 \subset \mathbf{R}^3$ is the unit sphere of $\mathbf{R}^3$, show that the velocity field along great circles, parametrized by arc length, is a parallel field. A similar argument holds for $S^n \subset \mathbf{R}^{n+1}$.

5. In Euclidean space, the parallel transport of a vector between two points does not depend on the curve joining the two points. Show, by example, that this fact may not be true on an arbitrary Riemannian manifold.

6. Let M be a Riemannian manifold and let p be a point of M. Consider a constant curve $f\colon I \to M$ given by $f(t) = p$, for all $t \in I$. Let V be a vector field along f (that is, V is a differentiable mapping of I into T_pM). Show that $\frac{DV}{dt} = \frac{dV}{dt}$,

that is to say, the covariant derivative coincides with the usual derivative of $V: I \to T_pM$.

7. Let $S^2 \subset \mathbf{R}^3$ be the unit sphere, c an arbitrary parallel of latitude on S^2 and V_o a tangent vector to S^2 at a point of c. Describe geometrically the parallel transport of V_o along c.
Hint: Consider the cone C tangent to S^2 along c and show that the parallel transport of V_o along c is the same, whether taken relative to S^2 or to C.

8. Consider the upper half-plane

$$\mathbf{R}^2_+ = \{(x, y) \in \mathbf{R}^2; y > 0\}$$

with the metric given by $g_{11} = g_{22} = \frac{1}{y^2}$, $g_{12} = 0$ (metric of Lobatchevski's non-euclidean geometry).
  a) Show that the Christoffel symbols of the Riemannian connection are: $\Gamma^1_{11} = \Gamma^2_{12} = \Gamma^1_{22} = 0$, $\Gamma^2_{11} = \frac{1}{y}$, $\Gamma^1_{12} = \Gamma^2_{22} = -\frac{1}{y}$.
  b) Let $v_o = (0, 1)$ be a tangent vector at point $(0, 1)$ of $\mathbf{R}^{2+}$ (v_o is a unit vector on the y-axis with origin at $(0, 1)$). Let $v(t)$ be the parallel transport of v_o along the curve $x = t$, $y = 1$. Show that $v(t)$ makes an angle t with the direction of the y-axis, measured in the clockwise sense.

Hint: The field $v(t) = (a(t), b(t))$ satisfies the system (2) which defines a parallel field and which, in this case, simplifies to

$$\begin{cases} \frac{da}{dt} + \Gamma^1_{12} b = 0, \\ \frac{db}{dt} + \Gamma^2_{11} a = 0. \end{cases}$$

Taking $a = \cos\theta(t)$, $b = \sin\theta(t)$ and noting that along the given curve we have $y = 1$, we obtain from the equations above that $\frac{d\theta}{dt} = -1$. Since $v(0) = v_o$, this implies that $\theta(t) = \pi/2 - t$.

9. (*Pseudo-Riemannian Metrics*). A *pseudo-Riemannian* metric on a smooth manifold M is a choice, at every point $p \in M$, of a non-degenerate symmetric bilinear form $\langle\, ,\, \rangle$ (not necessarily positive definite) on T_pM which varies differentiably with p. Except for the fact that $\langle\, ,\, \rangle$ need not be positive definite, all of the definitions that have been presented up to now make sense for a pseudo-Riemannian metric. For example, an affine

connection on M compatible with a pseudo-Riemannian metric on M satisfies equation (4) ; if, in addition, (5) holds, the affine connection is said to be *symmetric.*

a) Show that the theorem of Levi-Civita extends to pseudo-Riemannian metrics. The connection so obtained is called the ***pseudo-Riemannian connection.***

b) Introduce a pseudo-Riemannian metric on $\mathbf{R}^{n+1}$ by using the quadratic form:

$$Q(x_o, \ldots, x_n) = -(x_o)^2 + (x_1)^2 + \ldots + (x_n)^2, \quad (x_o, \ldots, x_n) \in \mathbf{R}^{n+1}.$$

Show that the parallel transport corresponding to the Levi-Civita connection of this metric coincides with the usual parallel transport of $\mathbf{R}^{n+1}$ (this pseudo-Riemannian metric is called the *Lorentz metric*; for $n = 3$, it appears naturally in relativity).

CHAPTER 3

GEODESICS; CONVEX NEIGHBORHOODS

1. Introduction

After fixing the basic terminology, we pass to the study of two fundamental concepts of Riemannian geometry, namely, geodesics and curvature. This chapter introduces the notion of a geodesic as a curve with zero acceleration. In the next chapter, we initiate the study of curvature.

One of the objectives of the present chapter is to show that a geodesic minimizes arc length for points "sufficiently close" (in a sense to be made precise); in addition, if a curve minimizes arc length between any two of its points, it is a geodesic. To prove these facts we need various concepts and theorems which will be useful later.

In Section 2 we introduce the tangent bundle TM of a differentiable manifold M which allows us to reduce the local study of geodesics on M to the study of the trajectories of a vector field (the geodesic field) on TM. In Section 3, we introduce the exponential map of an open set in TM to M which is simply a way of "collecting" all of the geodesics of M into a unique differentiable mapping. This notation is extremely useful, and, permits us, for example, to apply the inverse function theorem to show that any point of M posseses a neighborhood W such that any two points of W can be joined by a unique geodesic which minimizes arc length (see Theorem 3.7).

The concept of a geodesic, as a curve that minimizes the distance between two nearby points, is rather old. For surfaces in $\mathbf{R}^3$, the geodesics can be characterized as those curves $c(s)$ (where s is arc length) for which the acceleration $c''(s)$ in $\mathbf{R}^3$ is perpendicular to the surface (therefore, the acceleration of c "from the viewpoint" of the surface is zero). Such a characterization was apparently known, at least for convex surfaces, in 1697 by Johann Bernoulli, and the equations of geodesics for surfaces of the form $f(x, y, z) = 0$ was considered by Euler in 1732. Nevertheless, it was only with the work of Gauss [Ga] in 1827 that the relationship between the geodesics and the curvature of a surface was established (Cf. Introduction to Chap. 1). This relationship is fundamental and will appear in various forms throughout this book.

2. The geodesic flow

In what follows, M will be a Riemannian manifold, together with its Riemannian connection.

2.1 DEFINITION. A parametrized curve $\gamma: I \to M$ is a *geodesic at* $t_o \in I$ if $\frac{D}{dt}(\frac{d\gamma}{dt}) = 0$ at the point t_o; if γ is a geodesic at t, for all $t \in I$, we say that γ is a *geodesic*. If $[a, b] \subset I$ and $\gamma: I \to M$ is a geodesic, the restriction of γ to $[a, b]$ is called a *geodesic segment joining* $\gamma(a)$ to $\gamma(b)$.

At times, by abuse of language, we refer to the image $\gamma(I)$, of a geodesic γ, as a geodesic.

If $\gamma: I \to M$ is a geodesic, then

$$\frac{d}{dt}\langle\frac{d\gamma}{dt}, \frac{d\gamma}{dt}\rangle = 2\langle\frac{D}{dt}\frac{d\gamma}{dt}, \frac{d\gamma}{dt}\rangle = 0,$$

that is, the length of the tangent vector $\frac{d\gamma}{dt}$ is constant. We assume, from now on, that $\left|\frac{d\gamma}{dt}\right| = c \neq 0$, that is, we exclude the geodesics which reduce to points. The arc length s of γ, starting from a fixed origin, say $t = t_o$, is then given by

$$s(t) = \int_{t_o}^{t} \left|\frac{d\gamma}{dt}\right| dt = c(t - t_o).$$

Therefore, the parameter of a geodesic is proportional to arc length. When the parameter is actually arc length, that is, $c = 1$, we say that the geodesic γ is *normalized*.

Now we are going to determine the local equations satisfied by a geodesic γ in a system of coordinates $(U, \mathbf{x})$ about $\gamma(t_o)$. In U, a curve γ

$$\gamma(t) = (x_1(t), \ldots, x_n(t)).$$

will be a geodesic if and only if

$$0 = \frac{D}{dt}\left(\frac{d\gamma}{dt}\right) = \sum_k \left(\frac{d^2 x_k}{dt^2} + \sum_{i,j} \Gamma^k_{ij} \frac{dx_i}{dt}\frac{dx_j}{dt}\right)\frac{\partial}{\partial x^k}.$$

Hence the second order system

$$\frac{d^2 x_k}{dt^2} + \sum_{i,j} \Gamma^k_{ij} \frac{dx_i}{dt}\frac{dx_j}{dt} = 0, \quad k = 1, \ldots, n, \tag{1}$$

yields the desired equations.

To study the system (1), it is convenient to consider the tangent bundle TM, which will also be useful in future situations.

TM is the set of pairs (q, v), $q \in M$, $v \in T_qM$. If $(U, \mathbf{x})$ is a system of coordinates on M, then any vector in T_qM, $q \in \mathbf{x}(U)$, can be written as $\sum_{i=1}^n y_i \frac{\partial}{\partial x_i}$. Taking $(x_1, \ldots, x_n, y_1, \ldots, y_n)$ as coordinates of (q, v) in TU, it is easy to show that we obtain a differentiable structure for TM (Cf. Example 4.1 of Chap. 0).

Observe that $TU = U \times \mathbf{R}^n$, that is, the tangent bundle is locally a product. In addition, the canonical projection $\pi: TM \to M$ given by $\pi(q, v) = q$ is differentiable.

Any differentiable curve $t \to \gamma(t)$ in M determines a curve $t \to (\gamma(t), \frac{d\gamma}{dt}(t))$ in TM. If γ is a geodesic then, on TU, the curve

$$t \to \left(x_1(t), \ldots, x_n(t), \frac{dx_1(t)}{dt}, \ldots, \frac{dx_n(t)}{dt}\right)$$

satisfies the system

$$\left\{ \begin{array}{ll} \frac{dx_k}{dt} = y_k & \\ \frac{dy_k}{dt} = -\sum_{i,j} \Gamma^k_{ij} y_i y_j & k = 1, \ldots, n \end{array} \right. \tag{1'}$$

in terms of coordinates $(x_1, \ldots, x_n, y_1, \ldots, y_n)$ on TU. Therefore the second order system (1) on U is equivalent to the first order system (1′) on TU.

Let us recall the following result from differential equations.

2.2 Theorem. *If X is a C^∞ vector field on the open set V in the manifold M and $p \in V$ then there exist an open set $V_o \subset V$, $p \in V_o$, a number $\delta > 0$, and a C^∞ mapping $\varphi: (-\delta, \delta) \times V_o \to V$ such that the curve $t \to \varphi(t, q)$, $t \in (-\delta, \delta)$, is the unique trajectory of X which at the instant $t = 0$ passes through the point q, for every $q \in V_o$.*

The mapping $\varphi_t: V_o \to V$ given by $\varphi_t(q) = \varphi(t, q)$ is called the *flow* of X on V.

2.3 LEMMA. *There exists a unique vector field G on TM whose trajectories are of the form $t \to (\gamma(t), \gamma'(t))$, where γ is a geodesic on M.*

Proof. We shall first prove the uniqueness of G, supposing its existence. Consider a system of coordinates $(U, \mathbf{x})$ on M. From the hypothesis, the trajectories of G on TU are given by $t \to (\gamma(t), \gamma'(t))$ where γ is a geodesic. It follows that $t \to (\gamma(t), \gamma'(t))$ is a solution of the system of differential equations (1′). From the uniqueness of the trajectories of such a system, we conclude that if G exists, then it is unique.

To prove the existence of G, define it locally by the system (1′). Using the uniqueness, we conclude that G is well-defined on TM. □

2.4 DEFINITION. The vector field G defined above is called the *geodesic field* on TM and its flow is called the *geodesic flow* on TM.

Applying Theorem 2.2 to the geodesic field G at the point $(p, 0) \in TM$, we obtain the following fact:

For each $p \in M$ there exist an open set $\mathcal{U}$ in TU, where $(U, \mathbf{x})$ is a system of coordinates at p and $(p, 0) \in \mathcal{U}$, a number $\delta > 0$ and a C^∞ mapping, $\varphi: (-\delta, \delta) \times \mathcal{U} \to TU$, such that $t \to \varphi(t, q, v)$ is the unique trajectory of G which satisfies the initial condition $\varphi(0, q, v) = (q, v)$, for each $(q, v) \in \mathcal{U}$.

It is possible to choose $\mathcal{U}$ in the form

$$\mathcal{U} = \{(q, v) \in TU;\ q \in V \text{ and } v \in T_qM \text{ with } |v| < \varepsilon_1\},$$

where $V \subset U$ is a neighborhood of $p \in M$. Putting $\gamma = \pi \circ \varphi$, where $\pi: TM \to M$ is the canonical projection, we can describe the previous result in the following way.

2.5 PROPOSITION. *Given $p \in M$, there exist an open set $V \subset M$, $p \in V$, numbers $\delta > 0$ and $\varepsilon_1 > 0$ and a C^∞ mapping*

$$\gamma\colon (-\delta,\delta) \times \mathcal{U} \to M, \quad \mathcal{U} = \{(q,v); q \in V, v \in T_qM, |v| < \varepsilon_1\},$$

such that the curve $t \to \gamma(t,q,v)$, $t \in (-\delta,\delta)$, is the unique geodesic of M which, at the instant $t = 0$, passes through q with velocity v, for each $q \in V$ and for each $v \in T_qM$ with $|v| < \varepsilon_1$.

Proposition 2.5 asserts that if $|v| < \varepsilon_1$, the geodesic $\gamma(t,q,v)$ exists in an interval $(-\delta,\delta)$ and is unique. Actually, it is possible to increase the velocity of a geodesic by decreasing its interval of definition, or vice-versa. This follows from the following lemma of homogeneity.

2.6 LEMMA. *(Homogeneity of a geodesic). If the geodesic $\gamma(t,q,v)$ is defined on the interval $(-\delta,\delta)$, then the geodesic $\gamma(t,q,av)$, $a \in \mathbf{R}$, $a > 0$, is defined on the interval $(-\frac{\delta}{a},\frac{\delta}{a})$ and*

$$\gamma(t,q,av) = \gamma(at,q,v).$$

Proof. Let $h\colon (-\frac{\delta}{a},\frac{\delta}{a}) \to M$ be a curve given by $h(t) = \gamma(at,q,v)$. Then $h(0) = q$ and $\frac{dh}{dt}(0) = av$. In addition, since $h'(t) = a\gamma'(at,q,v)$,

$$\frac{D}{dt}\left(\frac{dh}{dt}\right) = \nabla_{h'(t)}h'(t) = a^2\nabla_{\gamma'(at,q,v)}\gamma'(at,q,v) = 0,$$

where, for the first equality, we extend $h'(t)$ to a neighborhood of $h(t)$ in M. Therefore, h is a geodesic passing through q with velocity av at the instant $t = 0$. By uniqueness,

$$h(t) = \gamma(at,q,v) = \gamma(t,q,av). \quad \Box$$

Proposition 2.5, together with the lemma of homogeneity, permits us to make the interval of definition of a geodesic uniformly large in a neighborhood of p. More precisely, we have the following fact.

2.7 PROPOSITION. *Given $p \in M$, there exist a neighborhood V of p in M, a number $\varepsilon > 0$ and a C^∞ mapping $\gamma\colon (-2,2) \times \mathcal{U} \to M$, $\mathcal{U} = \{(q,w) \in TM; q \in V, w \in T_qM, |w| < \varepsilon\}$ such that $t \to \gamma(t,q,w)$, $t \in (-2,2)$, is the unique geodesic of M which, at the instant $t = 0$, passes through q with velocity w, for every $q \in V$ and for every $w \in T_qM$, with $|w| < \varepsilon$.*

Proof. The geodesic $\gamma(t,q,v)$ of Proposition 2.5 is defined for $|t|<\delta$ and for $|v|<\varepsilon_1$. From the lemma of homogeneity, $\gamma(t,q,\frac{\delta v}{2})$ is defined for $|t|<2$. Taking $\varepsilon<\frac{\delta\varepsilon_1}{2}$, we obtain that the geodesic $\gamma(t,q,w)$ is defined for $|t|<2$ and $|w|<\varepsilon$. □

2.8 REMARK. By an analogous argument, we can make the velocity of a geodesic uniformly large in a neighborhood of p.

Proposition 2.7 permits us to introduce the concept of the exponential map in the following manner. Let $p\in M$ and let $\mathcal{U}\subset TM$ be an open set given by Proposition 2.7. Then the map $\exp\colon\mathcal{U}\to M$ given by

$$\exp(q,v)=\gamma(1,q,v)=\gamma(|v|,q,\frac{v}{|v|}),\quad (q,v)\in\mathcal{U},$$

is called the *exponential map* on $\mathcal{U}$.

It is clear that exp is differentiable. In most of the applications, we shall utilize the restriction of exp to an open subset of the tangent space T_qM, that is, we define

$$\exp_q\colon B_\varepsilon(0)\subset T_qM\to M$$

by $\exp_q(v)=\exp(q,v)$. Here, and in what follows, we denote by $B_\varepsilon(0)$ an open ball with center at the origin 0 of T_qM and of radius ε. It is easy to verify that $\exp_q$ is differentiable and that $\exp_q(0)=q$.

Geometrically, $\exp_q(v)$ is a point of M obtained by going out the length equal to $|v|$, starting from q, along a geodesic which passes through q with velocity equal to $\frac{v}{|v|}$.

2.9 PROPOSITION. *Given $q\in M$, there exists an $\varepsilon>0$ such that $\exp_q\colon B_\varepsilon(0)\subset T_qM\to M$ is a diffeomorphism of $B_\varepsilon(0)$ onto an open subset of M.*

Proof. Let us calculate $d(\exp_q)_o$:

$$\begin{aligned}d(\exp_q)_o(v)&=\frac{d}{dt}(\exp_q(tv))\Big|_{t=0}=\frac{d}{dt}(\gamma(1,q,tv))\Big|_{t=0}\\&=\frac{d}{dt}(\gamma(t,q,v))\Big|_{t=0}=v.\end{aligned}$$

Hence $d(\exp_q)_o$ is the identity of T_qM, and it follows from the inverse function theorem that $\exp_q$ is a local diffeomorphism on a neighborhood of 0. □

2.10 EXAMPLE. Let $M = \mathbf{R}^n$. Since the covariant derivative coincides with the usual derivative, the geodesics are straight lines parametrized proportionally to arc length. The exponential is clearly the identity (with the usual identification of the tangent space of $\mathbf{R}^n$ at p with $\mathbf{R}^n$).

2.11 EXAMPLE. Let $M = S^n \subset \mathbf{R}^{n+1}$ be the unit sphere of dimension n. As we saw in Exercise 4(b) of Chapter 2, the great circles of S^n, parametrized by arc length, are geodesics. We are going to show that all of the geodesics of S^n are great circles parametrized proportionally to arc length. Indeed, given $p \in S^n$ and a unit vector $v \in T_pS^n$, the intersection with S^n of the plane that contains the origin of $\mathbf{R}^{n+1}$, the point p, and the vector v, is a great circle that can be parametrized as the geodesic through p with velocity v. From the uniqueness of Proposition 2.5, the statement follows.

Given a point $(p, v) \in TM$, the point $\exp_p v \in M$ is obtained by running along the geodesic $\gamma(t, p, \frac{v}{|v|})$ a length equal to $|v|$, starting from p. In the case at hand, it is clear that $\exp_p$ is defined over the entire tangent space, and can be described in the following way: $\exp_p$ transforms $B_\pi(0)$ injectively into $S^n - \{q\}$, where q is the antipodal point to p; the boundary of $B_\pi(0)$ is transformed to q; the open annulus $B_{2\pi}(0) - \overline{B_\pi(0)}$ is transformed injectively onto $S^n - \{p, q\}$; the boundary of $B_{2\pi}(0)$ collapses to p, etc. (Fig. 1).

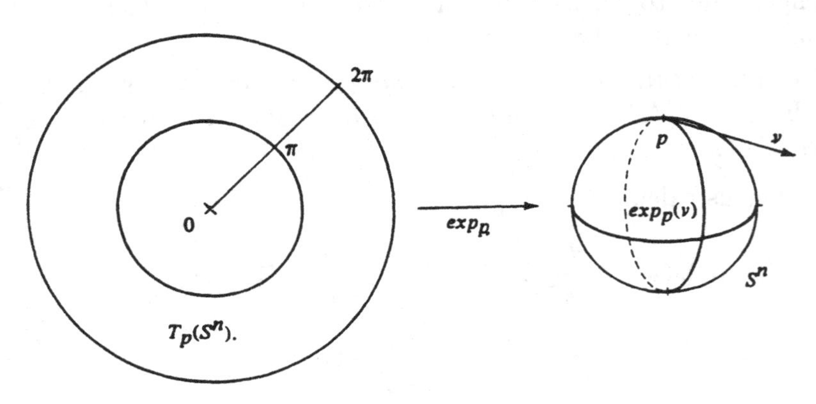

Figure 1

Observe that if we consider the Riemannian manifold $S^n - \{q\}$, $\exp_p$

will be defined only on $B_\pi(0) \subset T_p(S^n - \{q\})$.

3. Minimizing properties of geodesics

We now want to study certain minimizing properties of geodesics. For this it is necessary to consider some preliminary definitions and lemmas.

3.1 DEFINITION. A *piecewise differentiable curve* is a continuous mapping $c\colon [a,b] \to M$ of a closed interval $[a,b] \subset \mathbf{R}$ into M satisfying the following condition: there exists a partition $a = t_o < t_1 < \ldots < t_{k-1} < t_k = b$ of $[a,b]$ such that the restrictions $c\big|_{[t_i,t_{i+1}]}$, $i = 0,\ldots,k-1$, are differentiable. We say that c joins the points $c(a)$ and $c(b)$. $c(t_i)$ is called a *vertex* of c, and the angle formed by $\lim_{t\to t_i^+} c'(t)$ with $\lim_{t\to t_i^-} c'(t)$ is called the *vertex angle at* $c(t_i)$; *here* $\lim_{t\to t_i^+}$ ($\lim_{t\to t_i^-}$) *signifies that* t *approaches* t_i *through values above (below) that of* t_i.

The idea of parallel transport can easily be extended to piecewise differentiable curves: given $V_o \in T_{c(t)}M$, $t \in [t_i, t_{i+1}]$, extend V_o obtaining a parallel field $V(t)$, $t \in [t_i, t_{i+1}]$; taking $V(t_i)$ and $V(t_{i+1})$ as new initial values, we are able to extend $V(t)$ in a similar way over the interval $[t_{i-1}, t_{i+2}]$, and so on.

3.2 DEFINITION. A segment of the geodesic $\gamma\colon [a,b] \to M$ is called *minimizing* if $\ell(\gamma) \le \ell(c)$, where $\ell(\ )$ denotes the length of a curve and c is an arbitrary piecewise differentiable curve joining $\gamma(a)$ to $\gamma(b)$.

In the proof of Gauss' lemma, appearing in a moment, we use the following terminology.

3.3 DEFINITION. Let A be a connected set in $\mathbf{R}^2, U \subset A \subset \overline{U}$, U open, such that the boundary ∂A of A is a piecewise differentiable curve with vertex angles different from π. A *parametrized surface* in M is a differentiable mapping $s\colon A \subset \mathbf{R}^2 \to M$. (Observe that to say that s is differentiable on A means that there exists an open set $U \supset A$ to which s can be extended differentiably. The condition on the vertex angles of A is necessary to ensure that the differential of s does not depend on the given extension.)

A *vector field V along s* is a mapping which associates to each $q \in A$ a vector $V(q) \in T_{s(q)}M$, and which is differentiable in the following sense: if f is a differentiable function on M,then the mapping $q \to V(q)f$ is differentiable.

Let (u, v) be cartesian coordinates on $\mathbf{R}^2$. For v_o fixed, the mapping $u \to s(u, v_o)$, where u belongs to a connected component of $A \cap \{v = v_o\}$, is a curve in M, and $ds(\frac{\partial}{\partial u})$, which we indicate by $\frac{\partial s}{\partial u}$, is a vector field along this curve. This defines $\frac{\partial s}{\partial u}$ for all $(u, v) \in A$ and $\frac{\partial s}{\partial u}$ is a vector field along s. The vector field $\frac{\partial s}{\partial v}$ is defined analogously.

If V is a vector field along $s\colon A \to M$, let us define the *covariant derivative* $\frac{DV}{\partial u}$ and $\frac{DV}{\partial v}$ in the following way. $\frac{DV}{\partial u}(u, v_o)$ is the covariant derivative along the curve $u \to s(u, v_o)$ of the restriction of V to this curve. This defines $\frac{DV}{\partial u}(u, v)$ for all $(u, v) \in A$. $\frac{DV}{\partial v}$ is defined analogously.

3.4 LEMMA. *(symmetry). If M is a differentiable manifold with a symmetric connection and $s\colon A \to M$ is a parametrized surface then:*

$$\frac{D}{\partial v}\frac{\partial s}{\partial u} = \frac{D}{\partial u}\frac{\partial s}{\partial v}.$$

Proof. Let $\mathbf{x}\colon V \subset \mathbf{R}^n \to M$ be a system of coordinates in a neighborhood of a point of $s(A)$. We can write

$$\mathbf{x}^{-1} \circ s(u, v) = (x^1(u, v), \ldots, x^n(u, v)).$$

Therefore

$$\begin{aligned}
\frac{D}{\partial v}\left(\frac{\partial s}{\partial u}\right) &= \frac{D}{\partial v}\left(\sum_i \frac{\partial x^i}{\partial u}\frac{\partial}{\partial x^i}\right) \\
&= \sum_i \frac{\partial^2 x^i}{\partial v \partial u}\frac{\partial}{\partial x^i} + \sum_i \frac{\partial x^i}{\partial u}\nabla_{\Sigma_j (\partial x^j/\partial v)\partial/\partial x^j}\frac{\partial}{\partial x^i} \\
&= \sum_i \frac{\partial^2 x^i}{\partial v \partial u}\frac{\partial}{\partial x^i} + \sum_{i,j} \frac{\partial x^i}{\partial u}\frac{\partial x^j}{\partial v}\nabla_{\partial/\partial x^j}\frac{\partial}{\partial x^i}.
\end{aligned}$$

From the symmetry of the connection, $\nabla_{\partial/\partial x^j}\frac{\partial}{\partial x^i} = \nabla_{\partial/\partial x^i}\frac{\partial}{\partial x^j}$. Hence calculating $\frac{D}{\partial u}(\frac{\partial s}{\partial v})$ we obtain the same expression as above, which proves the lemma. □

In what follows we identify the tangent space to T_pM at $v \in T_pM$ with T_pM itself, and write $T_pM \approx T_v(T_pM)$.

3.5 LEMMA. *(Gauss). Let $p \in M$ and let $v \in T_pM$ such that $\exp_p v$ is defined. Let $w \in T_pM \approx T_v(T_pM)$. Then*

$$\langle (d\exp_p)_v(v), (d\exp_p)_v(w)\rangle = \langle v, w\rangle. \tag{2}$$

Proof. Let $w = w_T + w_N$, where w_T is parallel to v and w_N is normal to v. Since $d\exp_p$ is linear and, by the definition of $\exp_p$,

$$\langle (d\exp_p)_v(v), (d\exp_p)_v(w_T)\rangle = \langle v, w_T\rangle,$$

it suffices to prove (2) for $w = w_N$. It is clear that we can assume $w_N \neq 0$.

Since $\exp_p v$ is defined, there exists $\varepsilon > 0$ such that $\exp_p u$ is defined for

$$u = tv(s), \quad 0 \leq t \leq 1, \quad -\varepsilon < s < \varepsilon,$$

where $v(s)$ is a curve in T_pM with $v(0) = v$, $v'(0) = w_N$ and $|v(s)| =$const. We can, therefore, consider the parametrized surface

$$f: A \to M, \qquad A = \{(t, s); 0 \leq t \leq 1, -\varepsilon < s < \varepsilon\}$$

given by

$$f(t, s) = \exp_p tv(s).$$

Observe that the curves $t \to f(t, s_o)$ are geodesics (see Fig. 2).

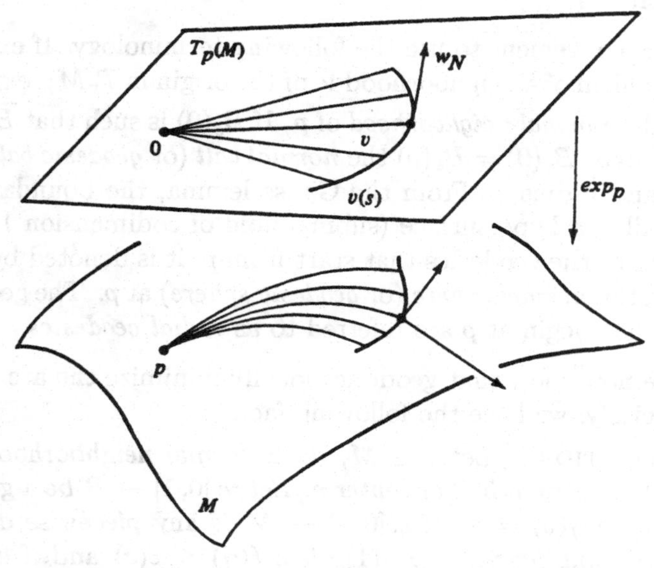

Figure 2

To prove (2) for $w = w_N$, observe first that

$$(3) \qquad \langle \frac{\partial f}{\partial s}, \frac{\partial f}{\partial t} \rangle (1,0) = \langle (d\exp_p)_v(w_N), (d\exp_p)_v(v) \rangle.$$

In addition, for all (t, s), we have

$$\frac{\partial}{\partial t}\langle \frac{\partial f}{\partial s}, \frac{\partial f}{\partial t} \rangle = \langle \frac{D}{\partial t}\frac{\partial f}{\partial s}, \frac{\partial f}{\partial t} \rangle + \langle \frac{\partial f}{\partial s}, \frac{D}{\partial t}\frac{\partial f}{\partial t} \rangle.$$

The last term of the expression above is zero, since $\frac{\partial f}{\partial t}$ is the tangent vector of a geodesic. From the symmetry of the connection, the first term of the sum is transformed into

$$\langle \frac{D}{\partial t}\frac{\partial f}{\partial s}, \frac{\partial f}{\partial t} \rangle = \langle \frac{D}{\partial s}\frac{\partial f}{\partial t}, \frac{\partial f}{\partial t} \rangle = \frac{1}{2}\frac{\partial}{\partial s}\langle \frac{\partial f}{\partial t}, \frac{\partial f}{\partial t} \rangle = 0.$$

It follows that $\langle \frac{\partial f}{\partial s}, \frac{\partial f}{\partial t} \rangle$ is independent of t. Since

$$\lim_{t \to 0} \frac{\partial f}{\partial s}(t, 0) = \lim_{t \to 0} (d\exp_p)_{tv} t w_N = 0,$$

we conclude that $\langle \frac{\partial f}{\partial s}, \frac{\partial f}{\partial t} \rangle (1,0) = 0$, which together with (3) proves the lemma. □

It is convenient to use the following terminology. If $\exp_p$ is a diffeomorphism of a neighborhood V of the origin in T_pM, $\exp_p V = U$ is called a *normal neighborhood* of p. If $B_\varepsilon(0)$ is such that $\overline{B_\varepsilon(0)} \subset V$, we call $\exp_p B_\varepsilon(0) = B_\varepsilon(p)$ the *normal ball* (or *geodesic ball*) with center p and radius ε. From the Gauss lemma, the boundary of a normal ball is a hypersurface (submanifold of codimension 1) in M orthogonal to the geodesics that start from p; it is denoted by $S_\varepsilon(p)$ and called the *normal sphere* (or *geodesic sphere*) at p. The geodesics in $B_\varepsilon(p)$ that begin at p are referred to as *radial geodesics*.

We now show that geodesics locally minimize the arc length. More precisely, we have the following fact.

3.6 PROPOSITION. *Let $p \in M$, U a normal neighborhood of p, and $B \subset U$ a normal ball of center p. Let $\gamma\colon [0,1] \to B$ be a geodesic segment with $\gamma(0) = p$. If $c\colon [0,1] \to M$ is any piecewise differentiable curve joining $\gamma(0)$ to $\gamma(1)$ then $\ell(\gamma) \le \ell(c)$ and if equality holds then $\gamma([0,1]) = c([0,1])$.*

Proof. Suppose initially that $c([0,1]) \subset B$. Since $\exp_p$ is a diffeomorphism on U, the curve $c(t)$, for $t \neq 0$, can be written uniquely as $\exp_p(r(t) \cdot v(t)) = f(r(t), t)$ where $t \to v(t)$ is a curve in T_pM with $|v(t)| = 1$ and $r\colon (0,1] \to \mathbf{R}$ is a positive piecewise differentiable function (we can suppose that if $t_1 \in (0,1]$ then $c(t_1) \neq p$; otherwise, ignore the interval $[0, t_1)$). It follows that, except for a finite number of points,

$$\frac{dc}{dt} = \frac{\partial f}{\partial r} r'(t) + \frac{\partial f}{\partial t}.$$

From the Gauss lemma, $\langle \frac{\partial f}{\partial r}, \frac{\partial f}{\partial t} \rangle = 0$. Since $\left| \frac{\partial f}{\partial r} \right| = 1$,

$$\left| \frac{dc}{dt} \right|^2 = |r'(t)|^2 + \left| \frac{\partial f}{\partial t} \right|^2 \geq |r'(t)|^2 \tag{1}$$

and so

$$\int_\varepsilon^1 \left| \frac{dc}{dt} \right| dt \geq \int_\varepsilon^1 |r'(t)|\, dt \geq \int_\varepsilon^1 r'(t) dt = r(1) - r(\varepsilon). \tag{2}$$

Taking $\varepsilon \to 0$ we obtain $\ell(c) \geq \ell(\gamma)$, because $r(1) = \ell(\gamma)$.

It is clear that if the inequality (1), or the second inequality (2), is strict then $\ell(c) > \ell(\gamma)$. If $\ell(c) = \ell(\gamma)$, then $\left| \frac{\partial f}{\partial t} \right| = 0$, that is, $v(t) = \text{const.}$, and $|r'(t)| = r'(t) > 0$. It follows that c is a monotonic reparametrization of γ, hence $c([0,1]) = \gamma([0,1])$.

If $c([0,1])$ is not contained in B, consider the first point $t_1 \in (0,1)$ for which $c(t_1)$ belongs to the boundary of B. If ρ is the radius of the geodesic ball B, we have:

$$\ell(c) \geq \ell_{[0,t_1]}(c) \geq \rho > \ell(\gamma). \quad \square$$

It should be noted that the proposition above is not global. If we consider a sufficiently large arc of a geodesic it can cease minimizing the arc length after awhile. For example the geodesics on the sphere which start at a point p are no longer minimizing after they pass through the antipode of p.

On the other hand, if a piecewise differentiable curve c is minimizing, we shall prove that c is a geodesic. For this, we need a refinement of Proposition 2.9, where we proved the existence of normal neighborhoods. We show below that for any $p \in M$ there exists a neighborhood W of p which is a normal neighborhood of each $q \in W$.

3.7 Theorem. *For any $p \in M$ there exist a neighborhood W of p and a number $\delta > 0$, such that, for every $q \in W$, $\exp_q$ is a diffeomorphism on $B_\delta(0) \subset T_qM$ and $\exp_q(B_\delta(0)) \supset W$, that is, W is a normal neighborhood of each of its points.*

Proof. Let $\varepsilon > 0$, V and $\mathcal{U}$ as in Proposition 2.7. Define $F: \mathcal{U} \to M \times M$ by $F(q,v) = (q, \exp_q v)$. Recall that $\mathcal{U} \subset TU$, where U is the domain of a system of coordinates $\mathbf{x}$ at p, with $V \subset \mathbf{x}(U)$. Consider, around $F(p,0) = (p,p) \in M \times M$ the system of coordinates $(U \times U; (\mathbf{x}, \mathbf{x}))$. Thus, the matrix of $dF_{(p,0)}$ is

$$\begin{pmatrix} I & I \\ 0 & I \end{pmatrix},$$

because $(d\exp_p)_o = I$. It follows that F is a local diffeomorphism in a neighborhood of $(p,0)$. This means that there exists a neighborhood $\mathcal{U}' \subset \mathcal{U}$ of $(p,0)$ in TM such that F maps $\mathcal{U}'$ diffeomorphically onto a neighborhood W' of (p,p) in $M \times M$. It is possible to choose $\mathcal{U}'$ of the form

$$\mathcal{U}' = \{(q,v); q \in V', v \in T_qM, |v| < \delta\},$$

where $V' \subset V$ is a neighborhood of p in M. Now choose a neighborhood $W \subset M$ of p such that $W \times W \subset W'$. We claim that W and δ, so obtained, satisfy the assertion of the Proposition.

Indeed, if $q \in W$ and $B_\delta(0) \subset T_qM$ then, since F is a diffeomorphism on $\mathcal{U}'$, we get

$$F(\{q\} \times B_\delta(0)) \supset \{q\} \times W.$$

From the definition of F, $\exp_q(B_\delta(0)) \supset W$. □

3.8 Remark. From the proposition above and the minimizing property of geodesics, it follows that given two points $q_1, q_2 \in W$ there exists a unique minimizing geodesic γ of length $< \delta$ joining q_1 to q_2. The proof shows, moreover, that γ depends differentiably on (q_1, q_2) in the following sense: given (q_1, q_2) there exists a unique $v \in T_{q_1}M$ (given by $F^{-1}(q_1, q_2) = (q_1, v)$) that depends differentiably on (q_1, q_2) and is such that $\gamma'(0) = v$.

It is customary to call W a *totally normal neighborhood* of $p \in M$.

3.9 COROLLARY. *If a piecewise differentiable curve $\gamma\colon [a,b] \to M$, with parameter proportional to arc length, has length less or equal to the length of any other piecewise differentiable curve joining $\gamma(a)$ to $\gamma(b)$ then γ is a geodesic. In particular, γ is regular.*

Proof. Let $t \in [a,b]$ and let W be a totally normal neighborhood of $\gamma(t)$. There exists a closed interval $I \subset [a,b]$, with non-empty interior, $t \in I$, such that $\gamma(I) \subset W$; the restriction $\gamma_I\colon I \to W$ is then a piecewise differentiable curve joining two points of a normal ball. From the hypothesis and from Proposition 3.6, $\ell(\gamma_I)$ is equal to the length of a radial geodesic joining these two points. Again, from Proposition 3.6, and from the fact that γ_I is parametrized proportionally to arc length, γ_I is a geodesic on I, and therefore at t. □

Using the corollary above, we can determine the geodesics in the Lobatchevski plane. It should be observed, and it is easy to verify, that the isometries of a Riemannian manifold take geodesics into geodesics.

3.10 EXAMPLE. Let G be the upper half-plane, that is, $G = \{(x,y) \in \mathbf{R}^2; y > 0\}$ with the Riemannian metric $g_{11} = g_{22} = \frac{1}{y^2}$, $g_{12} = g_{21} = 0$

We are going to show that the segment $\gamma\colon [a,b] \to G$, $a > 0$, of the y axis, given by $\gamma(t) = (0,t)$ is the image of a geodesic. In fact, for any arc $c\colon [a,b] \to G$ given by $c(t) = (x(t), y(t))$ with $c(a) = (0,a)$ and $c(b) = (0,b)$, we have that

$$\ell(c) = \int_a^b \left|\frac{dc}{dt}\right| dt = \int_a^b \sqrt{(\frac{dx}{dt})^2 + (\frac{dy}{dt})^2}\frac{dt}{y}$$
$$\geq \int_a^b \left|\frac{dy}{dt}\right| \frac{dt}{y} \geq \int_a^b \frac{dy}{y} = \ell(\gamma).$$

It follows that γ minimizes arc length for piecewise differentiable curves, and from Corollary 3.9, that the image of γ is a geodesic.

It is easy to see that the isometries of G (cf. Exercise 4 of Chap. 1)

$$z \to \frac{az+b}{cz+d}, \quad z = x + iy, \quad ad - bc = 1,$$

transform the $0y$ axis into (upper) semi-circles or rays $x = x_o$, $y > 0$. These curves are, therefore, geodesics of G. Indeed, they are all of the geodesics of G, since for each $p \in G$ and any direction in T_pG there passes such a circle with center on the $0x$ axis (Fig. 3; in the special case that a direction is normal to $0x$, the circle degenerates to a ray normal to $0x$).

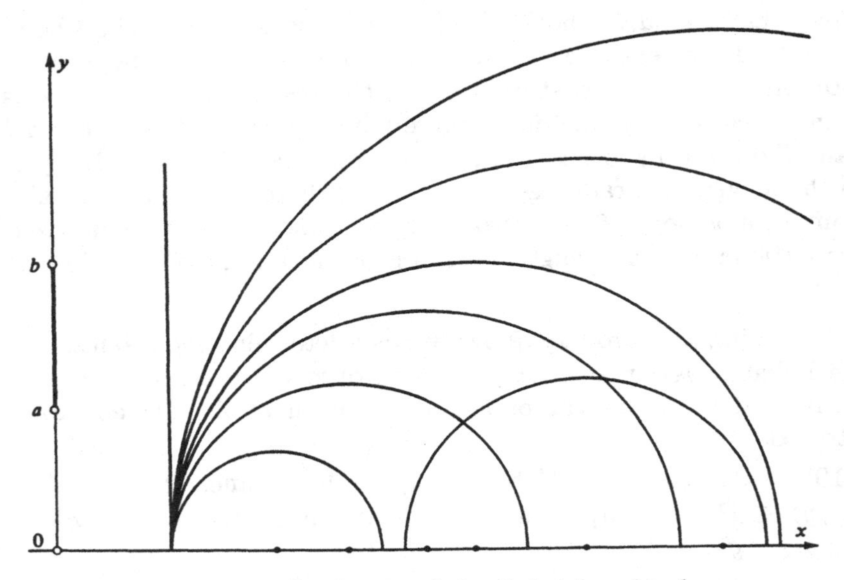

Figure 3. *Geodesics of the Lobatchevski plane.*

4. Convex neighborhoods

We saw in Theorem 3.7 (Cf. Remark 3.8) that any point $p \in M$ possesses a totally normal neighborhood, that is, a neighborhood W and a number $\delta > 0$ such that any two points $q_1, q_2 \in W$ can be joined by a minimizing geodesic of length $< \delta$. However, such a geodesic may not lie completely in W. We say that a subset $S \subset M$ is *strongly convex* if for any two points q_1, q_2 in the closure $\overline{S}$ of S there exists unique minimizing geodesic γ joining q_1 with q_2 whose interior is contained in S. Now we are going to prove that the radius of a totally normal ball can be chosen in such a way that the ball is strongly convex.

4.1 LEMMA. *For any $p \in M$ there exists a number $c > 0$ such that any geodesic in M that is tangent at $q \in M$ to the geodesic sphere*

$S_r(p)$ of radius $r < c$ stays out of the geodesic ball $B_r(p)$ for some neighborhood of q.

Proof. Let W be a totally normal neighborhood of p. Using the lemma of homogeneity, we can suppose, by conveniently restricting the interval of definition, that all of the geodesics of W have velocity one. We can, therefore, restrict ourselves to the unit tangent bundle $T_1 W$ given by

$$T_1 W = \{(q, v); q \in W, v \in T_q M, |v| = 1\}.$$

Let $\gamma: I \times T_1 W \to M$, $I = (-\varepsilon, \varepsilon)$, be the differentiable mapping such that $t \to \gamma(t, q, v)$ is the geodesic that at the instant $t = 0$ passes through q with velocity v, $|v| = 1$. Define $u(t, q, v) = \exp_p^{-1}(\gamma(t, q, v))$ and

$$F: I \times T_1 W \to R, \qquad F(t, q, v) = |u(t, q, v)|^2.$$

F measures the square of the "distance" from p to a point that is moving along the geodesic γ (Fig. 4). It is clear that u and F are differentiable, and that

$$\frac{\partial F}{\partial t} = 2\langle \frac{\partial u}{\partial t}, u\rangle,$$
$$\frac{\partial^2 F}{\partial t^2} = 2\langle \frac{\partial^2 u}{\partial t^2}, u\rangle + 2\left|\frac{\partial u}{\partial t}\right|^2.$$

Now let $r > 0$ be chosen so that

$$\exp_p B_r(0) = B_r(p) \subset W.$$

If a geodesic γ is tangent to the geodesic sphere $S_r(p)$ at the point $q = \gamma(0, q, v)$, then, from the Gauss lemma,

$$\langle \frac{\partial u}{\partial t}(0, q, v), u(0, q, v)\rangle = 0,$$

that is, $\frac{\partial F}{\partial t}(0, q, v) = 0$. If we show that, for r sufficiently small, the critical point $(0, q, v)$ of F is a strict minimum point, we have proven the lemma.

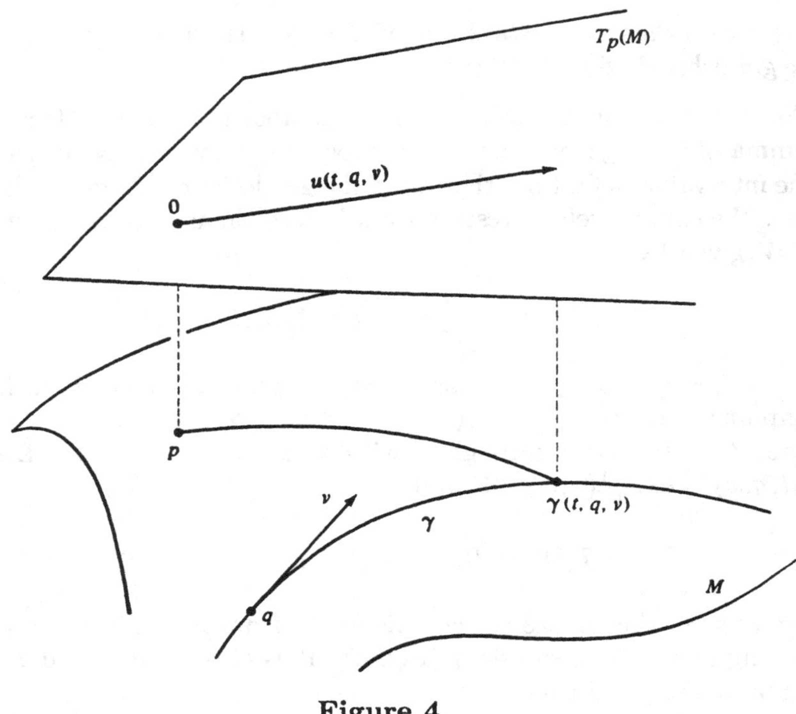

Figure 4

For this it suffices to observe that for $q = p$, we have $u(t,p,v) = tv$ and, therefore

$$\frac{\partial^2 F}{\partial t^2}(0,p,v) = 2\,|v|^2 = 2.$$

It follows that there exists a neighborhood $V \subset W$ of p such that $\frac{\partial^2 F}{\partial t^2}(0,q,v) > 0$, for all $q \in V$ and all $v \in T_qM$, $|v| = 1$. Let $c > 0$ be such that

$$\exp_p B_c(0) \subset V.$$

From what was previously proved, any geodesic in $B_c(p)$ that is tangent to the geodesic sphere of radius $r < c$ at the point $\gamma(0,q,v)$ produces a strict local minimum for F at $(0,q,v)$. It follows that, in a neighborhood of q, the points of γ stay outside of the ball $B_r(p)$. □

4.2 PROPOSITION. *(Convex neighborhoods). For any $p \in M$ there exists a number $\beta > 0$ such that the geodesic ball $B_\beta(p)$ is strongly convex.*

Proof. Let c be the number given in Lemma 4.1. Choose $\delta > 0$ and W in Theorem 3.7 in such a way that $\delta < \frac{c}{2}$. Take $\beta < \delta$ such that $\overline{B_\beta(p)} \subset W$. We shall prove that $B_\beta(p)$ is strongly convex. Let $q_1, q_2 \in \overline{B_\beta(p)}$ and let γ be the (unique) geodesic of length $< 2\delta < c$ joining q_1 to q_2. It is clear that γ is contained in $B_c(p)$ (Fig. 5).

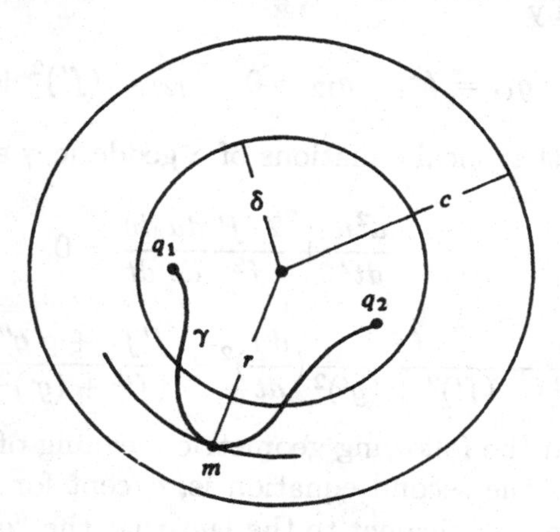

Figure 5

If the interior of γ is not contained in $B_\beta(p)$, then there exists a point m in the interior of γ where the maximum distance r from p to γ is attained. The points of γ in a neighborhood of m remain in the closure of $B_r(p)$. Since $m \in B_c(p)$ this contradicts Lemma 4.1 and proves the proposition. □

EXERCISES

1. (*Geodesics of a surface of revolution*). Denote by (u, v) the cartesian coordinates of $\mathbf{R}^2$. Show that the function $\varphi: U \subset \mathbf{R}^2 \to \mathbf{R}^3$ given by $\varphi(u, v) = (f(v) \cos u, f(v) \sin u, g(v))$,

$$U = \left\{(u, v) \in \mathbf{R}^2 : u_o < u < u_1; v_o < v < v_1\right\},$$

where f and g are differentiable functions, with $f'(v)^2 + g'(v)^2 \neq 0$ and $f(v) \neq 0$, is an immersion. The image $\varphi(U)$ is the surface generated by the rotation of the curve $(f(v), g(v))$ around the axis $0z$ and is called a *surface of revolution* S. The image by φ of the curves $u =$ constant and $v =$ constant are called *meridians* and *parallels*, respectively, of S.

a) Show that the induced metric in the coordinates (u, v) is given by

$$g_{11} = f^2, \quad g_{12} = 0, \quad g_{22} = (f')^2 + (g')^2.$$

b) Show that local equations of a geodesic γ are

$$\frac{d^2u}{dt^2} + \frac{2ff'}{f^2}\frac{du}{dt}\frac{dv}{dt} = 0,$$

$$\frac{d^2v}{dt^2} - \frac{ff'}{(f')^2 + (g')^2}(\frac{du}{dt})^2 + \frac{f'f'' + g'g''}{(f'^2 + (g')^2}(\frac{dv}{dt})^2 = 0.$$

c) Obtain the following geometric meaning of the equations above: the second equation is, except for meridians and parallels, equivalent to the fact that the "energy" $|\gamma'(t)|^2$ of a geodesic is constant along γ; the first equation signifies that if $\beta(t)$ is the oriented angle, $\beta(t) < \pi$, of γ with a parallel P intersecting γ at $\gamma(t)$, then

$$r \cos \beta = \text{const.},$$

where r is the radius of the parallel P (the equation above is called *Clairaut's relation*).

d) Use Clairaut's relation to show that a geodesic of the paraboloid

$$(f(v) = v, g(v) = v^2, 0 < v < \infty, -\varepsilon < u < 2\pi + \varepsilon),$$

which is not a meridian, intersects itself an infinite number of times(Fig. 6).

2. It is possible to introduce a Riemannian metric in the tangent bundle TM of a Riemannian manifold M in the following manner. Let $(p, v) \in TM$ and V, W be tangent vectors in TM at (p, v). Choose curves in TM

$$\alpha: t \to (p(t), v(t)), \;\; \beta: s \to (q(s), w(s)),$$

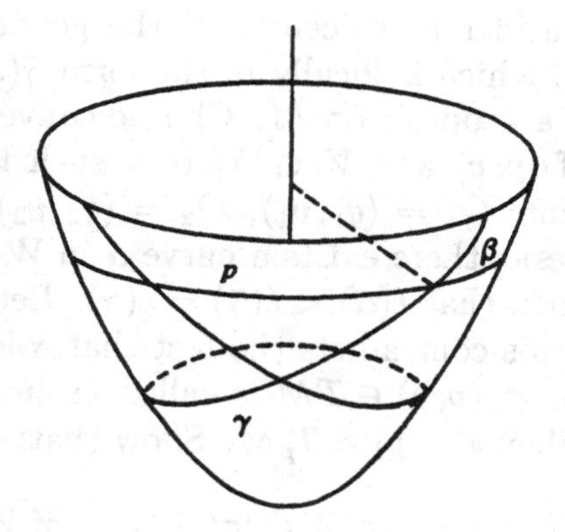

Figure 6. *Geodesics of a paraboloid.*

with $p(0) = q(0) = p$, $v(0) = w(0) = v$, and $V = \alpha'(0)$, $W = \beta'(0)$. Define an inner product on TM by

$$\langle V, W\rangle_{(p,v)} = \langle d\pi(V), d\pi(W)\rangle_p + \langle \frac{Dv}{dt}(0), \frac{Dw}{ds}(0)\rangle_p,$$

where $d\pi$ is the differential of $\pi: TM \to M$, $\pi(p,v) = p$.

a) Prove that this inner product is well-defined and introduces a Riemannian metric on TM.

b) A vector at $(p, v) \in TM$ that is orthogonal (for the metric above) to the fiber $\pi^{-1}(p) \approx T_pM$ is called a *horizontal vector*. A curve

$$t \to (p(t), v(t))$$

in TM is *horizontal* if its tangent vector is horizontal for all t. Prove that the curve

$$t \to (p(t), v(t))$$

is horizontal if and only if the vector field $v(t)$ is parallel along $p(t)$ in M.

c) Prove that the geodesic field is a horizontal vector field (i.e., it is horizontal at every point).

d) Prove that the trajectories of the geodesic field are geodesics on TM in the metric above.

Hint: Let $\bar{\alpha}(t) = (\alpha(t), v(t))$ be a curve in TM. Show that $\ell(\bar{\alpha}) \geq \ell(\alpha)$ and that the inequality is verified if v is parallel along α. Consider a trajectory of the geodesic flow passing through (p,v) which is locally of the form $\bar{\gamma}(t) = (\gamma(t), \gamma'(t))$, where $\gamma(t)$ is a geodesic on M. Choose convex neighborhoods $W \subset TM$ of (p,v) and $V \subset M$ of p such that $\pi(W) = V$. Take two points $Q_1 = (q_1, v_1)$, $Q_2 = (q_2, v_2)$ in $\bar{\gamma} \cap W$. If $\bar{\gamma}$ is not a geodesic, there exists a curve $\bar{\alpha}$ in W passing through Q_1 and Q_2 such that $\ell(\bar{\alpha}) < \ell(\bar{\gamma}) = \ell(\gamma)$. Let $\alpha = \pi(\bar{\alpha})$; since $\ell(\alpha) \leq \ell(\bar{\alpha})$, this contradicts the fact that γ is a geodesic.

e) A vector at $(p,v) \in TM$ is called *vertical* if it is tangent to the fiber $\pi^{-1}(p) \approx T_pM$. Show that:

$$\langle W, W\rangle_{(p,v)} = \langle d\pi(W), d\pi(W)\rangle_p, \quad \text{if } W \text{ is horizontal,}$$
$$\langle W, W\rangle_{(p,v)} = \langle W, W\rangle_p, \quad \text{if } W \text{ is vertical,}$$

where we are identifying the tangent space to the fiber with T_pM.

3. Let G be a Lie group, $\mathcal{G}$ its Lie algebra and let $X \in \mathcal{G}$ (see Example 2.6, Chap. 1). The trajectories of X determine a mapping $\varphi\colon (-\varepsilon, \varepsilon) \to G$ with $\varphi(0) = e$, $\varphi'(t) = X(\varphi(t))$.

a) Prove that $\varphi(t)$ is defined for all $t \in \mathbf{R}$ and that $\varphi(t+s) = \varphi(t)\cdot\varphi(s)$, ($\varphi\colon \mathbf{R} \to G$ is then called a *1-parameter subgroup* of G).

Hint: Let $\varphi(t_o) = y$, $t_o \in (-\varepsilon, \varepsilon)$. Show that, from the left invariance, $t \to y^{-1}\varphi(t)$, $t \in (-\varepsilon, \varepsilon)$, is also an integral curve of X passing through e for $t = t_o$. By uniqueness, $\varphi(t_o)^{-1}\varphi(t) = \varphi(t - t_o)$, hence φ can be extended out from t_o in an interval of radius ε. This shows that $\varphi(t)$ is defined for all $t \in \mathbf{R}$. In addition $\varphi(t_o)^{-1} = \varphi(-t_o)$ and, since t_o is arbitrary, we obtain $\varphi(t+s) = \varphi(t)\cdot\varphi(s)$.

b) Prove that if G has a bi-invariant metric $\langle\ ,\ \rangle$ then the geodesics of G that start from e are 1-parameter subgroups of G.

Hint: Use the relation (see Eq. (9) of Chap. 2)

$$\begin{aligned} 2\langle X, \nabla_Z Y\rangle &= Z\langle X, Y\rangle + Y\langle X, Z\rangle - X\langle Y, Z\rangle \\ &\quad + \langle Z, [X,Y]\rangle + \langle Y, [X,Z]\rangle - \langle X, [Y,Z]\rangle \end{aligned}$$

and the fact that the metric is left invariant to prove that $\langle X, \nabla_Y Y\rangle = \langle Y, [X,Y]\rangle$, where X, Y and Z are left invariant fields. Use also the fact that the bi-invariance of $\langle\, ,\rangle$ implies that

$$\langle [U,X],V\rangle = -\langle U,[V,X]\rangle, \qquad X,U,V \in \mathcal{G}.$$

It follows that $\nabla_Y Y = 0$, for all $Y \in \mathcal{G}$. Thus 1-parameter subgroups are geodesics. By uniqueness, geodesics are 1-parameter subgroups.

4. A subset A of a differentiable manifold M is *contractible* to a point $a \in A$ when the mapping id_A (identity on A) and $k_a: x \in A \to a \in A$ are homotopic (with base point a). A is *contractible* if it is contractible to one of its points.
   a) Show that a convex neighborhood in a Riemannian manifold M is a contractible subset (with respect to any of its points).
   b) Let M be a differentiable manifold. Show that there exists a covering $\{U_\alpha\}$ of M with the following properties:
      i) U_α is open and contractible, for each α.
      ii) If $U_{\alpha_1},\dots,U_{\alpha_r}$ are elements of the covering, then $\bigcap_1^r U_{\alpha_i}$ is contractible

5. Let M be a Riemannian manifold and $X \in \mathcal{X}(M)$. Let $p \in M$ and let $U \subset M$ be a neighborhood of p. Let $\varphi: (-\varepsilon,\varepsilon)\times U \to M$ be a differentiable mapping such that for any $q \in U$ the curve $t \to \varphi(t,q)$ is a trajectory of X passing through q at $t = 0$ (U and φ are given by the fundamental theorem for ordinary differential equations, Cf. Theorem 2.2). X is called a *Killing field* (or an *infinitesimal isometry*) if, for each $t_o \in (-\varepsilon,\varepsilon)$, the mapping $\varphi(t_o,\ ): U \subset M \to M$ is an isometry.
   Prove that:
   a) A vector vield v on $\mathbf{R}^n$ may be seen as a map $v: \mathbf{R}^n \to \mathbf{R}^n$; we say that the field is linear if v is a linear map. A linear field on $\mathbf{R}^n$, defined by a matrix A, is a Killing field if and only if A is anti-symmetric.
   b) Let X be a Killing field on M, $p \in M$, and let U be a normal neighborhood of p on M. Assume that p is a unique point of U that satisfies $X(p) = 0$. Then, in U, X is tangent to the geodesic spheres centered at p.
   c) Let X be a differentiable vector field on M and let $f: M \to N$ be an isometry. Let Y be a vector field on N defined

by $Y(f(p)) = df_p(X(p))$, $p \in M$. Then Y is a Killing field if and only if X is also a Killing vector field.

d) X is Killing $\Leftrightarrow \langle \nabla_Y X, Z\rangle + \langle \nabla_Z X, Y\rangle = 0$ for all $Y, Z \in \mathcal{X}(M)$ (the equation above is called the *Killing equation*).

Hint for $\Rightarrow$: By continuity, it suffices to prove the equation above for points $q \in U$ where $X(q) \neq 0$. If this is the case, let $S \subset U$ be a submanifold of U, passing through q, normal to $X(q) \neq 0$ at q, with $\dim S = \dim M - 1$. Let $(x_1, \ldots, x_{n-1})$ be coordinates in a neighborhood $V \subset S$ of q such that $(x_1, \ldots, x_{n-1}, t)$ are coordinates in a neighborhood $V \times (-\varepsilon, \varepsilon) \subset U$ and $X = \frac{\partial}{\partial t}$. Putting $X_i = \frac{\partial}{\partial x_i}$, obtain

$$\langle \nabla_{X_j} X, X_i\rangle + \langle \nabla_{X_i} X, X_j\rangle = X\langle X_i, X_j\rangle - \langle [X, X_i], X_j\rangle$$
$$- \langle [X, X_j], X_i\rangle = \frac{\partial}{\partial t}\langle X_i, X_j\rangle = 0,$$

where in the last equality the fact was used that X is a Killing field.

e) Let X be a Killing field on M with $X(q) \neq 0$, $q \in M$. Then there exists a system of coordinates $(x_1, \ldots, x_n)$ in a neighborhood of q, so that the coefficients g_{ij} of the metric in this system coordinates do not depend on x_n.

6. Let X be a Killing field (Cf. Exercise 5) on a connected Riemannian manifold M. Assume that there exists a point $q \in M$ such that $X(q) = 0$ and $\nabla_Y X(q) = 0$, for all $Y(q) \in T_qM$. Prove that $X \equiv 0$.

Hint: Show that, for all t, the local isometry $\varphi(t, \ ) \colon U \subset M \to M$ generated by the field X (Cf. Exercise 5) leaves the point q fixed and its differential at q , as a linear map of T_qM, is the identity. For this, observe that $d\varphi_t \colon T_qM \to T_qM$ for all t. In addition, $[X, Y](q) = (\nabla_X Y - \nabla_Y X)(q) = 0$, by hypothesis. Since

$$0 = [Y, X](q) = \lim_{t\to 0} \frac{1}{t}[d\varphi_t - \mathrm{Id}](Y) = \left.\frac{d}{dt}(d\varphi_t)\right|_{t=0}$$

and $d\varphi_{s+t} = d\varphi_s \cdot d\varphi_t$, conclude that $d\varphi_t$ does not depend on t, and it is equal to Id. Now use the exponential map to show that such an isometry is the identity on M.

7. (*Geodesic frame*). Let M be a Riemannian manifold of dimension n and let $p \in M$. Show that there exists a neighborhood $U \subset M$ of p and n vector fields $E_1, \ldots, E_n \in \mathcal{X}(U)$, orthonormal at each point of U, such that, at p, $\nabla_{E_i} E_j(p) = 0$.
Such a family E_i, $i = 1, \ldots, n$, of vector fields is called a *(local) geodesic frame* at p.

8. Let M be a Riemannian manifold. Let $X \in \mathcal{X}(M)$ and $f \in \mathcal{D}(M)$. Define the *divergence* of X as a function $\operatorname{div} X \colon M \to \mathbf{R}$ given by $\operatorname{div} X(p)$ = trace of the linear mapping $Y(p) \to \nabla_Y X(p)$, $p \in M$, and the *gradient* of f as a vector field grad f on M defined by

$$\langle \operatorname{grad} f(p), v\rangle = df_p(v), \quad p \in M,\ v \in T_pM.$$

a) Let E_i, $i = 1, \ldots, n = \dim M$, be a geodesic frame at $p \in M$ (See Exercise 7). Show that:

$$\operatorname{grad} f(p) = \sum_{i=1}^{n} (E_i(f)) E_i(p),$$

$$\operatorname{div} X(p) = \sum_{i=1}^{n} E_i(f_i)(p), \quad \text{where} \quad X = \sum_i f_i E_i.$$

b) Suppose that $M = \mathbf{R}^n$, with coordinates $(x_1, \ldots, x_n)$ and $\frac{\partial}{\partial x_i} = (0, \ldots, 1, \ldots, 0) = e_i$. Show that:

$$\operatorname{grad} f = \sum_{i=1}^{n} \frac{\partial f}{\partial x_i} e_i,$$

$$\operatorname{div} X = \sum_i \frac{\partial f_i}{\partial x_i}, \quad \text{where} \quad X = \sum_i f_i e_i.$$

9. Let M be a Riemannian manifold. Define an operator $\Delta \colon \mathcal{D}(M) \to \mathcal{D}(M)$ (the *Laplacian* of M) by

$$\Delta f = \operatorname{div} \operatorname{grad} f, \qquad f \in \mathcal{D}(M).$$

a) Let E_i be a geodesic frame at $p \in M$, $i = 1, \ldots, n = \dim M$ (see Exercise 7). Prove that

$$\Delta f(p) = \sum_i E_i(E_i(f))(p).$$

Conclude that if $M = \mathbf{R}^n$, Δ coincides with the usual Laplacian, namely, $\Delta f = \sum_i \frac{\partial^2 f}{\partial x_i^2}$.

b) Show that

$$\Delta(f \cdot g) = f\Delta g + g\Delta f + 2\langle \operatorname{grad} f, \operatorname{grad} g\rangle.$$

10. Let $f\colon [0,1] \times [0,a] \to M$ be a parametrized surface such that for all $t_o \in [0,a]$, the curve $s \to f(s,t_o)$, $s \in [0,1]$, is a geodesic parametrized by arc length, which is orthogonal to the curve $t \to f(0,t)$, $t \in [0,a]$, at the point $f(0,t_o)$. Prove that, for all $(s_o,t_o) \in [0,1] \times [0,a]$, the curves $s \to f(s,t_o)$, $t \to f(s_o,t)$ are orthogonal.
Hint: Differentiate $\langle \frac{\partial f}{\partial s}, \frac{\partial f}{\partial t}\rangle$ with respect to s, obtaining

$$\begin{aligned}\frac{d}{ds}\langle \frac{\partial f}{\partial s}, \frac{\partial f}{\partial t}\rangle &= \langle \frac{D}{ds}\frac{\partial f}{\partial s}, \frac{\partial f}{\partial t}\rangle + \langle \frac{\partial f}{\partial s}, \frac{D}{\partial t}\frac{\partial f}{\partial s}\rangle \\ &= \frac{1}{2}\frac{d}{dt}\langle \frac{\partial f}{\partial s}, \frac{\partial f}{\partial s}\rangle = 0,\end{aligned}$$

where we used the symmetry of the connection and the fact that $\frac{D}{ds}\frac{\partial f}{\partial s} = 0$.

†11. Let M be an oriented Riemannian manifold. Let ν be a differential form of degree $n = \dim M$ defined in the following way:

$$\begin{aligned}\nu(v_1,\dots,v_n)(p) &= \pm\sqrt{\det(\langle v_i, v_j\rangle)} \\ &= \text{orient. vol. } \{v_1,\dots,v_n\}, \quad p \in M,\end{aligned}$$

where $v_1,\dots,v_n \in T_p(M)$ are linearly independent, and the oriented volume is affected by the sign + or - depending on whether or not the basis $\{v_1,\dots,v_n\}$ belongs to the orientation of M; ν is called the *volume element* of M. For a vector field $X \in \mathcal{X}(M)$ define the *interior product* $i(X)\nu$ of X with ν as the $(n-1)$-form:

$$i(X)\nu(Y_2,\dots,Y_n) = \nu(X,Y_2,\dots,Y_n),\ Y_2,\dots,Y_n \in \mathcal{X}(M).$$

Prove that

$$d(i(X)\nu) = \operatorname{div} X \nu.$$

Hint: Let $p \in M$ and let E_i be a geodesic frame at p. Write X as a sum, $X = \sum f_i E_i$ and let ω_i be differential forms of degree one defined on a neighborhood of p by $\omega_i(E_j) = \delta_{ij}$. Show that $\omega_i \wedge \ldots \wedge \omega_n$ is a volume form ν on M. Next put $\theta_i = \omega_1 \wedge \ldots \wedge \hat{\omega}_i \wedge \ldots \wedge \omega_n$, where $\hat{\omega}_i$ signifies that the factor $\hat{\omega}_i$ is not present. Prove that $i(X)\nu = \sum_i (-1)^{i+1} f_i \theta_i$. It then follows that

$$d(i(X)\nu) = \sum_i (-1)^{i+1} df_i \wedge \theta_i + \sum_i (-1)^{i+1} f_i \wedge d\theta_i$$
$$= (\sum_i E_i(f_i))\nu + \sum_i (-1)^{i+1} f_i \wedge d\theta_i.$$

But $d\theta_i = 0$ at p, since

$$d\omega_k(E_i, E_j) = E_i\omega_k(E_j) - E_j\omega_k(E_i) - \omega_k([E_i, E_j])$$
$$= \omega_k(\nabla_{E_i} E_j - \nabla_{E_j} E_i).$$

Therefore

$$d(i(X)\nu)(p) = (\sum_i E_i(f_i)(p))\nu = \operatorname{div} X(p)\nu$$

and since p is arbitrary, this completes the proof.
Remark. The result obtained implies that the notion of the divergence of X makes sense on an oriented differentiable manifold on which a "volume element" has been chosen, that is, an n-form ν which takes positive values on positive bases.

†12. (*Theorem of E. Hopf*). Let M be a compact orientable Riemannian manifold which is also connected. Let f be a differentiable function on M with $\Delta f \geq 0$. Then $f = \text{const.}$ In particular, the harmonic functions on M, that is, those for which $\Delta f = 0$, are constant.
Hint: Take grad $f = X$. Using Stokes theorem and the result of exercise 11, obtain

$$\int_M \Delta f \nu = \int_M \operatorname{div} X \nu = \int_M d(i(X)\nu) = \int_{\partial M} i(X)\nu = 0.$$

Since $\Delta f \geq 0$, we have $\Delta f = 0$. Using again Stokes theorem on $f^2/2$, and the result of exercise 9(b), we obtain

$$0 = \int_M \Delta(f^2/2)\nu = \int_M f\Delta f\nu + \int_M |\text{grad}\, f|^2\, \nu$$

$$= \int_M |\text{grad}\, f|^2\, \nu,$$

which together with the connectedness of M, implies that $f =$ const..

†13. Let M be a Riemannian manifold and $X \in \mathcal{X}(M)$. Let $p \in M$ such that $X(p) \neq 0$. Choose a coordinate system $(t, x_2, \dots, x_n)$ in a neighborhood U of p such that $\frac{\partial}{\partial t} = X$. Show that if $\nu = g\ dt \wedge dx_2 \wedge \dots \wedge dx_n$ is a volume element of M, then

$$i(X)\nu = g\, dx_2 \wedge \dots \wedge dx_n.$$

Conclude from this, using the result of Exercise 11, that

$$\text{div} X = \frac{1}{g}\frac{\partial g}{\partial t}.$$

This proves that div$X$ intuitively measures the degree of variation of the volume element of M along the trajectories of X.

14. (*Liouville's Theorem*). Prove that if G is the geodesic field on TM then div$G = 0$. Conclude from this that the geodesic flow preserves the volume of TM.

Hint: Let $p \in M$ and consider a system $(u_1, \dots, u_n)$ of *normal coordinates* at p. Such coordinates are defined in a normal neighborhood U of p by considering an orthonormal basis $\{e_i\}$ of T_pM and taking $(u_1, \dots, u_n)$, $q = \exp_p(\sum_i u_i e_i)$, $i = 1, \dots, n$, as coordinates of q. In such a coordinate system, $\Gamma_{ij}^k(p) = 0$, since the geodesics that pass through p are given by linear equations. Therefore if $X = \sum x_i \frac{\partial}{\partial u_i}$, then div $X(p) = \sum \frac{\partial x_i}{\partial u_i}$.

Now let (u_i) be normal coordinates in a neighborhood $U \subset M$ around $p \in M$ and let (u_i, v_j), $v = \sum_j v_j \frac{\partial}{\partial u_j}$, $i, j = 1, \dots, n$, be coordinates on TM. Calculate the volume element of the natural metric of TM at (q, v), $q \in U$, $v \in T_qM$, and show that it is the volume element of the product metric on $U \times U$ at the point (q, q) (See Exercise 2(e)). Since the divergence of G only depends on the volume element (see Exercise 11), and G is horizontal, we can calculate $\operatorname{div} G$ in the product metric. Observe that in the coordinates (u_i, v_j) we have

$$G(u_i) = v_i, \quad G(v_j) = -\sum_{ik} \Gamma^j_{ik} v_i v_k, \quad k = 1, \ldots, n.$$

Since the Christoffel symbols of the product metric on $U \times U$ vanish at (p, p), we obtain finally, at p,

$$\operatorname{div} G = \sum_i \frac{\partial v_i}{\partial u_i} - \sum_j \frac{\partial}{\partial v_j} \Big(\sum_{ik} \Gamma^j_{ik} v_i v_k\Big) = 0.$$

CHAPTER 4

CURVATURE

1. Introduction

The notion of curvature in a Riemannian manifold was introduced by Riemann (See Riemann [Ri]) in a rather geometric manner, which we are now going to describe. Let p be a point of a Riemannian manifold M and let $\sigma \subset T_pM$ be a two dimensional subspace of the tangent space T_pM of M at p. Consider the set of geodesics that start at p and are tangent to σ. The segments of such geodesics in a normal neighborhood $U \subset M$ of p determine a submanifold of dimension two $S \subset M$ (with our present notation, S is the image of $\exp_p$ restricted to $\sigma \cap \exp_p^{-1}(U)$). S has a metric induced from the inclusion. Since Gauss had proved that the curvature of a surface can be expressed in terms of its metric, so Riemann could speak of the curvature of S at p, and indicate it by $K(p,\sigma)$, (nowadays, $K(p,\sigma)$ is called the sectional curvature of M at p with respect to σ). This was the curvature considered by Riemann in [Ri]. It is a natural generalization of the Gaussian curvature of surfaces and it is clear that if $M = R^n$, $K(p,\sigma) = 0$ for all p and all σ.

Riemann did not indicate a way to calculate the sectional curvature starting with the metric of M; that was done a few years later by Christoffel (see Christoffel [Cf]; Cf. also Eq. (2) of this chapter). Indeed, all the work of Riemann contains just one formula, namely, an expression for the metric for which $K(p,\sigma)$ is constant, for all p and σ, and even this formula was presented without proof. (The formula of Riemann will be presented in Exercise 1(c) of Chap. 8.)

As frequently happens in mathematics, a "workable" formulation of the concept of curvature required a long time for its development. When such a formulation finally appeared it had the advantage of being easy to use to prove theorems but it had the disadvantage of being so far removed from the initial intuitive concept that it looked as if it were some kind of arbitrary creation.

This chapter presents a definition of curvature that, intuitively, measures the amount that a Riemannian manifold deviates from being Euclidean (Cf. Def. 2.1). In Chapter 6, we are going to show that the notion of sectional curvature (Cf. Def. 3.2) obtained by starting with this definition of curvature generalizes the notion of Gaussian curvature for surfaces, and coincides with the concept introduced by Riemann.

2. Curvature

2.1 DEFINITION. The *curvature* R of a Riemannian manifold M is a correspondence that associates to every pair $X, Y \in \mathcal{X}(M)$ a mapping $R(X,Y): \mathcal{X}(M) \to \mathcal{X}(M)$ given by

$$R(X,Y)Z = \nabla_Y \nabla_X Z - \nabla_X \nabla_Y Z + \nabla_{[X,Y]} Z, \quad Z \in \mathcal{X}(M),$$

where ∇ is the Riemannian connection of M.

Observe that if $M = R^n$, then $R(X,Y)Z = 0$ for all $X, Y, Z \in \mathcal{X}(R^n)$. In fact, if the vector field Z is given by $Z = (z_1, \ldots, z_n)$, with the components of Z coming from the natural coordinates of R^n, we obtain

$$\nabla_X Z = (Xz_1, \ldots, Xz_n),$$

hence

$$\nabla_Y \nabla_X Z = (YXz_1, \ldots, YXz_n),$$

which implies that

$$R(X,Y)Z = \nabla_Y \nabla_X Z - \nabla_X \nabla_Y Z + \nabla_{[X,Y]} Z = 0,$$

as was stated. We are able, therefore, to think of R as a way of measuring how much M deviates from being Euclidean.

Another way of viewing definition 2.1 is to consider a system of coordinates $\{x_i\}$ around $p \in M$. Since $\left[\frac{\partial}{\partial x_i}, \frac{\partial}{\partial x_j}\right] = 0$, we obtain

$$R\left(\frac{\partial}{\partial x_i}, \frac{\partial}{\partial x_j}\right)\frac{\partial}{\partial x_k} = (\nabla_{\partial/\partial x_j} \nabla_{\partial/\partial x_i} - \nabla_{\partial/\partial x_i} \nabla_{\partial/\partial x_j})\frac{\partial}{\partial x_k},$$

that is, the curvature measures the non-commutativity of the covariant derivative.

These interpretations, are, however, more or less formal. In this chapter we advise the reader to get used to the formal properties of curvature, postponing until Chapter 6 the proof of a more geometric interpretation of curvature. Let us remark also that a frequently encountered definition of curvature in the literature differs from definition 2.1 by a sign.

2.2 PROPOSITION. *The curvature R of a Riemannian manifold has the following properties:*

(i) R is bilinear in $\mathcal{X}(M) \times \mathcal{X}(M)$, that is,

$$R(fX_1 + gX_2, Y_1) = fR(X_1, Y_1) + gR(X_2, Y_1),$$
$$R(X_1, fY_1 + gY_2) = fR(X_1, Y_1) + gR(X_1, Y_2),$$

$f, g \in \mathcal{D}(M), \quad X_1, X_2, Y_1, Y_2 \in \mathcal{X}(M).$

(ii) For any $X, Y \in \mathcal{X}(M)$, the curvature operator $R(X,Y)$: $\mathcal{X}(M) \to \mathcal{X}(M)$ is linear, that is,

$$R(X,Y)(Z+W) = R(X,Y)Z + R(X,Y)W,$$
$$R(X,Y)fZ = fR(X,Y)Z,$$

$f \in \mathcal{D}(M), \quad Z, W \in \mathcal{X}(M).$

Proof. Let us verify (ii) only, leaving (i) as an exercise for the reader. The first part of (ii) is obvious. As for the second, we have

$$\nabla_Y \nabla_X (fZ) = \nabla_Y (f\nabla_X Z + (Xf)Z) = f\nabla_Y \nabla_X Z + (Yf)(\nabla_X Z) + (Xf)(\nabla_Y Z) + (Y(Xf))Z.$$

Therefore,

$$\nabla_Y \nabla_X (fZ) - \nabla_X \nabla_Y (fZ) = f(\nabla_Y \nabla_X - \nabla_X \nabla_Y)Z + ((YX - XY)f)Z,$$

hence

$$R(X,Y)fZ = f\nabla_Y \nabla_X Z - f\nabla_X \nabla_Y Z + ([Y,X]f)Z + f\nabla_{[X,Y]}Z + ([X,Y]f)Z = fR(X,Y)Z. \quad \square$$

2.3 REMARK. An analysis of the proof above shows that the necessity of the appearance of the term $\nabla_{[X,Y]}Z$ in the definition of the curvature is connected to the fact that we want the mapping $R(X,Y)\colon \mathcal{X}(M) \to \mathcal{X}(M)$ to be linear (see the next Rem. 2.6).

2.4 Proposition. *(Bianchi Identity).*

$$R(X,Y)Z + R(Y,Z)X + R(Z,X)Y = 0.$$

Proof. From the symmetry of the Riemannian connection, we have,

$$R(X,Y)Z+R(Y,Z)X+R(Z,X)Y = \nabla_Y\nabla_X Z-\nabla_X\nabla_Y Z+\nabla_{[X,Y]}Z$$
$$+\nabla_Z\nabla_Y X - \nabla_Y\nabla_Z X + \nabla_{[Y,Z]}X + \nabla_X\nabla_Z Y - \nabla_Z\nabla_X Y + \nabla_{[Z,X]}Y$$
$$= \nabla_Y[X,Z]+\nabla_Z[Y,X]+\nabla_X[Z,Y]-\nabla_{[X,Z]}Y-\nabla_{[Y,X]}Z-\nabla_{[Z,Y]}X$$
$$= [Y,[X,Z]] + [Z,[Y,X]] + [X,[Z,Y]] = 0,$$

where the last equality follows from the Jacobi identity for vector fields. □

From now on, we shall write $\langle R(X,Y)Z,T\rangle = (X,Y,Z,T)$.

2.5 Proposition. (a) $(X,Y,Z,T)+(Y,Z,X,T)+(Z,X,Y,T) = 0$
(b) $(X,Y,Z,T) = -(Y,X,Z,T)$
(c) $(X,Y,Z,T) = -(X,Y,T,Z)$
(d) $(X,Y,Z,T) = (Z,T,X,Y)$.

Proof.
(a) is just the Bianchi identity again;
(b) follows directly from Definition 2.1;
(c) is equivalent to $(X,Y,Z,Z) = 0$, whose proof follows:

$$(X,Y,Z,Z) = \langle\nabla_Y\nabla_X Z - \nabla_X\nabla_Y Z + \nabla_{[X,Y]}Z, Z\rangle.$$

But

$$\langle\nabla_Y\nabla_X Z, Z\rangle = Y\langle\nabla_X Z, Z\rangle - \langle\nabla_X Z, \nabla_Y Z\rangle,$$

and

$$\langle\nabla_{[X,Y]}Z, Z\rangle = \frac{1}{2}[X,Y]\langle Z,Z\rangle.$$

Hence

$$(X,Y,Z,Z) = Y\langle\nabla_X Z, Z\rangle - X\langle\nabla_Y Z, Z\rangle + \frac{1}{2}[X,Y]\langle Z,Z\rangle$$
$$= \frac{1}{2}Y(X\langle Z,Z\rangle) - \frac{1}{2}X(Y\langle Z,Z\rangle)$$
$$+ \frac{1}{2}[X,Y]\langle Z,Z\rangle = -\frac{1}{2}[X,Y]\langle Z,Z\rangle$$
$$+ \frac{1}{2}[X,Y]\langle Z,Z\rangle = 0,$$

which proves (c).

In order to prove (d), we use (a), and write:

$$(X,Y,Z,T)+(Y,Z,X,T)+(Z,X,Y,T)=0,$$

$$(Y,Z,T,X)+(Z,T,Y,X)+(T,Y,Z,X)=0,$$

$$(Z,T,X,Y)+(T,X,Z,Y)+(X,Z,T,Y)=0,$$

$$(T,X,Y,Z)+(X,Y,T,Z)+(Y,T,X,Z)=0.$$

Summing the equations above, we obtain

$$2(Z,X,Y,T)+2(T,Y,Z,X)=0$$

and, therefore,

$$(Z,X,Y,T)=(Y,T,Z,X). \quad \Box$$

It is convenient to express what was seen above in a coordinate system $(U,\mathbf{x})$ based at the point $p\in M$. Let us indicate, as usual, $\frac{\partial}{\partial x_i}=X_i$. We put

$$R(X_i,X_j)X_k=\sum_\ell R^\ell_{ijk}X_\ell.$$

Thus R^ℓ_{ijk} are the components of the curvature R in $(U,\mathbf{x})$. If

$$X=\sum_i u^iX_i,\ Y=\sum_j v^jX_j,\ Z=\sum_k w^kX_k,$$

we obtain, from the linearity of R,

$$R(X,Y)Z=\sum_{i,j,k,\ell}R^\ell_{ijk}u^iv^jw^kX_\ell. \tag{1}$$

To express R^ℓ_{ijk} in terms of the coefficients Γ^k_{ij} of the Riemannian connection, we write,

$$\begin{aligned}R(X_i,X_j)X_k&=\nabla_{X_j}\nabla_{X_i}X_k-\nabla_{X_i}\nabla_{X_j}X_k\\&=\nabla_{X_j}(\sum_\ell\Gamma^\ell_{ik}X_\ell)-\nabla_{X_i}(\sum_\ell\Gamma^\ell_{jk}X_\ell),\end{aligned}$$

which by a direct calculation yields

$$(2) \qquad R^s_{ijk} = \sum_\ell \Gamma^\ell_{ik}\Gamma^s_{j\ell} - \sum_\ell \Gamma^\ell_{jk}\Gamma^s_{i\ell} + \frac{\partial}{\partial x_j}\Gamma^s_{ik} - \frac{\partial}{\partial x_i}\Gamma^s_{jk}.$$

Putting

$$\langle R(X_i, X_j)X_k, X_s\rangle = \sum_\ell R^\ell_{ijk} g_{\ell s} = R_{ijks},$$

we can write the identities of Proposition 2.5 as:

$$R_{ijks} + R_{jkis} + R_{kijs} = 0$$

$$R_{ijks} = -R_{jiks}$$

$$R_{ijks} = -R_{ijsk}$$

$$R_{ijks} = R_{ksij}.$$

2.6 REMARK. The equation (1), which depends on the linearity of the operator R, shows that the value of $R(X,Y)Z$ at the point p depends uniquely on the values of X, Y, Z at p and the values of the functions R^ℓ_{ijk} at p. Observe that this contrasts with the behavior of the covariant derivative (See Rem. 2.3, Chap. 2), the reason being that the covariant derivative is not linear in all of its arguments. In general, entities, such as the curvature, that are linear, are called tensors on M (more details will be given in Section 5).

3. Sectional curvature

Closely related to the curvature operator is the sectional (or Riemannian) curvature that we are now going to define.

In what follows it is convenient to use the following notation. Given a vector space V, we denote by $|x \wedge y|$ the expression

$$\sqrt{|x|^2\,|y|^2 - \langle x, y\rangle^2},$$

which represents the area of a two-dimensional parallelogram determined by the pair of vectors $x, y \in V$.

3.1 PROPOSITION. *Let $\sigma \subset T_pM$ be a two-dimensional subspace of the tangent space T_pM and let $x, y \in \sigma$ be two linearly independent vectors. Then*

$$K(x,y) = \frac{(x,y,x,y)}{|x \wedge y|^2}$$

does not depend on the choice of the vectors $x, y \in \sigma$.

Proof. To avoid calculating, we observe that we can pass from the basis $\{x,y\}$ of σ to any other basis $\{x', y'\}$ by iterating the following elementary transformations:

(a) $\{x,y\} \to \{y,x\}$,
(b) $\{x,y\} \to \{\lambda x, y\}$,
(c) $\{x,y\} \to \{x + \lambda y, y\}$.

It is easy to see that $K(x,y)$ is invariant by such transformations and that completes the proof. □

3.2 DEFINITION. Given a point $p \in M$ and a two-dimensional subspace $\sigma \subset T_pM$, the real number $K(x,y) = K(\sigma)$, where $\{x,y\}$ is any basis of σ, is called the *sectional curvature* of σ at p.

Besides the fact that the sectional curvature has interesting geometrical interpretations, its importance comes from the fact that knowledge of $K(\sigma)$, for all σ, determines the curvature R completely. This is a purely algebraic fact:

3.3 LEMMA. *Let V be a vector space of dimension ≥ 2, provided with an inner product $\langle\ ,\ \rangle$. Let $R: V \times V \times V \to V$ and $R': V \times V \times V \to V$ be tri-linear mappings such that conditions (a), (b), (c) and (d) of Proposition 2.5 are satisfied by*

$$(x,y,z,t) = \langle R(x,y)z, t\rangle, \quad (x,y,z,t)' = \langle R'(x,y)z, t\rangle.$$

If x, y are two linearly independent vectors, we may write,

$$K(\sigma) = \frac{(x,y,x,y)}{|x \wedge y|^2}, \qquad K'(\sigma) = \frac{(x,y,x,y)'}{|x \wedge y|^2},$$

where σ is the bi-dimensional subspace generated by x and y. If for all $\sigma \subset V$, $K(\sigma) = K'(\sigma)$, then $R = R'$.

Proof. It suffices to prove that $(x, y, z, t) = (x, y, z, t)'$ for any $x, y, z, t \in V$. Observe first that, by hypothesis, we have $(x, y, x, y) = (x, y, x, y)'$, for all $x, y \in V$. Then

$$(x+z, y, x+z, y) = (x+z, y, x+z, y)',$$

hence

$$(x, y, x, y) + 2(x, y, z, y) + (z, y, z, y) = (x, y, x, y)' + 2(x, y, z, y)' + (z, y, z, y)'$$

and, therefore

$$(x, y, z, y) = (x, y, z, y)',$$

for all $x, y, z \in V$.

Using what we have just proved, we obtain

$$(x, y+t, z, y+t) = (x, y+t, z, y+t)',$$

hence

$$(x, y, z, t) + (x, t, z, y) = (x, y, z, t)' + (x, t, z, y)',$$

which can be written further as

$$(x, y, z, t) - (x, y, z, t)' = (y, z, x, t) - (y, z, x, t)'.$$

It follows that, the expression $(x, y, z, t) - (x, y, z, t)'$ is invariant by cyclic permutations of the first three elements. Therefore, by (a) of Proposition 2.5, we have

$$3[(x, y, z, t) - (x, y, z, t)'] = 0,$$

hence

$$(x, y, z, t) = (x, y, z, t)'$$

for all $x, y, z, t \in V$. □

The Riemannian manifolds that have constant sectional curvature played a fundamental role in the development of Riemannian Geometry. We shall treat these manifolds in more detail in Chapter 8 of this book. At the moment, we wish only to show how the lemma above allows us to obtain a characterization of such manifolds by means of the components $R_{ijk\ell}$ of the curvature in an orthonormal basis. This follows from the lemma below.

3.4 LEMMA. *Let M be a Riemannian manifold and p a point of M. Define a tri-linear mapping $R': T_pM \times T_pM \times T_pM \to T_pM$ by*

$$\langle R'(X,Y,W), Z\rangle = \langle X, W\rangle\langle Y, Z\rangle - \langle Y, W\rangle\langle X, Z\rangle,$$

for all $X, Y, W, Z \in T_pM$. Then M has constant sectional curvature equal to K_o if and only if $R = K_oR'$, where R is the curvature of M.

Proof. Assume that $K(p,\sigma) = K_o$ for all $\sigma \subset T_pM$, and set $\langle R'(X,Y,W), Z\rangle = (X,Y,W,Z)'$. Observe that R' satisfies the properties (a), (b), (c) and (d) of Proposition 2.5. Since

$$(X,Y,X,Y)' = \langle X, X\rangle\langle Y, Y\rangle - \langle X, Y\rangle^2,$$

we have that, for all pairs of vectors $X, Y \in T_pM$,

$$R(X,Y,X,Y) = K_o(|X|^2\,|Y|^2 - \langle X, Y\rangle^2) = K_oR'(X,Y,X,Y).$$

Lemma 3.3 implies that, for all X, Y, W, Z,

$$R(X,Y,W,Z) = K_oR'(X,Y,W,Z),$$

hence $R = K_oR'$. The converse is immediate.

3.5 COROLLARY. *Let M be a Riemannian manifold, p a point of M and $\{e_1, \ldots, e_n\}$, $n = \dim M$, an orthonormal basis of T_pM. Define $R_{ijk\ell} = \langle R(e_i, e_j)e_k, e_\ell\rangle$, $i, j, k, \ell = 1, \ldots, n$. Then $K(p,\sigma) = K_o$ for all $\sigma \subset T_pM$, if and only if*

$$R_{ijk\ell} = K_o(\delta_{ik}\delta_{j\ell} - \delta_{i\ell}\delta_{jk}),$$

where

$$\delta_{ij} = \begin{cases} 1, & \text{if } i = j \\ 0, & i \neq j. \end{cases}$$

In other words, $K(p,\sigma) = K_o$ for all $\sigma \subset T_pM$ if and only if $R_{ijij} = -R_{ijji} = K_o$ for all $i \neq j$, and $R_{ijk\ell} = 0$ in the other cases.

4. Ricci curvature and scalar curvature

Certain combinations of sectional curvature appear with such frequency that they deserve special names.

Let $x = z_n$ be a unit vector in T_pM; we take an orthonormal basis $\{z_1, z_2, \ldots, z_{n-1}\}$ of the hyperplane in T_pM orthogonal to x and consider the following averages:

$$\mathrm{Ric}_p(x) = \frac{1}{n-1}\sum_i \langle R(x, z_i)x, z_i\rangle, \quad i = 1, 2, \ldots, n-1,$$

$$K(p) = \frac{1}{n}\sum_j \mathrm{Ric}_p(z_j) = \frac{1}{n(n-1)}\sum_{ij}\langle R(z_i, z_j)z_i, z_j\rangle,$$
$$j = 1, \ldots, n.$$

We are going to prove that the expressions above do not depend on the choice of the corresponding orthonormal basis; these expressions are called the *Ricci curvature* in the direction x and the *scalar curvature* at p, respectively.

To prove these facts, we give an intrinsic characterization of the expressions above. First, define a bilinear form on T_pM as follows: let $x, y \in T_pM$ and put

$$Q(x, y) = \text{trace of the mapping} \quad z \mapsto R(x, z)y.$$

Q is obviously bilinear. Choosing x a unit vector and then completing it to an orthonormal basis $\{z_1, \ldots, z_{n-1}, z_n = x\}$ of T_pM we have

$$\begin{aligned} Q(x, y) &= \sum_i \langle R(x, z_i)y, z_i\rangle \\ &= \sum_i \langle R(y, z_i)x, z_i\rangle = Q(y, x), \end{aligned}$$

that is, Q is symmetric and $Q(x, x) = (n-1)\,\mathrm{Ric}_p(x)$; this proves that $\mathrm{Ric}_p(x)$ is intrinsically defined.

On the other hand, the bilinear form Q on T_pM corresponds to a linear self-adjoint mapping K, given by

$$\langle K(x), y\rangle = Q(x, y).$$

Taking an orthonormal basis $\{z_1, \ldots, z_n\}$, we have

$$\text{Trace of } \quad K = \sum_j \langle K(z_j), z_j\rangle = \sum_j Q(z_j, z_j)$$
$$= (n-1)\sum_j \mathrm{Ric}_p(z_j) = n(n-1)K(p),$$

which proves the statement.

The bilinear form $\frac{1}{n-1}Q$ is, at times, called *the Ricci tensor.*

As usual we should express what was done above in a coordinate system (x_i). Let $X_i = \frac{\partial}{\partial x_i}$, $g_{ij} = \langle X_i, X_j\rangle$, and g^{ij} the inverse matrix of g_{ij} (i.e., $\sum_k g_{ik}g^{k\ell} = \delta_i^\ell$). Then the coefficients of the bilinear form $\frac{1}{n-1}Q$ in the basis $\{X_i\}$ are given by

$$\frac{1}{n-1}R_{ik} = \frac{1}{n-1}\sum_j R_{ijk}^j = \frac{1}{n-1}\sum_{sj} R_{ijks}g^{sj}.$$

We observe now that if $A: T_pM \to T_pM$ is a linear self-adjoint mapping and $B: T_pM \times T_pM \to \mathbf{R}$ is the associated bilinear form, i.e., $B(X,Y) = \langle A(X), Y\rangle$, then the trace of A is equal to $\sum_{ik} B(X_i, X_k)g^{ik}$. Thus, the scalar curvature in the coordinate system (x_i) is given by

$$K = \frac{1}{n(n-1)}\sum_{ik} R_{ik}g^{ik}.$$

To conclude this section, we are going to establish a relation that will be useful in the future.

Let $f: A \subset R^2 \to M$ be a parametrized surface (Cf. Def. 3.3, Chap. 3) and let (s,t) be the usual coordinates of R^2. Let $V = V(s,t)$ be a vector field along f. For each (s,t), it is possible to define $R(\frac{\partial f}{\partial s}, \frac{\partial f}{\partial t})V$ in an obvious manner.

4.1 LEMMA.

$$\frac{D}{\partial t}\frac{D}{\partial s}V - \frac{D}{\partial s}\frac{D}{\partial t}V = R(\frac{\partial f}{\partial s}, \frac{\partial f}{\partial t})V. \tag{3}$$

Proof. The proof is a long calculation. Choose a system of coordinates $(U, \mathbf{x})$ based at $p \in M$. Let $V = \sum_i v^i X_i$, where $v^i = v^i(s,t)$ e $X_i = \frac{\partial}{\partial x_i}$. Then

$$\frac{D}{\partial s}V = \frac{D}{\partial s}(\sum_i v^i X_i) = \sum_i v^i \frac{D}{\partial s}X_i + \sum_i \frac{\partial v^i}{\partial s}X_i,$$

and

$$\frac{D}{\partial t}\left(\frac{D}{\partial s}V\right) = \sum_i v^i \frac{D}{\partial t}\frac{D}{\partial s}X_i + \sum_i \frac{\partial v^i}{\partial t}\frac{D}{\partial s}X_i + \sum_i \frac{\partial v^i}{\partial s}\frac{D}{\partial t}X_i + \sum_i \frac{\partial^2 v^i}{\partial t \partial s}X_i.$$

Therefore, interchanging the roles of s and t in the expression above, and subtracting, we obtain

$$\frac{D}{\partial t}\frac{D}{\partial s}V - \frac{D}{\partial s}\frac{D}{\partial t}V = \sum v^i\left(\frac{D}{\partial t}\frac{D}{\partial s}X_i - \frac{D}{\partial s}\frac{D}{\partial t}X_i\right).$$

Let us now calculate $\frac{D}{\partial t}\frac{D}{\partial s}X_i$. Put

$$f(s,t) = (x_1(s,t), \ldots, x_n(s,t)).$$

Then $\frac{\partial f}{\partial s} = \sum_j \frac{\partial x_j}{\partial s}X_j$ and $\frac{\partial f}{\partial t} = \sum_k \frac{\partial x_k}{\partial t}X_k$. Thus, we have

$$\frac{D}{\partial s}X_i = \nabla_{\Sigma_j(\partial x_j/\partial s)X_j}(X_i) = \sum_j \frac{\partial x_j}{\partial s}\nabla_{X_j}X_i$$

and

$$\begin{aligned}\frac{D}{\partial t}\frac{D}{\partial s}X_i &= \frac{D}{\partial t}\left(\sum_j \frac{\partial x_j}{\partial s}\nabla_{X_j}X_i\right)\\ &= \sum_j \frac{\partial^2 x_j}{\partial t \partial s}\nabla_{X_j}X_i + \sum_j \frac{\partial x_j}{\partial s}\nabla_{\Sigma_k(\partial x_k/\partial t)X_k}(\nabla_{X_j}X_i)\\ &= \sum_j \frac{\partial^2 x_j}{\partial t \partial s}\nabla_{X_j}X_i + \sum_{jk}\frac{\partial x_j}{\partial s}\frac{\partial x_k}{\partial t}\nabla_{X_k}\nabla_{X_j}X_i,\end{aligned}$$

or

$$\left(\frac{D}{\partial t}\frac{D}{\partial s} - \frac{D}{\partial s}\frac{D}{\partial t}\right)X_i = \sum_{jk}\frac{\partial x_j}{\partial s}\frac{\partial x_k}{\partial t}(\nabla_{X_k}\nabla_{X_j}X_i - \nabla_{X_j}\nabla_{X_k}X_i).$$

Joining everything together, we finally get

$$\begin{aligned}\left(\frac{D}{\partial t}\frac{D}{\partial s} - \frac{D}{\partial s}\frac{D}{\partial t}\right)V &= \sum_{ijk} v^i \frac{\partial x_j}{\partial s}\frac{\partial x_k}{\partial t}R(X_j, X_k)X_i\\ &= R\left(\frac{\partial f}{\partial s}, \frac{\partial f}{\partial t}\right)V. \quad \square\end{aligned}$$

5. Tensors on Riemannian manifolds

The notion of curvature is a particular case of the idea of a tensor, which is a useful object in differential geometry. We present here a rapid introduction to the study of tensors on a Riemannian manifold. The idea of a tensor is a natural generalization of the idea of a vector field, an important point being that, analogously to vector fields, tensors can be differentiated covariantly.

For what follows it is useful to observe that $\mathcal{X}(M)$ is a module over $\mathcal{D}(M)$, that is, $\mathcal{X}(M)$ has a linear structure when we take as "scalars" the elements of $\mathcal{D}(M)$.

5.1 DEFINITION. A *tensor T of order r* on a Riemannian manifold is a multilinear mapping

$$T\colon \underbrace{\mathcal{X}(M) \times \cdots \times \mathcal{X}(M)}_{r \text{ factors}} \to \mathcal{D}(M).$$

This means that given $Y_1, \ldots, Y_r \in \mathcal{X}(M)$, $T(Y_1, \ldots, Y_r)$, is a differentiable function on M, and that T is linear in each argument, that is,

$$T(Y_1, \ldots, fX + gY, \ldots, Y_r) = fT(Y_1, \ldots, X, \ldots, Y_r) + gT(Y_1, \ldots, Y, \ldots, Y_r),$$

for all $X, Y \in \mathcal{X}(M)$, $f, g \in \mathcal{D}(M)$.

A tensor T is a pointwise object in a sense that we now explain. Fix a point $p \in M$ and let U be a neighborhood of p in M on which it is possible to define vector fields $E_1 \ldots, E_n \in \mathcal{X}(M^n)$, in such a fashion that at each $q \in U$, the vectors $\{E_i(q)\}$, $i = 1, \ldots, n$, form a basis of T_qM; we say, in this case, that $\{E_i\}$ is a *moving frame* on U. Let

$$Y_1 = \sum_{i_1} y_{i_1} E_{i_1}, \ldots, Y_r = \sum_{i_r} y_{i_r} E_{i_r}, \quad i_1, \ldots, i_r = 1, \ldots, n,$$

be the restrictions to U of the vector fields $Y_1, \ldots, Y_r$, expressed in the moving frame $\{E_i\}$. By linearity,

$$T(Y_1, \ldots, Y_r) = \sum_{i_1, \ldots, i_r} y_{i_1} \cdots y_{i_r}\, T(E_{i_1}, \ldots, E_{i_r}).$$

The functions $T(E_{i_1}, \ldots, E_{i_r}) = T_{i_1 \ldots i_r}$ on U are called the *components* of T in the frame $\{E_i\}$.

The expression above implies that the value of $T(Y_1, \ldots, Y_r)$ at a point $p \in M$ depends only on the values at p of the components of T, and the values of $Y_1, \ldots, Y_r$ at p. It is in this sense that we say that T is a pointwise object.

5.2 EXAMPLE. The *curvature tensor*

$$R: \mathcal{X}(M) \times \mathcal{X}(M) \times \mathcal{X}(M) \times \mathcal{X}(M) \to \mathcal{D}(M)$$

is defined by

$$R(X, Y, Z, W) = \langle R(X,Y)Z, W\rangle, \qquad X, Y, Z, W \in \mathcal{X}(M).$$

It is easy to verify that R is a tensor of order 4, whose components in the frame $\left\{X_i = \frac{\partial}{\partial x_i}\right\}$ associated with the system of coordinates (x_i) is

$$R(X_i, X_j, X_k, X_\ell) = R_{ijk\ell}.$$

5.3 EXAMPLE. The "metric tensor" $G: \mathcal{X}(M) \times \mathcal{X}(M) \to \mathcal{D}(M)$ is defined by $G(X,Y) = \langle X, Y\rangle$, $X, Y \in \mathcal{X}(M)$. G is a tensor of order 2 and its components in the frame $\{X_i\}$ are the coefficients g_{ij} of the Riemannian metric in the given system of coordinates.

5.4 EXAMPLE. The Riemannian connection ∇ defined by:

$$\nabla: \mathcal{X}(M) \times \mathcal{X}(M) \times \mathcal{X}(M) \to \mathcal{D}(M)$$

$$\nabla(X, Y, Z) = \langle \nabla_X Y, Z\rangle, \quad X, Y, Z \in \mathcal{X}(M),$$

is *not* a tensor, because ∇ is not linear with respect to the argument Y.

5.5 REMARK. It is possible to define the notion of a tensor on a differentiable manifold which does not have a Riemannian metric. In this case, it is necessary to distinguish the covariant tensors (which we have already defined) from the contravariant tensors (which can be defined in an analogous manner, using in place of $\mathcal{X}(M)$ its dual $\mathcal{X}^*(M)$). On a Riemannian manifold this is unnecessary, because the Riemannian metric associates to each $X \in \mathcal{X}(M)$ a unique element $\omega \in \mathcal{X}^*(M)$ given by

$$\omega(Y) = \langle X, Y\rangle, \quad \text{for all} \quad Y \in \mathcal{X}(M).$$

Such a correspondence allows us to identify the contravariant tensors with the covariant tensors. For reasons of economy, we restrict ourselves to covariant tensors.

5.6 REMARK. For various reasons, it is convenient to identify the field $X \in \mathcal{X}(M)$ with the tensor $X\colon \mathcal{X}(M) \to \mathcal{D}(M)$ given by $X(Y) = \langle X, Y\rangle$, for all $Y \in \mathcal{X}(M)$.

It is possible to covariantly differentiate tensors. We will show in a moment that the following definition is fairly natural.

5.7 DEFINITION. Let T be a tensor of order r. The *covariant differential* ∇T of T is a tensor of order $(r+1)$ given by

$$\nabla T(Y_1,\dots,Y_r,Z) = Z(T(Y_1,\dots,Y_r)) - T(\nabla_Z Y_1,\dots,Y_r) - \cdots - T(Y_1,\dots,Y_{r-1},\nabla_Z Y_r).$$

For each $Z \in \mathcal{X}(M)$, the *covariant derivative* $\nabla_Z T$ *of* T *relative to* Z is a tensor of order r given by

$$\nabla_Z T(Y_1,\dots,Y_r) = \nabla T(Y_1,\dots,Y_r,Z).$$

We are going to show that, in a convenient frame, the definition of the covariant derivative of a tensor T relative to $Z \in \mathcal{X}(M)$ becomes quite natural. For this, let $p \in M$ and let $\alpha\colon (-\varepsilon,\varepsilon) \to M$ be a differentiable curve with $\alpha(0) = p$, $\alpha'(t) = Z(\alpha(t))$. Let $\{e_1,\dots,e_n\}$ be a basis of T_pM and let $e_i(t)$ be the parallel transport of e_i along $\alpha = \alpha(t)$, for $i = 1,\dots,n$. Let $T_{i_1\dots i_r}(t)$ be the components, in the basis $\{e_i(t)\}$, of the restriction $T(\alpha(t))$ of T to the curve α. Then, by the definition of $\nabla_Z T$,

$$(\nabla_Z T)(e_{i_1}(t),\dots,e_{i_r}(t)) = \frac{d}{dt}T_{i_1\dots i_r}(t) - T(\nabla_Z e_{i_1}(t),\dots,e_{i_r}(t)) - \cdots - T(e_{i_1}(t),\dots,\nabla_Z e_{i_r}(t)).$$

Since $\nabla_Z e_i(t) = 0$, we have, by linearity,

$$(\nabla_Z T)_{i_1\dots i_r} = (\nabla_Z T)(e_{i_1}(t),\dots,e_{i_r}(t)) = \frac{d}{dt}T_{i_1\dots i_r}.$$

In other words, in this frame, the components of the covariant derivative of T are the usual derivatives of the components of T.

5.8 EXAMPLE. The covariant differential of the metric tensor is the zero tensor. Indeed, for all $X, Y, Z \in \mathcal{X}(M)$,

$$\nabla G(X,Y,Z) = Z\langle X,Y\rangle - \langle \nabla_Z X, Y\rangle - \langle X, \nabla_Z Y\rangle = 0,$$

because ∇ is the Riemannian connection.

5.9 EXAMPLE. Let $X \in \mathcal{X}(M)$. Identify X with the tensor that associates to the vector field $Y \in \mathcal{X}(M)$ the function $\langle X, Y\rangle$ (See Rem. 5.6). The covariant derivative of the tensor X relative to the vector field $Z \in \mathcal{X}(M)$ is such that, for all $Y \in \mathcal{X}(M)$,

$$\begin{aligned} \nabla_Z X(Y) = \nabla X(Y,Z) &= Z(X(Y)) - X(\nabla_Z Y) \\ &= Z\langle X,Y\rangle - \langle X, \nabla_Z Y\rangle = \langle \nabla_Z X, Y\rangle. \end{aligned}$$

We conclude thus that the tensor $\nabla_Z X$ can be identified with the vector field $\nabla_Z X$. This justifies the notation adopted, and shows that the covariant derivative of tensors is a generalization of the covariant derivative of vector fields.

EXERCISES

1. Let G be a Lie group with a bi-invariant metric $\langle\ ,\ \rangle$. Let $X, Y, Z \in \mathcal{X}(G)$ be unit left invariant vector fields on G.
   a) Show that $\nabla_X Y = \frac{1}{2}[X,Y]$.
   Hint: Use the symmetry of the connection and the fact that $\nabla_X X = 0$ (Cf. Exercise 3 of Chap. 3).
   b) Conclude from (a) that $R(X,Y)Z = \frac{1}{4}[[X,Y],Z]$.
   c) Prove that, if X and Y are orthonormal, the sectional curvature $K(\sigma)$ of G with respect to the plane σ generated by X and Y is given by

$$K(\sigma) = \frac{1}{4}\,\|[X,Y]\|^2.$$

   Therefore, *the sectional curvature $K(\sigma)$ of a Lie group with bi-invariant metric is non-negative and is zero if and only if σ*

is generated by vectors X, Y which commute, that is, such that $[X, Y] = 0$.

2. Let X be a Killing field (See Exercise 5 of Chap. 3) on a Riemannian manifold M. Define a mapping $A_X: \mathcal{X}(M) \to \mathcal{X}(M)$ by $A_X(Z) = \nabla_Z X$, $Z \in \mathcal{X}(M)$. Consider the function $f: M \to R$ given by $f(q) = \langle X, X \rangle_q$, $q \in M$. Let $p \in M$ be a critical point of f (that is, $df_p = 0$). Prove that for any $Z \in \mathcal{X}(M)$, at p,
   a) $\langle A_X(Z), X \rangle(p) = 0$.
   b) $\langle A_X(Z), A_X(Z) \rangle(p) = \frac{1}{2} Z_p(Z\langle X, X \rangle) + \langle R(X, Z)X, Z \rangle$.

   Hint for (b): Put $S = \frac{1}{2} ZZ\langle X, X \rangle + \langle R(X, Y)X, Z \rangle(p)$. Using the Killing equation $\langle \nabla_Z X, X \rangle + \langle \nabla_X X, Z \rangle = 0$ (cf. Exercise 5 of Chap. 3), we obtain

$$S = \langle \nabla_{[X,Z]} X, Z \rangle - \langle \nabla_X X, \nabla_Z Z \rangle - \langle \nabla_X \nabla_Z, Z \rangle.$$

   Using the Killing equation again, we obtain

$$\begin{aligned} S &= -\langle \nabla_Z X, \nabla_X Z \rangle + \langle \nabla_Z X, \nabla_Z X \rangle \\ &\quad + \langle \nabla_Z X, \nabla_X Z \rangle - \langle \nabla_X X, \nabla_Z Z \rangle \\ &= \langle \nabla_Z X, \nabla_Z X \rangle - \langle \nabla_X X, \nabla_Z Z \rangle. \end{aligned}$$

   Because of the Killing equation at p, $\nabla_X X(p) = 0$, and we conclude the assertion.

3. Let M be a compact Riemannian manifold of even dimension whose sectional curvature is positive. Prove that every Killing field X on M has a singularity (i.e., there exists $p \in M$ such that $X(p) = 0$).

   Hint: Let $f: M \to R$ be the function $f(q) = \langle X, X \rangle(q)$, $q \in M$, and let $p \in M$ be a minimum point of of f (Cf. the previous Exercise). Suppose that $X(p) \neq 0$. Define a linear mapping $A: T_pM \to T_pM$ by $A(y) = A_X Y = \nabla_Y X$, where Y is an extension of $y \in T_pM$. Let $E \subset T_pM$ be orthogonal to $X(p)$. Use the previous exercise to show that $A: E \to E$ is an anti-symmetric isomorphism. This implies that $\dim E = \dim M - 1$ is even, which is a contradiction; thus $X(p) = 0$.

4. Let M be a Riemannian manifold with the following property: given any two points $p, q \in M$, the parallel transport from p to

q does not depend on the curve that joins p to q. Prove that the curvature of M is identically zero, that is, for all $X, Y, Z \in \mathcal{X}(M)$, $R(X,Y)Z = 0$.

Hint: Consider a parametrized surface $f: U \subset R^2 \to M$, where

$$U = \{(s,t) \in R^2; -\varepsilon < t < 1+\varepsilon, -\varepsilon < s < 1+\varepsilon, \varepsilon > 0\}$$

and $f(s,0) = f(0,0)$, for all s. Let $V_o \in T_{f(0,0)}(M)$ and define a field V along f by: $V(s,0) = V_o$ and, if $t \neq 0$, $V(s,t)$ is the parallel transport of V_o along the curve $t \to f(s,t)$. Then, from Lemma 4.1,

$$\frac{D}{\partial s}\frac{D}{\partial t}V = 0 = \frac{D}{\partial t}\frac{D}{\partial s}V + R(\frac{\partial f}{\partial t}, \frac{\partial f}{\partial s})V.$$

Since parallel transport does not depend on the curve chosen, $V(s,1)$ is the parallel transport of $V(0,1)$ along the curve $s \to f(s,1)$, hence $\frac{D}{\partial s}V(s,1) = 0$. Thus,

$$R_{f(0,1)}(\frac{\partial f}{\partial t}(0,1), \frac{\partial f}{\partial s}(0,1))V(0,1) = 0.$$

Use the arbitrariness of f and V_o to conclude what is required.

5. Let γ: $[0,\ell] \to M$ be a geodesic and let $X \in \mathcal{X}(M)$ be such that $X(\gamma(0)) = 0$. Show that

$$\nabla_{\gamma'}(R(\gamma', X)\gamma')(0) = (R(\gamma', X')\gamma')(0),$$

where $X' = \frac{DX}{dt}$.

Hint Let R be the curvature tensor of Example 5.2. Observe that, for all $Z \in \mathcal{X}(M)$, and $t = 0$,

$$\begin{aligned} 0 &= (\nabla_{\gamma'}R)(\gamma', X, \gamma', Z) \\ &= \frac{d}{dt}\langle R(\gamma', X)\gamma', Z\rangle - \langle R(\gamma', X')\gamma', Z\rangle - \langle R(\gamma', X)\gamma', Z'\rangle \\ &= \langle \nabla_{\gamma'}(R(\gamma', X)\gamma'), Z\rangle - \langle R(\gamma', X')\gamma', Z\rangle. \end{aligned}$$

6. (*Locally symmetric spaces*). Let M be a Riemannian manifold. M is a *locally symmetric* space if $\nabla R = 0$, where R is

the curvature tensor of M. (The geometric significance of this condition will be given in Exercise 14 of Chap. 8).

a) Let M be a locally symmetric space and let $\gamma\colon [0,\ell) \to M$ be a geodesic of M. Let X, Y, Z be parallel vector fields along γ. Prove that $R(X,Y)Z$ is a parallel field along γ.

b) Prove that if M is locally symmetric, connected, and has dimension two, then M has constant sectional curvature.

c) Prove that if M has constant (sectional) curvature, then M is a locally symmetric space

7. Prove the *2nd Bianchi Identity*:

$$\nabla R(X,Y,Z,W,T) + \nabla R(X,Y,W,T,Z) + \nabla R(X,Y,T,Z,W) = 0$$

for all $X,Y,Z,W,T \in \mathcal{X}(M)$.
Hint: Since the objects involved are all tensors, it suffices to prove the equality at a point $p \in M$. Choose a geodesic frame $\{e_i\}$ based at p (See Exercise 7 of Chap. 3). In this frame $\nabla_{e_i} e_j(p) = 0$, hence

$$\begin{aligned}\nabla R(e_i,e_j,e_k,e_\ell,e_h) &= e_h\langle R(e_i,e_j)e_k,e_\ell\rangle = e_h\langle R(e_k,e_\ell)e_i,e_j\rangle \\ &= \langle \nabla_{e_h}\nabla_{e_\ell}\nabla_{e_k}e_i - \nabla_{e_h}\nabla_{e_k}\nabla_{e_\ell}e_i + \nabla_{e_h}\nabla_{[e_k,e_\ell]}e_i, e_j\rangle.\end{aligned}$$

Therefore, using the Jacobi identity for the bracket, we find

$$\begin{aligned}\nabla R(e_i,e_j,e_k,e_\ell,e_h) &+ \nabla R(e_i,e_j,e_\ell,e_h,e_k) \\ &+ \nabla R(e_i,e_j,e_h,e_k,e_\ell) = R(e_\ell,e_h,\nabla_{e_k}e_i,e_j) \\ &+ R(e_h,e_k,\nabla_{e_\ell}e_i,e_j) + R(e_k,e_\ell,\nabla_{e_h}e_i,e_j) = 0,\end{aligned}$$

since each one of the summands vanishes at p. The general case follows by linearity.

8. (*Schur's Theorem*). Let M^n be a connected Riemannian manifold with $n \geq 3$. Suppose that M is *isotropic*, that is, for each $p \in M$, the sectional curvature $K(p,\sigma)$ does not depend on $\sigma \subset T_pM$. Prove that M has constant sectional curvature, that is, $K(p,\sigma)$ also does not depend on p.
Hint: Define a tensor R' of order 4 by

$$R'(W,Z,X,Y) = \langle W,X\rangle\langle Z,Y\rangle - \langle Z,X\rangle\langle W,Y\rangle.$$

If $K(p,\sigma) = K$ does not depend on σ, by Lemma 3.4, $R = KR'$. Therefore, for all $U \in \mathcal{X}(M)$, $\nabla_U R = (UK)R'$. Using the 2nd Bianchi identity (see Exercise 7):

$$\nabla R(W,Z,X,Y,U) + \nabla R(W,Z,Y,U,X) + \nabla R(W,Z,U,X,Y) = 0,$$

we obtain, for all $X,Y,W,Z,U \in \mathcal{X}(M)$,

$$\begin{aligned} 0 = (UK)(\langle W,X\rangle\langle Z,Y\rangle - \langle Z,X\rangle\langle W,Y\rangle) \\ + (XK)(\langle W,Y\rangle\langle Z,U\rangle - \langle Z,Y\rangle\langle W,U\rangle) \\ + (YK)(\langle W,U\rangle\langle Z,X\rangle - \langle Z,U\rangle\langle W,X\rangle). \end{aligned}$$

Fix $p \in M$. Because $n \geq 3$, it is possible, fixing X at p, to choose Y and Z at p such that $\langle X,Y\rangle = \langle Y,Z\rangle = \langle Z,X\rangle = 0$, $\langle Z,Z\rangle = 1$. Put $U = Z$ at p. The relation above yields, for all W,

$$\langle (XK)Y - (YK)X, W\rangle = 0.$$

Since X and Y are linearly independent at p, we conclude that $XK = 0$ for all $X \in T_pM$. Thus $K = \text{const}$.

9. Prove that the scalar curvature $K(p)$ at $p \in M$ is given by

$$K(p) = \frac{1}{\omega_{n-1}} \int_{S^{n-1}} \mathrm{Ric}_p(x)\, dS^{n-1},$$

where ω_{n-1} is the area of the sphere S^{n-1} in T_pM and dS^{n-1} is the area elements on S^{n-1}.

Hint: Use the following general argument on quadratic forms. Consider an orthonormal basis $e_1, \dots, e_n$ in T_pM such that if $x = \sum_{i=1}^n x_i e_i$,

$$\mathrm{Ric}_p(x) = \sum \lambda_i x_i^2, \quad \lambda_i \quad \text{real}.$$

Because $|x| = 1$, the vector $(x_1, \dots, x_n) = \nu$ is a unit normal vector on S^{n-1}. Denoting $V = (\lambda_1 x_1, \dots, \lambda_n x_n)$, and using Stokes Theorem, we obtain

$$\begin{aligned} \frac{1}{\omega_{n-1}} \int_{S^{n-1}} \left(\sum \lambda_i x_i^2\right) dS^{n-1} &= \frac{1}{\omega_{n-1}} \int_{S^{n-1}} \langle V, \nu\rangle\, dS^{n-1} \\ &= \frac{1}{\omega_{n-1}} \int_{B^n} \operatorname{div} V\, dB^n, \end{aligned}$$

where B^n is the unit ball whose boundary is $S^{n-1} = \partial B^n$. Noting that $\operatorname{vol} B^n/\omega_n = 1/n$, we conclude that

$$\frac{1}{\omega_{n-1}} \int_{S^{n-1}} \operatorname{Ric}_p(x) dS^{n-1} = \frac{1}{n} \operatorname{div} V = \frac{\sum \lambda_i}{n}$$
$$= \frac{\sum \operatorname{Ric}_p(e_i)}{n} = K(p).$$

10. (Einstein manifolds). A Riemannian manifold M^n is called an *Einstein manifold* if, for all $X, Y \in \mathcal{X}(M)$, $\operatorname{Ric}(X,Y) = \lambda \langle X, Y \rangle$, where $\lambda: M \to \mathbf{R}$ is a real valued function. Prove that:
    a) If M^n is connected and Einstein, with $n \geq 3$, then λ is constant on M.
    b) If M^3 is a connected Einstein manifold then M^3 has constant sectional curvature.

    Hint for (a): Consider a geodesic orthonormal frame $\{e_i\}$, $i = 1, \ldots, n \geq 3$, at a point $p \in M$ (see Exercise 7 of Chap. 3). The 2nd Bianchi identity (see Exercise 7) at p can be written

$$(*) \qquad e_s(R_{hijk}) + e_j(R_{hiks}) + e_k(R_{hish}) = 0,$$

where R_{hijk} are the components of the curvature tensor in this frame and we take into account that $\nabla_{e_i} e_j(p) = 0$. Observe that $\langle e_i, e_k \rangle = g_{ik} = \delta_{ik} = \delta^{ik}$. Multiplying (*) by $\delta_{ik}\delta_{hj}$ and summing on i, k, h, j, we obtain: for the first part,

$$\sum_{ikjh} \delta_{hj}\delta_{ik} e_s(R_{hijk}) = e_s(\sum_{ikjh} \delta_{hj}\delta_{ik} R_{hijk})$$
$$= e_s(\sum_{hj} \delta_{hj} R_{hj}) = e_s(\sum_{hj} \delta_{hj}(\lambda \delta_{hj})) = n e_s(\lambda);$$

for the second part,

$$\sum_{ikjh} \delta_{hj}\delta_{ik} e_j(R_{hiks}) = -\sum_{jh} \delta_{hj} e_j(\sum_{ik} \delta_{ik} R_{hisk})$$
$$= \sum_{jh} \delta_{hj} e_j(\lambda \delta_{hs}) = -e_s(\lambda);$$

and for the third part,

$$\sum_{ikjh} \delta_{hj}\delta_{ik}e_k(R_{hiks}) = -e_s(\lambda).$$

Therefore, (*) implies that, for all s, $(n-2)e_s(\lambda) = 0$. From the arbitrariness of p, λ is constant on M.

CHAPTER 5

JACOBI FIELDS

1. Introduction

In this chapter we shall derive a first relation between the two basic concepts introduced previously, namely, geodesics and curvature. As we shall see (Cf. Rem. 2.11), the curvature $K(p,\sigma)$, $\sigma \subset T_pM$, determines how fast the geodesics, that start from p and are tangent to σ, spread apart. In order to formalize precisely this velocity of the deviation of the geodesics, it is necessary to introduce the so-called Jacobi fields. Jacobi fields are vector fields along geodesics, defined by means of a differential equation that arise naturally in the study of the exponential mapping (Cf. Sec. 2). Besides furnishing the relation mentioned above, the Jacobi fields allow us to obtain a simple characterization of the singularities of the exponential map (Cf. Prop. 3.5).

2. The Jacobi equation

Let M be a Riemannian manifold and let $p \in M$. In the proof of the Gauss Lemma we saw that if $\exp_p$ is defined at $v \in T_pM$, and if $w \in T_v(T_pM)$, then

$$(d\exp_p)_v w = \frac{\partial f}{\partial s}(1,0),$$

where f is a parametrized surface given by

$$f(t,s) = \exp_p tv(s), \quad 0 \le t \le 1, \quad -\varepsilon \le s \le \varepsilon,$$

and $v(s)$ is a curve in T_pM with $v(0) = v$, $v'(0) = w$.

We would like to obtain information on $|(d\exp_p)_v(w)|$. One of the reasons for this is that $|(d\exp_p)_v(w)|$ denotes, intuitively, the rate of spreading of the geodesics $t \to \exp_p tv(s)$ which start from p. As we shall see below, such spreading is associated with the value of the sectional curvature at p with respect to the plane generated by v and w. Another reason is that if we have $|(d\exp_p)_v(w)| = 0$ with $w \neq 0$, then v will be a critical point of $\exp_p$.

It is convenient to extend our objective slightly and study the field

$$(d\exp_p)_{tv}(tw) = \frac{\partial f}{\partial s}(t,0)$$

along the geodesic $\gamma(t) = \exp_p(tv)$, $0 \leq t \leq 1$.

The basic remark is that $\frac{\partial f}{\partial s}$ satisfies a differential equation. In fact, since γ is a geodesic, we have for all (t,s), $\frac{D}{\partial t}\frac{\partial f}{\partial t} = 0$. Thus, from Lemma 4.1 of Chap. 4,

$$\begin{aligned} 0 = \frac{D}{\partial s}\left(\frac{D}{\partial t}\frac{\partial f}{\partial t}\right) &= \frac{D}{\partial t}\frac{D}{\partial s}\frac{\partial f}{\partial t} - R\left(\frac{\partial f}{\partial s}, \frac{\partial f}{\partial t}\right)\frac{\partial f}{\partial t} \\ &= \frac{D}{\partial t}\frac{D}{\partial t}\frac{\partial f}{\partial s} + R\left(\frac{\partial f}{\partial t}, \frac{\partial f}{\partial s}\right)\frac{\partial f}{\partial t}. \end{aligned}$$

Putting $\frac{\partial f}{\partial s}(t,0) = J(t)$, we obtain the fact that J satisfies the equation

$$\frac{D^2 J}{dt^2} + R(\gamma'(t), J(t))\gamma'(t) = 0. \tag{1}$$

The equation above is called the *Jacobi equation*. Since it appears in a variety of situations, it is useful to make a separate study of it. We start with a definition.

2.1 DEFINITION. Let $\gamma\colon [0,a] \to M$ be a geodesic in M. A vector field J along γ is said to be a *Jacobi field* if it satisfies the Jacobi equation (1), for all $t \in [0,a]$.

A Jacobi field is determined by its initial conditions $J(0)$, $\frac{DJ}{dt}(0)$. Indeed, let $e_1(t), \ldots, e_n(t)$ be parallel, orthonormal fields along γ. We shall write:

$$J(t) = \sum_i f_i(t)e_i(t), \quad a_{ij} = \langle R(\gamma'(t), e_i(t))\gamma'(t), e_j(t)\rangle,$$

$$i, j = 1, \ldots, n = \dim M.$$

Then

$$\frac{D^2 J}{dt^2} = \sum_i f_i''(t) e_i(t),$$

and

$$\begin{aligned} R(\gamma', J)\gamma' &= \sum_j \langle R(\gamma', J)\gamma', e_j \rangle e_j \\ &= \sum_{ij} f_i \langle R(\gamma', e_i)\gamma', e_j \rangle e_j = \sum_{ij} f_i a_{ij} e_j. \end{aligned}$$

Therefore, the equation (1) is equivalent to the system

$$f_j''(t) + \sum_i a_{ij}(t) f_i(t) = 0, \quad j = 1, \ldots, n,$$

which is a linear system of the second order. Hence, given the initial conditions $J(0)$, $\frac{DJ}{dt}(0)$, there exists a C^∞ solution of the system, defined on $[0, a]$. There exist, therefore, $2n$ linearly independent Jacobi fields along γ.

2.2 REMARK. It is worth noting that $\gamma'(t)$ and $t\gamma'(t)$ are Jacobi fields along γ. The first field has derivative zero and vanishes nowhere; the second field is zero if and only if $t = 0$. Due to these facts, we shall consider Jacobi fields along γ that are normal to γ'.

2.3 EXAMPLE. (*Jacobi fields on manifolds of constant curvature*). Let M be a Riemannian manifold of constant sectional curvature K, and let $\gamma\colon [0, \ell] \to M$ be a normalized geodesic on M. Further let J be a Jacobi field along γ, normal to γ'. We claim that from the fact that $|\gamma'| = 1$ and from Lemma 3.4 of Chapter 4, it follows that

$$R(\gamma', J)\gamma' = KJ.$$

Indeed, for all vector fields T along γ we have

$$\begin{aligned} \langle R(\gamma', J)\gamma', T \rangle &= K\{\langle \gamma', \gamma' \rangle \langle J, T \rangle - \langle \gamma', T \rangle \langle J, \gamma' \rangle\} \\ &= K \langle J, T \rangle, \end{aligned}$$

which was asserted.

As a result, the Jacobi equation can be written as

$$\frac{D^2 J}{dt^2} + KJ = 0. \tag{2}$$

Let $w(t)$ be a parallel field along γ with $\langle \gamma'(t), w(t)\rangle = 0$ and $|w(t)| = 1$. It is easy to verify that

$$J(t) = \begin{cases} \frac{\sin(t\sqrt{K})}{\sqrt{K}} w(t), & \text{if } K > 0, \\ tw(t), & \text{if } K = 0, \\ \frac{\sinh(t\sqrt{-K})}{\sqrt{-K}} w(t), & \text{if } K < 0, \end{cases}$$

is a solution of (2) with initial conditions $J(0) = 0,\ J'(0) = w(0)$.

As we saw previously, given $p \in M$, $v \in T_pM$, and $w \in T_v(T_pM)$, we can construct a Jacobi field along the geodesic $\gamma\colon [0,1] \to M$, given by $\gamma(t) = \exp_p tv$. For that, we consider the parametrized surface given by $f(t,s) = \exp_p tv(s)$, where $v(s)$ is a curve in T_pM with $v(0) = v$, $v'(0) = w$, and take $J(t) = \frac{\partial f}{\partial s}(t,0)$. Notice that $J(0) = 0$.

We are going to show that this is essentially the only way of constructing Jacobi fields along $\gamma(t)$ with $J(0) = 0$. More precisely, we have the following proposition.

2.4 PROPOSITION. *Let $\gamma\colon [0,a] \to M$ be a geodesic and let J be a Jacobi field along γ with $J(0) = 0$. Put $\frac{DJ}{dt}(0) = w$ and $\gamma'(0) = v$. Consider w as an element of $T_{av}(T_{\gamma(0)}M)$ and construct a curve $v(s)$ in $T_{\gamma(0)}M$ with $v(0) = av$, $v'(0) = w$. Put $f(t,s) = \exp_p(\frac{t}{a}v(s))$, $p = \gamma(0)$, and define a Jacobi field $\bar{J}$ by $\bar{J}(t) = \frac{\partial f}{\partial s}(t,0)$. Then $\bar{J} = J$ on $[0,a]$.*

Proof. For $s = 0$, we have

$$\frac{D}{dt}\frac{\partial f}{\partial s} = \frac{D}{\partial t}((d\exp_p)_{tv}(tw)) = \frac{D}{\partial t}(t(d\exp_p)_{tv}(w))$$
$$= (d\exp_p)_{tv}(w) + t\frac{D}{\partial t}((d\exp_p)_{tv}(w)).$$

Therefore, for $t = 0$,

$$\frac{D\bar{J}}{dt}(0) = \frac{D}{\partial t}\frac{\partial f}{\partial s}(0,0) = (d\exp_p)_o(w) = w.$$

Since $J(0) = \bar{J}(0) = 0$ and $\frac{DJ}{dt}(0) = \frac{D\bar{J}}{dt}(0) = w$, we conclude, from the uniqueness theorem, that $J = \bar{J}$. □

2.5 COROLLARY. *Let $\gamma\colon [0,a] \to M$ be a geodesic. Then a Jacobi field J along γ with $J(0) = 0$ is given by*

$$J(t) = (d\exp_p)_{t\gamma'(0)}(tJ'(0)), \qquad t \in [0,a].$$

2.6 REMARK. It is possible to obtain a construction analogous to Proposition 2.4 for Jacobi fields that do not satisfy the condition $J(0) = 0$. Since we shall not use this fact, we leave its proof as an exercise (Exercise 2).

Now we are going to relate the rate of spreading of the geodesics that start from $p \in M$ with the curvature at p. From now on, for simplicity of notation, we shall put $\frac{DJ}{dt} = J'$, $\frac{D^2J}{dt^2} = J''$, etc.

2.7 PROPOSITION. *Let $p \in M$ and $\gamma\colon [0,a] \to M$ be a geodesic with $\gamma(0) = p$, $\gamma'(0) = v$. Let $w \in T_v(T_pM)$ with $|w| = 1$ and let J be a Jacobi field along γ given by*

$$J(t) = (d\exp_p)_{tv}(tw), \qquad 0 \leq t \leq a.$$

Then the Taylor expansion of $|J(t)|^2$ about $t = 0$ is given by

$$|J(t)|^2 = t^2 - \frac{1}{3}\langle R(v,w)v, w\rangle t^4 + R(t), \tag{3}$$

where $\lim_{t\to 0} \frac{R(t)}{t^4} = 0$.

Proof. Since $J(0) = 0$ and $J'(0) = w$, we have, for the first three coefficients:

$$\begin{aligned}
&\langle J, J\rangle(0) = 0,\\
&\langle J, J\rangle'(0) = 2\langle J, J'\rangle(0) = 0,\\
&\langle J, J\rangle''(0) = 2\langle J', J'\rangle(0) + 2\langle J'', J\rangle(0) = 2.
\end{aligned}$$

On the other hand, since $J''(0) = -R(\gamma', J)\gamma'(0) = 0$, we have

$$\langle J, J\rangle'''(0) = 6\langle J', J''\rangle(0) + 2\langle J''', J\rangle(0) = 0.$$

Now we need the following fact:

$$\nabla_{\gamma'}(R(\gamma', J)\gamma')(0) = R(\gamma', J')\gamma'(0). \tag{4}$$

To prove (4), note that for any W, we have at $t = 0$,

$$\begin{aligned}\langle \frac{D}{dt}(R(\gamma', J)\gamma'), W\rangle &= \frac{d}{dt}\langle R(\gamma', W)\gamma', J\rangle - \langle R(\gamma', J)\gamma', W'\rangle \\ &= \langle \frac{D}{dt}(R(\gamma', W)\gamma'), J\rangle + \langle R(\gamma', W)\gamma', J'\rangle \\ &= \langle R(\gamma', J')\gamma', W\rangle\end{aligned}$$

which implies (4).

It follows from (4) and from the Jacobi equation that $J'''(0) = -R(\gamma', J')\gamma'(0)$. Therefore,

$$\begin{aligned}\langle J, J\rangle''''(0) &= 8\langle J', J'''\rangle(0) + 6\langle J'', J''\rangle(0) + 2\langle J'''', J\rangle(0) \\ &= -8\langle J', R(\gamma', J')\gamma'\rangle(0) = -8\langle R(v, w)v, w\rangle.\end{aligned}$$

Putting together the calculation above, we obtain (3). □

2.8 REMARK. The expression (4) can also be obtained using the (covariant) derivation of tensors described in Section 5 of Chap. 4 (see exercise 5 of Chap. 4).

2.9 COROLLARY. *If $\gamma: [0, \ell] \to M$ is parametrized by arc length, (i.e., $|v| = 1$) and $\langle w, v\rangle = 0$, the expression $\langle R(v, w)v, w\rangle$ is the sectional curvature at p with respect to the plane σ generated by v and w. Therefore, in this situation,*

$$|J(t)|^2 = t^2 - \frac{1}{3}K(p, \sigma)t^4 + R(t). \tag{5}$$

2.10 COROLLARY. *With the same conditions as in the previous corollary,*

$$|J(t)| = t - \frac{1}{6}K(p, \sigma)t^3 + \tilde{R}(t), \qquad \lim_{t\to 0}\frac{\tilde{R}(t)}{t^3} = 0. \tag{6}$$

2.11 REMARK. The expression (6) essentially contains the relation between geodesics and curvature, mentioned in the beginning of this chapter. Indeed, considering the parametrized surface

$$f(t, s) = \exp_p tv(s), \quad t \in [0, \delta], \quad s \in (-\varepsilon, \varepsilon),$$

where δ is chosen so small that $\exp_p tv(s)$ is defined, and $v(s)$ is a curve in T_pM with $|v(s)| = 1$, $v(0) = v$, $v'(0) = w$, we see that the rays $t \to tv(s)$, $t \in [0, \delta]$, that start from the origin 0 of T_pM, deviate from the ray $t \to tv(0)$ with the velocity

$$\left|\left(\frac{\partial}{\partial s} tv(s)\right)(0)\right| = |tw| = t.$$

On the other hand, (6) tells us that the geodesics $t \to \exp_p(tv(s))$ deviate from the geodesic $\gamma(t) = \exp_p tv(0)$ with a velocity that differs from t by a term of the third order in t, given by $-\frac{1}{6}K(p,\sigma)t^3$. This tells us that, locally, the geodesics spread apart less than the rays in T_pM, if $K_p(\sigma) > 0$, and that they spread apart more than the rays in T_pM, if $K_p(\sigma) < 0$. Actually, for t small, the value $K(p,\sigma)t^3$ furnishes an approximation for the extent of this spread with an error of order t^3.

3. Conjugate points

Now we are going to turn to the relationship between the singularities of the exponential map and Jacobi fields. Before doing this, we require some definitions.

3.1 DEFINITION. Let $\gamma\colon [0, a] \to M$ be a geodesic. The point $\gamma(t_o)$ is said to be *conjugate* to $\gamma(0)$ along γ, $t_o \in (0, a]$, if there exists a Jacobi field J along γ, not identically zero, with $J(0) = 0 = J(t_o)$. The maximum number of such linearly independent fields is called the *multiplicity* of the conjugate point $\gamma(t_o)$.

Observe that if $\gamma(t_o)$ is conjugate to $\gamma(0)$, then $\gamma(0)$ is conjugate to $\gamma(t_o)$.

3.2 REMARK. If the dimension of M is n, there exist exactly n linearly independent Jacobi fields along the geodesic $\gamma\colon [0, a] \to M$, which are zero at $\gamma(0)$. This follows from the fact, easily checked, that the Jacobi fields $J_1, \dots, J_k$ with $J_i(0) = 0$ are linearly independent if and only if $J_1'(0), \dots, J_k'(0)$ are linearly independent. In addition, the Jacobi field $J(t) = t\gamma'(t)$ never vanishes for $t \neq 0$ (see Rem. 2.2). From this we deduce that the multiplicity of a conjugate point never exceeds $n - 1$.

3.3 EXAMPLE. Let $S^n = \{x \in R^{n+1}; |x| = 1\}$. In this example, we assume a fact which will be proved in the next chapter, namely, the sectional curvatures of S^n are all equal to one. The Jacobi field on S^n given in example 2.3, that is, $J(t) = (\sin t)w(t)$, satisfies the condition $J(0) = J(\pi) = 0$. Therefore, along any geodesic γ of S^n, the antipodal point $\gamma(\pi)$ to $\gamma(0)$ is conjugate to $\gamma(0)$. It is trivial to verify that there exists $n-1$ such fields that are linearly independent, that is, the multiplicity of $\gamma(\pi)$ as a conjugate point of $\gamma(0)$ is $n-1$.

3.4 DEFINITION. The set of (first) conjugate points to the point $p \in M$, for all the geodesics that start at p, is called the *conjugate locus* of p and is denoted by $C(p)$.

On S^n, $C(p) = \{-p\}$, for all p. The case of S^n, however, is not typical. A more typical example is given by the ellipsoid, where $C(p)$ is, in general, a curve with four singular points (see Fig. 4 of Chap. 13); Cf. Braunmühl, A., "Geodätische Linien auf dreiachsigen Flächen 2-Grades", Math. Ann., 20 (1882), 557-586.

The following proposition relates conjugate points with the singularities of the exponential map.

3.5 PROPOSITION. *Let $\gamma\colon [0,a] \to M$ be a geodesic and put $\gamma(0) = p$. The point $q = \gamma(t_o)$, $t_o \in (0,a]$, is conjugate to p along γ if and only if $v_o = t_o\gamma'(0)$ is a critical point of $\exp_p$. In addition, the multiplicity of q as a conjugate point of p is equal to the dimension of the kernel of the linear map $(d\exp_p)_{v_o}$.*

Proof. The point $q = \gamma(t_o)$ is a conjugate point of p along γ if and only if there exists a non-zero Jacobi field J along γ with $J(0) = J(t_o) = 0$. Let $v = \gamma'(0)$ and $w = J'(0)$. From Corollary 2.5, $J(t) = (d\exp_p)_{tv}(tw)$, $t \in [0,a]$. Observe that J is non-zero if and only if $w \neq 0$. Therefore, $q = \gamma(t_o)$ is conjugate to p if and only if

$$0 = J(t_o) = (d\exp_p)_{t_o v}(t_o w), \quad w \neq 0,$$

that is, if and only if, $t_o v$ is a critical point of $\exp_p$. The first assertion is therefore proved.

The multiplicity of q is equal to the number of linearly independent Jacobi fields $J_1, \dots, J_k$ which are zero at 0 and at t_o. As is easy to verify, the fields $J_1, \dots, J_k$ are linearly independent if and only if $J_1'(0), \dots, J_k'(0)$ are linearly independent in T_pM. From the construction above, the multiplicity of q is equal to the dimension of the kernel of $(d\exp_p)_{t_0 v}$. □

We conclude this chapter by presenting some properties of Jacobi fields that will be useful later on.

3.6 PROPOSITION. *Let J be a Jacobi field along the geodesic $\gamma\colon [0,a] \to M$. Then*

$$\langle J(t), \gamma'(t)\rangle = \langle J'(0), \gamma'(0)\rangle t + \langle J(0), \gamma'(0)\rangle, \quad t \in [0,a].$$

Proof. Omitting the t for the sake of notation, we have from the Jacobi equation,

$$\langle J', \gamma'\rangle' = \langle J'', \gamma'\rangle = -\langle R(\gamma', J)\gamma', \gamma'\rangle = 0.$$

Therefore, $\langle J', \gamma'\rangle = \langle J'(0), \gamma'(0)\rangle$. In addition,

$$\langle J, \gamma'\rangle' = \langle J', \gamma'\rangle = \langle J'(0), \gamma'(0)\rangle.$$

Integrating this last equation in t, we obtain finally

$$\langle J, \gamma'\rangle = \langle J'(0), \gamma'(0)\rangle t + \langle J(0), \gamma'(0)\rangle. \quad \square$$

3.7 COROLLARY. *If $\langle J, \gamma'\rangle(t_1) = \langle J, \gamma'\rangle(t_2)$, $t_1, t_2 \in [0,a]$, $t_1 \neq t_2$, then $\langle J, \gamma'\rangle$ does not depend on t; in particular, if $J(0) = J(a) = 0$, then $\langle J, \gamma'\rangle(t) \equiv 0$.*

3.8 COROLLARY. *Suppose that $J(0) = 0$. Then $\langle J'(0), \gamma'(0)\rangle = 0$ if and only if $\langle J, \gamma'\rangle(t) \equiv 0$; in particular, the space of Jacobi fields J with $J(0) = 0$ and $\langle J, \gamma'\rangle(t) \equiv 0$ has dimension equal to $n-1$.*

3.9 PROPOSITION. *Let $\gamma\colon [0,a] \to M$ be a geodesic. Let $V_1 \in T_{\gamma(0)}M$ and $V_2 \in T_{\gamma(a)}M$. If $\gamma(a)$ is not conjugate to $\gamma(0)$ there exists a unique Jacobi field J along γ, with $J(0) = V_1$ and $J(a) = V_2$.*

Proof. Let $\mathcal{J}$ be the space of Jacobi fields J with $J(0) = 0$. Define a mapping $\Theta\colon \mathcal{J} \to T_{\gamma(a)}M$ by

$$\Theta(J) = J(a), \qquad J \in \mathcal{J}.$$

Since $\gamma(a)$ is not conjugate to $\gamma(0)$, Θ is injective. Indeed, if $J_1 \neq J_2$ with $J_1(a) = J_2(a)$, we should have $J_1 - J_2$, a non-zero

Jacobi field, with $(J_1 - J_2)(0) = 0$, which contradicts the fact that $\gamma(a)$ is not conjugate to $\gamma(0)$.

Since Θ is linear, it follows from injectivity and the fact that $\dim \mathcal{J} = \dim T_{\gamma(a)}M$ that Θ is an isomorphism. Hence there exists $\bar{J}_1 \in \mathcal{J}$ with $\bar{J}_1(0) = 0$ and $\bar{J}_1(a) = V_2$.

By an analogous argument, there exists a Jacobi field $\bar{J}_2$ along γ with $\bar{J}_2(a) = 0$, $\bar{J}_2(0) = V_1$. The desired field is now given by $J = \bar{J}_1 + \bar{J}_2$. Uniqueness is clear. □

3.10 COROLLARY. *Let $\gamma\colon [0,a] \to M$ be a geodesic in M, $\dim M = n$, and let $\mathcal{J}^{\perp}$ be the space of Jacobi fields with $J(0) = 0$, $J'(0) \perp \gamma'(0)$. Let $\{J_1, \dots, J_{n-1}\}$ be a basis of $\mathcal{J}^{\perp}$. If $\gamma(t)$, $t \in (0,a]$, is not conjugate to $\gamma(0)$, then $\{J_1(t), \dots, J_{n-1}(t)\}$ is a basis for the orthogonal complement $\{\gamma'(t)\}^{\perp} \subset T_{\gamma(t)}M$ of $\gamma'(t)$.*

EXERCISES

1. Let M be a Riemannian manifold with sectional curvature identically zero. Show that, for every $p \in M$, the mapping $\exp_p\colon B_\varepsilon(0) \subset T_pM \to B_\varepsilon(p)$ is an isometry, where $B_\varepsilon(p)$ is a normal ball at p.

2. Let M be a Riemannian manifold, $\gamma\colon [0,1] \to M$ a geodesic, and J a Jacobi field along γ. Prove that there exists a parametrized surface $f(t,s)$, where $f(t,0) = \gamma(t)$ and the curves $t \to f(t,s)$ are geodesics, such that $J(t) = \frac{\partial f}{\partial s}(t,0)$.
Hint: Choose a curve $\lambda(s)$, $s \in (-\varepsilon, \varepsilon)$ in M such that $\lambda(0) = \gamma(0)$, $\lambda'(0) = J(0)$. Along λ choose a vector field $W(s)$ with $W(0) = \gamma'(0)$, $\frac{DW}{ds}(0) = \frac{DJ}{dt}(0)$. Define $f(s,t) = \exp_{\lambda(s)} tW(s)$ and verify that $\frac{\partial f}{\partial s}(0,0) = \frac{d\lambda}{ds}(0) = J(0)$ and
$$\frac{D}{dt}\frac{\partial f}{\partial s}(0,0) = \frac{D}{ds}\frac{\partial f}{\partial t}(0,0) = \frac{DW}{ds}(0) = \frac{DJ}{dt}(0).$$

3. Let M be a Riemannian manifold with non-positive sectional curvature. Prove that, for all p, the conjugate locus $C(p)$ is empty.

Hint: Assume the existence of a non-trivial Jacobi field along the geodesic $\gamma\colon [0,a] \to M$, with $\gamma(0) = p$, $J(0) = J(a) = 0$. Use the Jacobi equation to show that $\frac{d}{dt}\langle \frac{DJ}{dt}, J\rangle \geq 0$. Conclude that $\langle \frac{DJ}{dt}, J\rangle \equiv 0$. Since $\frac{d}{dt}\langle J, J\rangle = 2\langle \frac{DJ}{dt}, J\rangle \equiv 0$, we have $\|J\|^2 = \text{const.} = 0$, a contradiction.

4. Let $b < 0$ and let M be a manifold with constant negative sectional curvature equal to b. Let $\gamma\colon [0,\ell] \to M$ be a normalized geodesic, and let $v \in T_{\gamma(\ell)}M$ such that $\langle v, \gamma'(\ell)\rangle = 0$ and $|v| = 1$. Since M has negative curvature, $\gamma(\ell)$ is not conjugate to $\gamma(0)$ (see Exercise 3). Show that the Jacobi field J along γ determined by $J(0) = 0$, $J(\ell) = v$ is given by

$$J(t) = \frac{\sinh(t\sqrt{-b})}{\sinh(\ell\sqrt{-b})} w(t),$$

where $w(t)$ is the parallel transport along γ of the vector

$$w(0) = \frac{u_o}{|u_o|}, \qquad u_o = (d\exp_p)^{-1}_{\ell\gamma'(0)}(v)$$

and where u_o is considered as a vector $T_{\gamma(0)}M$ by the identification $T_{\gamma(0)}M \approx T_{\ell\gamma'(0)}(T_{\gamma(0)}M)$.
Hint: From example 2.3, the Jacobi field J_1 along γ satisfying $J_1(0) = 0$, $J_1'(0) = \frac{u_o}{|u_o|}$, is given by

$$J_1(t) = \frac{\sinh t\sqrt{-b}}{\sqrt{-b}} w(t).$$

In addition, from Corollary 2.5,

$$J_1(\ell) = (d\exp_p)_{\ell\gamma'(0)}(\ell w(0)).$$

It follows that

$$J(\ell) = v = (d\exp_p)_{\ell\gamma'(0)}(u_o) = J_1(\ell)\frac{|u_o|}{\ell}.$$

Therefore,

$$J(t) = J_1(t)\frac{|u_o|}{\ell} = \frac{\sinh t\sqrt{-b}}{\sqrt{-b}} w(t)\frac{|u_o|}{\ell}.$$

In addition, since

$$1 = |v| = |J(\ell)| = \frac{\sinh \ell\sqrt{-b}}{\sqrt{-b}} \frac{|u_o|}{\ell},$$

we have

$$\frac{|u_o|}{\ell} = \left(\frac{\sinh \ell\sqrt{-b}}{\sqrt{-b}}\right)^{-1},$$

which implies what was asserted.

5. *Jacobi fields and conjugate points on locally symmetric spaces* (Cf. Exercise 6 of Chap. 4).
Let $\gamma\colon [0,\infty) \to M$ be a geodesic in a locally symmetric space M and let $v = \gamma'(0)$ be its velocity at $p = \gamma(0)$. Define a linear transformation $K_v\colon T_pM \to T_pM$ by

$$K_v(x) = R(v,x)v, \qquad x \in T_pM.$$

a) Prove that K_v is self-adjoint.
b) Choose an orthonormal basis $\{e_1, \ldots, e_n\}$ of T_pM that diagonalizes K_v, that is,

$$K_v(e_i) = \lambda_i e_i, \qquad i = 1, \ldots, n.$$

Extend the e_i to fields along γ by parallel transport. Show that, for all t,

$$K_{\gamma'(t)}(e_i(t)) = \lambda_i e_i(t),$$

where λ_i does not depend on t.
Hint: Use Exercise 6(a), of Chap. 4.
c) Let $J(t) = \sum_i x_i(t)e_i(t)$ be a Jacobi field along γ. Show that the Jacobi equation is equivalent to the system

$$\frac{d^2x_i}{dt^2} + \lambda_i x_i = 0, \qquad i = 1, \ldots, n.$$

d) Show that the conjugate points of p along γ are given by $\gamma(\pi k/\sqrt{\lambda_i})$, where k is a positive integer and λ_i is a positive eigenvalue of K_v.

6. Let M be a Riemannian manifold of dimension two (in this case we say that M is a surface). Let $B_\delta(p)$ be a normal ball around the point $p \in M$ and consider the parametrized surface

$$f(\rho, \theta) = \exp_p \rho v(\theta), \quad 0 < \rho < \delta, \quad -\pi < \theta < \pi,$$

where $v(\theta)$ is a circle of radius δ in T_pM parametrized by the central angle θ.

   a) Show that (ρ, θ) are coordinates in an open set $U \subset M$ formed by the open ball $B_\delta(p)$ minus the ray $\exp_p(-\rho v(0))$ $0 < \rho < \delta$. Such coordinates are called *polar coordinates* at p.

   b) Show that the coefficients g_{ij} of the Riemannian metric in these coordinates are:

$$g_{12} = 0, \quad g_{11} = \left|\frac{\partial f}{\partial \rho}\right|^2 = |v(\theta)|^2 = 1, \quad g_{22} = \left|\frac{\partial f}{\partial \theta}\right|^2.$$

   c) Show that, along the geodesic $f(\rho, 0)$, we have

$$(\sqrt{g_{22}})_{\rho\rho} = -K(p)\rho + R(\rho), \quad \text{where} \quad \lim_{\rho \to 0} \frac{R(\rho)}{\rho} = 0$$

   and $K(p)$ is the sectional curvature of M at p.

   d) Prove that

$$\lim_{\rho \to 0} \frac{(\sqrt{g_{22}})_{\rho\rho}}{\sqrt{g_{22}}} = -K(p).$$

This last expression is the value of the Gaussian curvature of M at p given in polar coordinates (Cf., for example, M. do Carmo [dC 2] p. 288). This fact from the theory of surfaces, and (d) shows that, in dimension two, the sectional curvature coincides with the Gaussian curvature. In the next chapter, we shall give a more direct proof of this fact.

7. Let M be a Riemannian manifold of dimension two. Let $p \in M$ and let $V \subset T_pM$ be a neighborhood of the origin where $\exp_p$ is a diffeomorphism. Let $S_r(0) \subset V$ be a circle of radius r centered at the origin, and let L_r be the length of the curve $\exp_p(S_r)$ in M. Prove that the sectional curvature at $p \in M$ is given by

$$K(p) = \lim_{r \to 0} \frac{3}{\pi} \frac{2\pi r - L_r}{r^3}.$$

Hint: Use Exercise 6.

8. Let $\gamma\colon [0,a] \to M$ be a geodesic and let X be a Killing field on M.
   a) Show that the restriction $X(\gamma(s))$ of X to $\gamma(s)$ is a Jacobi field along γ.
   b) Use item (a) to show that (Cf. Exercise 6 of Chap. 3) if M is connected and there exists $p \in M$ with $X(p) = 0$ and $\nabla_Y X(p) = 0$, for all $Y(p) \in T_pM$, then $X = 0$ on M.

CHAPTER 6

ISOMETRIC IMMERSIONS

1. Introduction

In this chapter we shall consider the following situation. Let $f: M \to \overline{M}$ be a differentiable immersion of a manifold M of dimension n into a Riemannian manifold $\overline{M}$ of dimension equal to $k = n + m$. The Riemannian metric of $\overline{M}$ induces, in a natural manner, a Riemannian metric on M: if $v_1, v_2 \in T_pM$, define $\langle v_1, v_2 \rangle = \langle df_p(v_1), df_p(v_2) \rangle$. In this situation, f becomes an isometric immersion of M into $\overline{M}$. We should like to study the relationship between the geometry of M and that of $\overline{M}$.

As always, the motivation for this study comes from the classical case of surfaces S in $\mathbf{R}^3$. For the purpose of motivation, we can restrict ourselves to the case that the surface S is the graph $\{(x, y, z) \in \mathbf{R}^3; z = f(x, y)\}$ of a differentiable function f with $f(0,0) = 0$ and $f_x(0,0) = f_y(0,0) = 0$ (this last condition means that S is tangent to the x, y plane). In this case, we know from Calculus, that the behavior of S in a neighborhood of the origin $0 \in \mathbf{R}^3$ is strongly influenced by the quadratic form

$$II(x, y) = f_{xx}(0)x^2 + 2f_{xy}(0)xy + f_{yy}(0)y^2$$

defined in the x, y plane. II is called the second fundamental form of S at the point 0 and, for instance, the Gaussian curvature of S at 0 is given by

$$K = f_{xx}f_{yy} - f_{xy}^2. \tag{1}$$

Our first objective is to generalize the notion of the second fundamental form to the case $f: M \to \overline{M}$. Since the codimension m can be larger than 1, the quadratic form so defined must take values in a vector space of dimension $m \geq 1$.

As we shall see, the relations between the Riemannian metrics of M and $\overline{M}$ can be expressed by means of the second fundamental form. Among these relations, the most important is probably the Gauss formula (see Theorem 2.5) which generalizes (1) and yields the difference between the curvatures of M and $\overline{M}$ with the help of expressions involving the second fundamental form. Since the curvatures are defined intrinsically, the Gauss formula generalizes the fundamental theorem of Gauss, mentioned in the Introduction to Chapter 1, which was the point of departure for Riemannian Geometry.

Using the Gauss formula, we shall give a geometric interpretation of the sectional curvature that is essentially the definition of curvature used by Riemann (Cf. Chap. 4, Introduction).

In the last section, we introduce the equations of Codazzi and Ricci, which, together with Gauss' equation, form the fundamental equations of the local theory of isometric immersions.

2. The second fundamental form

Let $f: M^n \to \overline{M}^{n+m=k}$ be an immersion. Then, for each $p \in M$, there exists a neighborhood $U \subset M$ of p such that $f(U) \subset \overline{M}$ is a submanifold of $\overline{M}$. This means that there exists a neighborhood $\overline{U} \subset \overline{M}$ of $f(p)$ and a diffeomorphism $\varphi: \overline{U} \to V \subset \mathbf{R}^k$ to an open set V of $\mathbf{R}^k$, such that φ maps $f(U) \cap \overline{U}$ diffeomorphically onto an open set of a subspace of $\mathbf{R}^n \subset \mathbf{R}^k$ (See Fig. 1). To simplify the notation, we shall identify U with $f(U)$ and each vector $v \in T_qM$, $q \in U$, with $df_q(v) \in T_{f(q)}\overline{M}$. We shall use such identifications to extend, for example, a local vector field (that is, defined on U) on M to a local vector field (that is, defined on $\overline{U}$) on $\overline{M}$; if U is sufficiently small, such an extension is always possible, as is easily seen using the diffeomorphism φ.

For each $p \in M$, the inner product on $T_p\overline{M}$ splits $T_p\overline{M}$ into the direct sum

$$T_p\overline{M} = T_pM \oplus (T_pM)^{\perp},$$

where $(T_pM)^{\perp}$ is the orthogonal complement of T_pM in $T_p\overline{M}$.

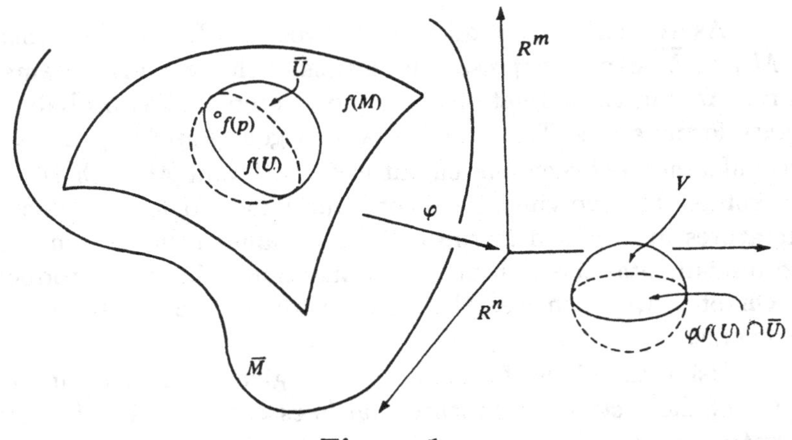

Figure 1

If $v \in T_p\overline{M}$, $p \in M$, we can write

$$v = v^T + v^N, \quad v^T \in T_pM, \quad v^N \in (T_pM)^\perp.$$

We call v^T the *tangential component* of v and v^N the *normal component* of v. Such a splitting is clearly differentiable, in the sense that the mappings

$$(p, v) \to (p, v^T) \quad \text{and} \quad (p, v) \to (p, v^N)$$

of $T\overline{M}$ into $T\overline{M}$ are differentiable.

The Riemannian connection on $\overline{M}$ will be denoted by $\overline{\nabla}$. If X and Y are local vector fields on M, and $\overline{X}$, $\overline{Y}$ are local extensions to $\overline{M}$, define

$$\nabla_X Y = (\overline{\nabla}_{\overline{X}}\overline{Y})^T.$$

It is easy to verify that this is the Riemannian connection relative to the metric induced on M (Cf. Exercise 3 of Chap. 2).

We want to define the second fundamental form of the immersion $f: M \to \overline{M}$. To do this, it is convenient to introduce beforehand the following definition. If X, Y are local vector fields on M,

$$B(X, Y) = \overline{\nabla}_{\overline{X}}\overline{Y} - \nabla_X Y$$

is a local vector field on $\overline{M}$ normal to M. $B(X, Y)$ does not depend on the extensions $\overline{X}$, $\overline{Y}$. Indeed, if $\overline{X}_1$ is another extension of X, we have

$$(\overline{\nabla}_{\overline{X}}\overline{Y} - \nabla_X Y) - (\overline{\nabla}_{\overline{X}_1}\overline{Y} - \nabla_X Y) = \overline{\nabla}_{\overline{X}-\overline{X}_1}\overline{Y},$$

which vanishes on M, because $\overline{X} - \overline{X}_1 = 0$ on M; using what was proved, it follows that if $\overline{Y}_1$ is another extension of Y,

$$(\overline{\nabla}_{\overline{X}}\overline{Y} - \nabla_X Y) - (\overline{\nabla}_{\overline{X}}\overline{Y}_1 - \nabla_X Y) = \overline{\nabla}_X(\overline{Y} - \overline{Y}_1) = 0,$$

because $\overline{Y} - \overline{Y}_1 = 0$ on M.

Therefore $B(X,Y)$ is well-defined. In what follows, let us denote by $\mathcal{X}(U)^{\perp}$ the differentiable vector fields on U that are normal to $f(U) \approx U$.

2.1 PROPOSITION. *If $X, Y \in \mathcal{X}(U)$, the mapping $B: \mathcal{X}(U) \times \mathcal{X}(U) \to \mathcal{X}(U)^{\perp}$ given by*

$$B(X,Y) = \overline{\nabla}_{\overline{X}}\overline{Y} - \nabla_X Y$$

is bilinear and symmetric.

Proof. From the properties of linearity of a connection, it is immediate that B is additive in X and Y and that

$$B(fX,Y) = fB(X,Y),\ f \in \mathcal{D}(U).$$

It remains to show that

$$B(X, fY) = fB(X,Y),\ f \in \mathcal{D}(U).$$

Denoting the extension of f to $\overline{U}$ by $\overline{f}$, we have

$$\begin{aligned} B(X, fY) &= \overline{\nabla}_{\overline{X}}(\overline{f}\overline{Y}) - \nabla_X(fY) \\ &= \overline{f}\overline{\nabla}_{\overline{X}}\overline{Y} - f\nabla_X Y + \overline{X}(\overline{f})\overline{Y} - X(f)Y. \end{aligned}$$

Since $f = \overline{f}$, $\overline{X}(\overline{f}) = X(f)$, and $Y = \overline{Y}$ on M, we conclude that the last two cancel, leaving,

$$B(X, fY) = fB(X,Y)$$

as we claimed.

To show that B is symmetric, we use the symmetry of the Riemannian connection, obtaining

$$B(X,Y) = \overline{\nabla}_{\overline{X}}\overline{Y} - \nabla_X Y = \overline{\nabla}_{\overline{Y}}\overline{X} + [\overline{X}, \overline{Y}] - \nabla_Y X - [X,Y].$$

Since, $[\overline{X},\overline{Y}] = [X,Y]$ on M, we conclude that $B(X,Y) = B(Y,X)$. □

Because B is bilinear, we see by writing B in a system of coordinates, that the value $B(X,Y)(p)$ depends only on the values $X(p)$ and $Y(p)$.

Now we are in a position to define the second fundamental form. Let $p \in M$ and $\eta \in (T_pM)^\perp$. The mapping $H_\eta\colon T_pM \times T_pM \to \mathbf{R}$ given by

$$H_\eta(x,y) = \langle B(x,y), \eta\rangle, \quad x,y \in T_pM,$$

is, by Proposition 2.1, a symmetric bilinear form.

2.2 DEFINITION. The quadratic form $I\!I_\eta$ defined on T_pM by

$$I\!I_\eta(x) = H_\eta(x,x)$$

is called *the second fundamental form* of f at p along the normal vector η.

Sometimes the expression *second fundamental form* is also used to designate the mapping B which at every point $p \in M$ is a symmetric bilinear mapping, taking values in $(T_pM)^\perp$.

Observe that the bilinear mapping H_η is associated to a linear self-adjoint operator $S_\eta\colon T_pM \to T_pM$ by

$$\langle S_\eta(x), y\rangle = H_\eta(x,y) = \langle B(x,y), \eta\rangle.$$

The following proposition expresses the linear operator associated to the second fundamental form in terms of the covariant derivative.

2.3 PROPOSITION. *Let $p \in M$, $x \in T_pM$ and $\eta \in (T_pM)^\perp$. Let N be a local extension of η normal to M. Then*

$$S_\eta(x) = -(\overline{\nabla}_x N)^T.$$

Proof. Let $y \in T_pM$ and let X, Y be local extensions of x, y, respectively, which are tangent to M. Then $\langle N, Y\rangle = 0$, and therefore

$$\begin{aligned}\langle S_\eta(x), y\rangle &= \langle B(X,Y)(p), N\rangle = \langle \overline{\nabla}_X Y - \nabla_X Y, N\rangle(p) \\ &= \langle \overline{\nabla}_X Y, N\rangle(p) = -\langle Y, \overline{\nabla}_X N\rangle(p) = \langle -\overline{\nabla}_x N, y\rangle,\end{aligned}$$

for all $y \in T_pM$. $\square$

2.4 EXAMPLE. Consider the particular case in which the codimension of the immersion is 1, i.e., $f: M^n \to \overline{M}^{n+1}$; $f(M) \subset \overline{M}$ is then called a *hypersurface*. (Observe that a hypersurface can have self-intersections).

Let $p \in M$ and $\eta \in (T_pM)^\perp$, $|\eta| = 1$. Since $S_\eta : T_pM \to T_pM$ is symmetric, there exists an orthonormal basis of eigenvectors $\{e_1, \ldots, e_n\}$ of T_pM with real eigenvalues $\lambda_1, \ldots, \lambda_n$, i.e., $S_\eta(e_i) = \lambda_i e_i$, $1 \leq i \leq n$. If M and $\overline{M}$ are both orientable and oriented (i.e., orientations are chosen on M and $\overline{M}$) then the vector η is uniquely determined if we require that both $\{e_1, \ldots, e_n\}$ is a basis in the orientation of M, and $\{e_1, \ldots, e_n, \eta\}$ is a basis in the orientation of $\overline{M}$. In this case, we say that the e_i are *principal directions* and that the $\lambda_i = k_i$ are *principal curvatures* of f. The symmetric functions of $\lambda_1, \ldots, \lambda_n$ are invariants of the immersion. For example: $\det(S_\eta) = \lambda_1 \ldots \lambda_n$ is called the *Gauss-Kronecker curvature* of f and $\frac{1}{n}(\lambda_1 + \cdots + \lambda_n)$ is called the *mean curvature* of f.

An important case occurs when $\overline{M} = \mathbf{R}^{n+1}$. Here, we can give an interesting geometric interpretation of S_η. To begin with, let N be a local extension of η, which is a unit vector field normal to M. Let $S_1^n = \{x \in \mathbf{R}^{n+1}; |x| = 1\}$ be the unit sphere in $\mathbf{R}^{n+1}$ and define the *Gauss spherical mapping*, $g: M^n \to S_1^n$, by translating the origin of the field N to the origin of $\mathbf{R}^{n+1}$ and taking

$$g(q) = \text{endpoint of the translation of } N(q).$$

Since T_qM and $T_{g(q)}S_1^n$ are parallel, we can identify them, and see that $dg_q : T_qM \to T_qM$ is given by

$$dg_q(x) = \frac{d}{dt}(N \circ c(t))_{t=0} = \overline{\nabla}_x N = (\overline{\nabla}_x N)^T = -S_\eta(x),$$

where $c: (-\varepsilon, \varepsilon) \to M$ is a curve with $c(0) = q$, $c'(0) = x$, and we have used the fact that $\langle N, N \rangle = 1$ to guarantee that $\overline{\nabla}_x N = (\overline{\nabla}_x N)^T$. It follows that $-S_\eta$ is the derivative of the Gauss spherical mapping.

The Gauss mapping has important topological implications. As an example, we shall prove the following fact: Let M^n, $n \geq 2$, be a connected, compact, orientable manifold. If there exists an immersion $f\colon M^n \to \mathbf{R}^{n+1}$ with non-vanishing Gauss-Kronecker curvature at every point of M, then M is diffeomorphic to S_1^n. The proof depends on the properties of covering spaces (see for example, M. do Carmo [dC 2] §5.6). The fact that $\det(dg_p) \neq 0$, for the Gauss mapping, tells us that $g\colon M \to S_1^n$ is a local diffeomorphism. Since M is compact, g is a covering map, and because S_1^n is simply connected, g is a diffeomorphism.

We now relate the curvature of M with the curvature of $\overline{M}$ and the second fundamental forms. If $x, y \in T_pM \subset T_p\overline{M}$, are linearly independent, denote by $K(x,y)$ and $\overline{K}(x,y)$ the sectional curvatures of M and $\overline{M}$, respectively, in the plane generated by x and y. Another proof of the next theorem will come up in Section 3 of this chapter.

2.5 THEOREM. *(Gauss). Let $p \in M$ and let x, y be orthonormal vectors in T_pM. Then*

$$(1) \qquad K(x,y) - \overline{K}(x,y) = \langle B(x,x), B(y,y)\rangle - |B(x,y)|^2.$$

Proof. Let X, Y be local orthogonal extensions of x, y, respectively, which are tangent to M; we denote the local extensions to $\overline{M}$ of X, Y by $\overline{X}$, $\overline{Y}$. Then

$$\begin{aligned} K(x,y) &- \overline{K}(x,y) \\ &= \langle \nabla_Y \nabla_X X - \nabla_X \nabla_Y X - (\overline{\nabla}_{\overline{Y}} \overline{\nabla}_{\overline{X}} \overline{X} - \overline{\nabla}_{\overline{X}} \overline{\nabla}_{\overline{Y}} \overline{X}), Y\rangle(p) \\ &\quad + \langle \nabla_{[X,Y]} X - \overline{\nabla}_{[\overline{X},\overline{Y}]} \overline{X}, Y\rangle(p). \end{aligned}$$

Observe, first of all, that the last term is zero, because

$$\langle \nabla_{[X,Y]} X - \overline{\nabla}_{[\overline{X},\overline{Y}]} \overline{X}, Y\rangle(p) = -\langle (\overline{\nabla}_{[\overline{X},\overline{Y}]} \overline{X})^N, Y\rangle(p) = 0.$$

On the other hand, if we denote by $E_1, \ldots, E_m$, $m = \dim \overline{M} - \dim M$, local orthonormal fields which are normal to M, we have

$$B(X,Y) = \sum_i H_i(X,Y) E_i, \quad H_i = H_{E_i}, \quad i = 1, \ldots, m.$$

Therefore, at p,

$$\overline{\nabla}_{\overline{Y}}\overline{\nabla}_{\overline{X}}\overline{X} = \overline{\nabla}_{\overline{Y}}(\sum_i H_i(X,X)E_i + \nabla_X X)$$
$$= \sum_i \left\{H_i(X,X)\overline{\nabla}_{\overline{Y}}E_i + \overline{Y}H_i(X,X)E_i\right\} + \overline{\nabla}_{\overline{Y}}\nabla_X X.$$

Hence, at p,

$$\text{(2)}\quad \langle \overline{\nabla}_{\overline{Y}}\overline{\nabla}_{\overline{X}}\overline{X}, Y\rangle = -\sum_i H_i(X,X)H_i(Y,Y) + \langle \nabla_Y\nabla_X X, Y\rangle.$$

Similarly,

$$\text{(3)}\quad \langle \overline{\nabla}_{\overline{X}}\overline{\nabla}_{\overline{Y}}\overline{X}, Y\rangle = -\sum_i H_i(X,Y)H_i(X,Y) + \langle \nabla_X\nabla_Y X, Y\rangle.$$

Using (2) and (3), we obtain (1). □

2.6 REMARK. In the case of a hypersurface $f\colon M^n \to \overline{M}^{n+1}$, the Gauss formula (1) has a very simple expression. Let $p \in M$ and $\eta \in (T_pM)^{\perp}$. Let $\{e_1,\dots,e_n\}$ be an orthonormal basis of T_pM in which $S_\eta = S$ is diagonal, that is, $S(e_i) = \lambda_i e_i$, $i = 1,\dots,n$, where $\lambda_1,\dots,\lambda_n$ are the eigenvalues of S. Then $H(e_i,e_i) = \lambda_i$ and $H(e_i,e_j) = 0$, if $i \neq j$. Therefore (1) can be written

$$\text{(4)}\qquad K(e_i,e_j) - \overline{K}(e_i,e_j) = \lambda_i\lambda_j.$$

2.7 REMARK. In the case in which $M = M^2 \subset \overline{M} = \mathbf{R}^3$, the product $\lambda_1\lambda_2$ of the principal curvatures coincides with Gaussian curvature of the surface. In this case, the previous Remark shows that the Gaussian curvature coincides with the sectional curvature of the surface, and implies the famous Theorem Egregium of Gauss, which asserts that the Gaussian curvature of $M^2 \subset \mathbf{R}^3$ is an invariant under isometries.

2.8 EXAMPLE. (*curvature of S^n*). We are going to show that the sectional curvature of the unit sphere $S^n \subset \mathbf{R}^{n+1}$ is a constant equal to 1.

For this, orient S^n by the inward pointing unit normal $N(x) = -x \in \mathbf{R}^{n+1}$, $|x| = 1$. The Gauss mapping is then equal to $(-i)$, where i is the identity of S^n. It follows then that the self-adjoint operator associated to H_N has all of its eigenvalues equal to 1. This means that for all $p \in S^n$, every $v \in T_pS^n$ is an eigenvector. Using the expression (4), we conclude that all sectional curvatures of S^n are equal to 1, as we have claimed.

An immersion $f: M \to \overline{M}$ is said to be *geodesic* at $p \in M$ if for every $\eta \in (T_pM)^\perp$ the second fundamental form H_η is identically zero at p. An immersion f is called *totally geodesic* if it is geodesic for all $p \in M$. The reason for this terminology is given by the next proposition.

2.9 PROPOSITION. *An immersion $f: M \to \overline{M}$ is geodesic at $p \in M$ if and only if every geodesic γ of M starting from p is a geodesic of $\overline{M}$ at p.*

Proof. Let $\gamma(0) = p$ and $\gamma'(0) = x$. Let N be a local extension, normal to M, of a vector η normal to M at p, and let X be a local extension of $\gamma'(t)$ to a tangent field on M. Since $\langle X, N\rangle = 0$, we obtain, at p,

$$\begin{aligned} H_\eta(x,x) &= \langle S_\eta(x), x\rangle = -\langle \overline{\nabla}_X N, X\rangle \\ &= -X\langle N, X\rangle + \langle N, \overline{\nabla}_X X\rangle = \langle N, \overline{\nabla}_X X\rangle. \end{aligned}$$

It follows that f is geodesic at p if and only if, for all $x \in T_pM$, the geodesic γ of M that is tangent to x at p satisfies the condition: $\overline{\nabla}_X X(p)$ does not have a normal component. Therefore f is geodesic at p if and only if every geodesic γ of M starting from p is a geodesic of $\overline{M}$ at p. □

Proposition 2.9 allows us to get what is probably the best geometric interpretation of sectional curvature. Let M be a Riemannian manifold and let p be a point of M. Let $B \subset T_pM$ be an open ball in T_pM on which $\exp_p$ is a diffeomorphism, and let $\sigma \subset T_pM$ be a subspace of dimension two. Then $\exp_p(\sigma \cap B) = S$ is a submanifold of dimension two of M passing through p. Intuitively, S is the surface formed by "small" geodesics that start from p and are tangent to σ at p. By Proposition 2.9, S is geodesic at p, hence the second fundamental forms of the inclusion $i: S \subset M$ vanish at p. As a submanifold of M, S has an induced Riemannian metric whose

Gaussian curvature at p will be denoted by K_S. It follows from the Gauss formula that

$$K_S(p) = K(p, \sigma).$$

In other words, the sectional curvature $K(p, \sigma)$ is the Gaussian curvature, at p, of a small surface formed by geodesics of M that start from p and are tangent to σ. This was exactly the way in which Riemann defined sectional curvature in [Ri].

Examples of totally geodesic submanifolds are rare. In the case that $\overline{M} = \mathbf{R}^n$, the linear subspaces and their translates are evidently totally geodesic submanifolds. In the case that $\overline{M} = S^n \subset \mathbf{R}^{n+1}$, the intersections $\sum$ of linear subspaces of $\mathbf{R}^{n+1}$ with S^n are totally geodesic submanifolds. This comes from the fact that for every $p \in \sum$, the geodesics of S^n that start from p and are tangent to $\sum$ are geodesics of $\sum$.

It was proven by E. Cartan that if a Riemannian manifold M has the property that, for every $p \in M$ and every two dimensional subspace of $\sigma \subset T_pM$, there exists a totally geodesic submanifold of M tangent to σ, then M has constant sectional curvature.

A much weaker condition than that of being totally geodesic is the condition of being minimal.

2.10 DEFINITION. An immersion $f: M \to \overline{M}$ is called *minimal* if for every $p \in M$ and every $\eta \in (T_pM)^{\perp}$ the trace of $S_\eta = 0$.

Choosing an orthonormal frame $E_1, \ldots, E_m$ of vectors in $\mathcal{X}(U)^{\perp}$, where U is a neighborhood of p in which f is an embedding, we can write, at p,

$$B(x, y) = \sum_i H_i(x, y) E_i, \quad x, y \in T_pM, \quad i = 1, \ldots, m,$$

where $H_i = H_{E_i}$. It is not hard to verify that the normal vector given by

$$H = \frac{1}{n} \sum_i (\text{trace } S_i) E_i,$$

where $S_i = S_{E_i}$, does not depend on the chosen frame E_i. The vector H is called the *mean curvature vector* of f. It is clear that f is minimal if and only if $H(p) = 0$, for all $p \in M$.

The reason for the use of the word minimal in this context is that such immersions minimize the volume, in the induced metric, in the same way that geodesics minimize arc length. More precisely, if $M \subset \overline{M}$ is a minimal submanifold and $D \subset M$ is a sufficiently small domain of M with regular boundary ∂D, then the volume of D in the induced metric is less than or equal to the volume of any other submanifold of $\overline{M}$ with the same boundary.

We are not going to enter into details here. There exists a vast literature on the subject and the reader may consult Chern [Ch2], Lawson [La], or Osserman [Os] for further study. Even in dimension $n = 2$, the topic is quite active, particularly with regard to questions related to the Plateau problem (Cf. Chap. 2 of Lawson [La]). For consideration of some current problems, the reader may consult W. Meeks [Me] and M. do Carmo, *Minimal Surfaces: stability and finiteness*, International Congress of Mathematicians, Helsinki, 1978.

The theory of isometric immersions is, by itself, a vast dominion of Riemannian geometry, of which we have only presented the most elementary parts. The reader may find more information on the subject in M. Dajczer [Da] and in the references therein mentioned.

3. The fundamental equations

Given an isometric immersion $f: M^n \to \overline{M}^{n+m}$, we have at each $p \in M$ the decomposition

$$T_p\overline{M} = T_pM \oplus (T_pM)^\perp,$$

which varies differentiably with p. This means that, locally, the portion of the tangent bundle $T\overline{M}$ which sits over M can be decomposed into the direct sum of the *tangent bundle* TM and the *normal bundle* $TM^\perp$. In what follows, we shall systematically use Latin letters X, Y, Z, etc., to denote differentiable vector fields tangent to M and Greek letters ξ, η, ζ, etc., to indicate differentiable vector fields normal to M.

Given X and η, we saw that the tangent component of $\overline{\nabla}_X\eta$ is given by $(\overline{\nabla}_X\eta)^T = -S_\eta(X)$. We plan now to study the normal component of $\overline{\nabla}_X\eta$, which will be called the *normal connection* $\nabla^\perp$ of the immersion. Explicitly,

$$\text{(5)} \qquad \nabla^\perp_X\eta = (\overline{\nabla}_X\eta)^N = \overline{\nabla}_X\eta - (\overline{\nabla}_X\eta)^T = \overline{\nabla}_X\eta + S_\eta(X).$$

It is easy to verify that the normal connection $\nabla^\perp$ has all of the usual properties of a connection, that is, it is linear in X, additive in η, and

$$\nabla^\perp_X(f\eta) = f\nabla^\perp_X\eta + X(f)\eta, \quad f \in \mathcal{D}(M).$$

In a similar way as in the case of the tangent bundle, we can introduce for the normal connection $\nabla^\perp$, a notion of curvature in the normal bundle which is called the *normal curvature* $R^\perp$ of the immersion and is defined by

$$R^\perp(X,Y)\eta = \nabla^\perp_Y\nabla^\perp_X\eta - \nabla^\perp_X\nabla^\perp_Y\eta + \nabla^\perp_{[X,Y]}\eta.$$

Everything about immersions occurs as if the geometry decomposes into two geometries: the geometry of the tangent bundle and the geometry of the normal bundle. These geometries are related by the second fundamental form of the immersion by means of expressions that generalize the classical equations of Gauss and Codazzi in the theory of surfaces. The objective of this section is to establish these relations.

3.1 PROPOSITION. *The following equations are valid:*
(a) Gauss equation

$$\langle \overline{R}(X,Y)Z,T\rangle = \langle R(X,Y)Z,T\rangle - \langle B(Y,T),B(X,Z)\rangle + \langle B(X,T),B(Y,Z)\rangle.$$

(b) Ricci equation

$$\langle \overline{R}(X,Y)\eta,\zeta\rangle - \langle R^\perp(X,Y)\eta,\zeta\rangle = \langle [S_\eta,S_\zeta]X,Y\rangle,$$

where $[S_\eta,S_\zeta]$ *denotes the operator* $S_\eta \circ S_\zeta - S_\zeta \circ S_\eta$.

Proof. Observe that $\overline{\nabla}_X Y = \nabla_X Y + B(X,Y)$. Since

$$\begin{aligned}\overline{R}(X,Y)Z &= \overline{\nabla}_Y\overline{\nabla}_X Z - \overline{\nabla}_X\overline{\nabla}_Y Z + \overline{\nabla}_{[X,Y]}Z\\ &= \overline{\nabla}_Y(\nabla_X Z + B(X,Z)) - \overline{\nabla}_X(\nabla_Y Z + B(Y,Z))\\ &\quad + \nabla_{[X,Y]}Z + B([X,Y],Z),\end{aligned}$$

we have

$$\begin{aligned}\overline{R}(X,Y)Z = R(X,Y)Z &+ B(Y,\nabla_X Z) + \nabla^{\perp}_Y B(X,Z)\\ &- S_{B(X,Z)}Y - B(X,\nabla_Y Z) - \nabla^{\perp}_X B(Y,Z)\\ &+ S_{B(Y,Z)}X + B([X,Y],Z).\end{aligned} \tag{6}$$

Taking the inner product of (6) with T, since the normal terms vanish, we obtain finally

$$\begin{aligned}\langle\overline{R}(X,Y)Z,T\rangle &= \langle R(X,Y)Z,T\rangle - \langle S_{B(X,Z)}Y,T\rangle + \langle S_{B(Y,Z)}X,T\rangle\\ &= \langle R(X,Y)Z,T\rangle - \langle B(Y,T),B(X,Z)\rangle\\ &\quad + \langle B(X,T),B(Y,Z)\rangle\end{aligned}$$

which is the Gauss equation.

To get the Ricci equation, we calculate

$$\begin{aligned}\overline{R}(X,Y)\eta &= \overline{\nabla}_Y\overline{\nabla}_X\eta - \overline{\nabla}_X\overline{\nabla}_Y\eta + \overline{\nabla}_{[X,Y]}\eta\\ &= \overline{\nabla}_Y(\nabla^{\perp}_X\eta - S_\eta X) - \overline{\nabla}_X(\nabla^{\perp}_Y\eta - S_\eta Y)\\ &\quad + \nabla^{\perp}_{[X,Y]}\eta - S_\eta[X,Y]\\ &= R^{\perp}(X,Y)\eta - S_{\nabla^{\perp}_X\eta}Y - \nabla_Y(S_\eta X) - B(S_\eta X,Y)\\ &\quad + S_{\nabla^{\perp}_Y\eta}X + \nabla_X(S_\eta Y) + B(X,S_\eta Y) - S_\eta[X,Y].\end{aligned}$$

Multiplying the expression by ζ and observing that $\langle B(X,Y),\eta\rangle = \langle S_\eta X,Y\rangle$, we obtain

$$\begin{aligned}\langle\overline{R}(X,Y)\eta,\zeta\rangle &= \langle R^{\perp}(X,Y)\eta,\zeta\rangle - \langle B(S_\eta X,Y),\zeta\rangle + \langle B(X,S_\eta Y),\zeta\rangle\\ &= \langle R^{\perp}(X,Y)\eta,\zeta\rangle + \langle(S_\eta S_\zeta - S_\zeta S_\eta)X,Y\rangle\\ &= \langle R^{\perp}(X,Y)\eta,\zeta\rangle + \langle[S_\eta,S_\zeta]X,Y\rangle,\end{aligned}$$

which is Ricci's equation. □

3.2 REMARK. Theorem 2.5 of this chapter is a special case of Gauss' equation.

3.3 Remark. We say that the normal bundle of an immersion is *flat* if $R^\perp = 0$. Assume that the ambient space $\overline{M}$ has constant sectional curvature. Then the Ricci equation can be written as

$$\langle R^\perp(X,Y)\eta,\zeta\rangle = -\langle [S_\eta, S_\zeta]X, Y\rangle.$$

It follows that $R^\perp = 0$ if and only if $[S_\eta, S_\zeta] = 0$ for all η, ζ, that is, if and only if for all $p \in M$ there exists a basis of T_pM which diagonalizes all of the S_η simultaneously.

The Gauss and Ricci equations are algebraic expressions that relate the curvatures of the tangent and normal bundles, respectively, with the second fundamental form of the immersion. A non-algebraic relation is given by the Codazzi equation, for which we need to "differentiate" the second fundamental form considered as a tensor.

Given an isometric immersion, let us denote the space of differentiable vector fields normal to M by $\mathcal{X}(M)^\perp$. The second fundamental form of the immersion can then be considered as a tensor

$$B\colon \mathcal{X}(M) \times \mathcal{X}(M) \times \mathcal{X}(M)^\perp \to \mathbf{R}$$

defined by

$$B(X,Y,\eta) = \langle B(X,Y),\eta\rangle.$$

The definition of covariant derivative extends to tensors of this type in a natural way:

$$\begin{aligned}(\overline{\nabla}_X B)(Y,Z,\eta) = X(B(Y,Z,\eta)) &- B(\nabla_X Y, Z, \eta)\\ &- B(Y, \nabla_X Z, \eta) - B(Y,Z,\nabla^\perp_X \eta).\end{aligned}$$

3.4 Proposition. *(Codazzi's equation). With the notation above,*

$$\langle \overline{R}(X,Y)Z,\eta\rangle = (\overline{\nabla}_Y B)(X,Z,\eta) - (\overline{\nabla}_X B)(Y,Z,\eta).$$

Proof. Observe, to begin with, that

$$\begin{aligned}(\overline{\nabla}_X B)(Y,Z,\eta) &= X\langle B(Y,Z),\eta\rangle - \langle B(\nabla_X Y, Z),\eta\rangle\\ &\quad - \langle B(Y,\nabla_X Z),\eta\rangle - \langle B(Y,Z),\nabla^\perp_X\eta\rangle\\ &= \langle \nabla^\perp_X(B(Y,Z)),\eta\rangle - \langle B(\nabla_X Y, Z),\eta\rangle\\ &\quad - \langle B(Y,\nabla_X Z),\eta\rangle.\end{aligned}$$

Consider now expression (6) in the proof of Proposition 3.1, and multiply both sides of (6) by η. We obtain, taking into account the remark above,

$$\begin{aligned}\langle \overline{R}(X,Y)Z,\eta\rangle &= \langle B(Y,\nabla_X Z),\eta\rangle + \langle \nabla_Y^{\perp} B(X,Z),\eta\rangle \\ &\quad - \langle B(X,\nabla_Y Z),\eta\rangle - \langle \nabla_X^{\perp} B(Y,Z),\eta\rangle \\ &\quad + \langle B(\nabla_X Y,Z),\eta\rangle - \langle B(\nabla_Y X,Z)\eta\rangle \\ &= -(\overline{\nabla}_X B)(Y,Z,\eta) + (\overline{\nabla}_Y B)(X,Z,\eta)\end{aligned}$$

which is Codazzi's equation. □

3.5 Remark. If the ambient space $\overline{M}$ has constant sectional curvature, the Codazzi equation reduces to

$$(\overline{\nabla}_X B)(Y,Z,\eta) = (\overline{\nabla}_Y B)(X,Z,\eta).$$

If, in addition, the codimension of the immersion is 1, $\nabla_X^{\perp}\eta = 0$, hence

$$\begin{aligned}\overline{\nabla}_X B(Y,Z,\eta) &= X\langle S_\eta(Y),Z\rangle - \langle S_\eta(\nabla_X Y),Z\rangle - \langle S_\eta(Y),\nabla_X Z\rangle \\ &= \langle \nabla_X(S_\eta(Y)),Z\rangle - \langle S_\eta(\nabla_X Y),Z\rangle.\end{aligned}$$

Therefore, in this case, the Codazzi equation can be written

$$\nabla_X(S_\eta(Y)) - \nabla_Y(S_\eta(X)) = S_\eta([X,Y]).$$

The importance of the equations of Gauss, Codazzi, and Ricci is that, in the case in which the ambient space has constant sectional curvature, they play an analogous role to that of the compatibility equations in the local theory of surfaces (Cf. M. do Carmo [dC 2], pp. 235-236). Indeed, the compatibility equations in surface theory are only special cases of the Gauss and Codazzi equations obtained in this section. In the present case, it is possible to state an analogous theorem to the fundamental theorem of local surface theory (cf. loc. cit. p. 236). We refer the reader to the article of K. Tenenblat, "On isometric immersions of Riemannian manifolds", Boletim da Soc. Bras. de Mat. vol. 2 (1971), 23-36. A very nice discussion on the subject can be found in the article of H. Jacobowitz, "The Gauss-Codazzi equations", Tensor N.S., 39 (1982), 15-22.

EXERCISES

1. Let M_1 and M_2 be Riemannian manifolds, and consider the product $M_1 \times M_2$, with the product metric. Let ∇^1 be the Riemannian connection of M_1 and let ∇^2 be the Riemannian connection of M_2.
   a) Show that the Riemannian connection ∇ of $M_1 \times M_2$ is given by $\nabla_{Y_1+Y_2}(X_1 + X_2) = \overset{1}{\nabla}_{Y_1} X_1 + \overset{2}{\nabla}_{Y_2} X_2$, $X_1, Y_1 \in \mathcal{X}(M_1)$, $X_2, Y_2 \in \mathcal{X}(M_2)$.
   b) For every $p \in M_1$, the set $(M_2)_p = \{(p,q) \in M_1 \times M_2; q \in M_2\}$ is a submanifold of $M_1 \times M_2$, naturally diffeomorphic to M_2. Prove that $(M_2)_p$ is a totally geodesic submanifold of $M_1 \times M_2$.
   c) Let $\sigma(x,y) \subset T_{(p,q)}(M_1 \times M_2)$ be a plane such that $x \in T_pM_1$ and $y \in T_qM_2$. Show that $K(\sigma) = 0$.

2. Show that $\mathbf{x}\colon \mathbf{R}^2 \to \mathbf{R}^4$ given by

$$\mathbf{x}(\theta, \varphi) = \frac{1}{\sqrt{2}}(\cos\theta, \sin\theta, \cos\varphi, \sin\varphi), \quad (\theta, \varphi) \in \mathbf{R}^2$$

   is an immersion of $\mathbf{R}^2$ into the unit sphere $S^3(1) \subset \mathbf{R}^4$, whose image $\mathbf{x}(\mathbf{R}^2)$ is a torus T^2 with sectional curvature zero in the induced metric.

3. Let M be a Riemannian manifold and let $N \subset K \subset M$ be submanifolds of M. Suppose that N is totally geodesic in K and that K is totally geodesic in M. Prove that N is totally geodesic in M.

4. Let $N_1 \subset M_1$, $N_2 \subset M_2$ be totally geodesic submanifolds of the Riemannian manifolds M_1 and M_2, respectively. Prove that $N_1 \times N_2$ is a totally geodesic submanifold of the product $M_1 \times M_2$ with the product metric.

5. Prove that the sectional curvature of the Riemannian manifold $S^2 \times S^2$ with the product metric, where S^2 is the unit sphere in $\mathbf{R}^3$, is non-negative. Find a totally geodesic, flat torus, T^2, embedded in $S^2 \times S^2$.

6. Let G be a Lie group with a bi-invariant metric. Let H be a Lie group and let $h: H \to G$ be an immersion that is also a homomorphism of groups (that is, H is a Lie subgroup of G). Show that h is a totally geodesic immersion.

7. Show that if M is a totally geodesic submanifold of $\overline{M}$, then, for any tangent fields to M, ∇ and $\overline{\nabla}$ coincide.

8. (*The Clifford torus*). Consider the immersion $\mathbf{x}: \mathbf{R}^2 \to \mathbf{R}^4$ given in Exercise 2.

   a) Show that the vectors

$$e_1 = (-\sin\theta, \cos\theta, 0, 0), \quad e_2 = (0, 0, -\sin\varphi, \cos\varphi)$$

   form an orthonormal basis of the tangent space, and that the vectors $n_1 = \frac{1}{\sqrt{2}}(\cos\theta, \sin\theta, \cos\varphi, \sin\varphi)$, $n_2 = \frac{1}{\sqrt{2}}(-\cos\theta, -\sin\theta, \cos\varphi, \sin\varphi)$ form an orthonormal basis of the normal space.

   b) Use the fact that

$$\langle S_{n_k}(e_i), e_j\rangle = -\langle \overline{\nabla}_{e_i} n_k, e_j\rangle = \langle \overline{\nabla}_{e_i} e_j, n_k\rangle,$$

   where $\overline{\nabla}$ is the covariant derivative (that is, the usual derivative) of $\mathbf{R}^4$, and $i, j, k = 1, 2$, to establish that the matrices of S_{n_1} and S_{n_2} with respect to the basis $\{e_1, e_2\}$ are

$$S_{n_1} = \begin{pmatrix} -1 & 0 \\ 0 & -1 \end{pmatrix}$$

$$S_{n_2} = \begin{pmatrix} 1 & 0 \\ 0 & -1 \end{pmatrix}$$

   c) From Exercise 2, $\mathbf{x}$ is an immersion of the torus T^2 into $S^3(1)$ (the Clifford torus). Show that $\mathbf{x}$ is a minimal immersion.

9. Let $f: M^n \to \mathbf{R}^{m+n}$ be an immersion. Let $\eta \in (T_pM)^\perp$, $p \in M$ and $V = T_pM \oplus \mathbf{R}\eta \subset \mathbf{R}^{m+n}$, where $\mathbf{R}\eta = \{\lambda\eta \mid \lambda \in \mathbf{R}\}$. Let $\pi: \mathbf{R}^{m+n} \to \mathbf{R}^{m+n}$ be the orthogonal projection onto $T_pM \oplus \mathbf{R}\eta$. Since η is transversal to M at p, $\pi \mid U$ is an embedding, where U is a sufficiently small neighborhood of p in M. Let $M' = \pi(U) \subset T_pM \oplus \mathbf{R}\eta$ and let

$S'_\eta: T_pM' = T_pM \to T_pM'$ be the operator associated to the second fundamental form of M' at p in the direction of η. Show that $S'_\eta = S_\eta$, where $S_\eta: T_pM \to T_pM$ is the operator associated to the second fundamental form of M' at p in the direction of η.

Hint: Let N and N' be normal fields along U and $\pi(U)$, respectively, such that $N(p) = N'(p) = \eta$. Then if $X \in T_pM$, $S_\eta(X) = -(\overline{\nabla}_X N)^T$ and $S'_\eta(X) = -(\overline{\overline{\nabla}}_X N')^T$. Show that it is possible to choose N in such a way that $N' = d\pi(N)$ and observe that the restriction $d\pi \mid T_pM = \text{id}$. Hence, at p, $S'_\eta(X) = -(\overline{\overline{\nabla}}_X(d\pi N))^T = -d\pi(\overline{\nabla}_X N)^T = S_\eta(X)$.

10. Let $f: M^n \to \overline{M}^{n+k}$ be an isometric immersion and let $S_\eta: TM \to TM$ be the operator associated to the second fundamental form of f along the normal field η. Consider S_η as a tensor of order 2 given by $S_\eta(X,Y) = \langle S_\eta(X), Y\rangle$, $X, Y \in \mathcal{X}(M)$. Observe that saying the operator S_η is self-adjoint is equivalent to saying that the tensor S_η is symmetric, that is, $S_\eta(X,Y) = S_\eta(Y,X)$. Prove that for all $V \in \mathcal{X}(M)$, the tensor $\nabla_V S_\eta$ is symmetric.
Hint: Differentiating $\langle S_\eta X, Y\rangle = \langle X, S_\eta Y\rangle$ with respect to V, we obtain

$$\langle \nabla_V(S_\eta X), Y\rangle + \langle S_\eta X, \nabla_V Y\rangle = \langle \nabla_V X, S_\eta Y\rangle + \langle X, \nabla_V(S_\eta Y)\rangle.$$

Using the fact that

$$\langle((\nabla_V S_\eta)X), Y\rangle = \langle \nabla_V(S_\eta X), Y\rangle - \langle S_\eta(\nabla_V X), Y\rangle$$

and the previous expression, we obtain easily that

$$\langle(\nabla_V S_\eta)X, Y\rangle = \langle X, (\nabla_V S_\eta)Y\rangle.$$

11. Let $f: \overline{M}^{n+1} \to \mathbf{R}$ be a differentiable function. Define the *Hessian*, Hess f of f at $p \in \overline{M}$ as the linear operator

$$\text{Hess}\, f: T_p\overline{M} \to T_p\overline{M}, \quad (\text{Hess}\, f)Y = \overline{\nabla}_Y \operatorname{grad} f, \quad Y \in T_p\overline{M},$$

where $\overline{\nabla}$ is the Riemannian connection of $\overline{M}$. Let a be a regular value of f and let $M^n \subset \overline{M}^{n+1}$ be the hypersuperface in $\overline{M}$ defined by $M = \{p \in \overline{M}; f(p) = a\}$. Prove that:

a) The Laplacian $\overline{\Delta} f$ is given by

$$\overline{\Delta} f = \text{trace Hess} f.$$

b) If $X, Y \in \mathcal{X}(\overline{M})$, then

$$\langle (\text{Hess} f)Y, X\rangle = \langle Y, (\text{Hess} f)X\rangle.$$

Conclude that Hess f is self-adjoint, hence determines a symmetric bilinear form on $T_p\overline{M}$, $p \in \overline{M}$, given by $(\text{Hess} f)(X,Y) = \langle (\text{Hess} f)X, Y\rangle$, $X, Y \in T_p\overline{M}$.

c) The mean curvature H of $M \subset \overline{M}$ is given by

$$nH = -\operatorname{div}\left(\frac{\operatorname{grad} f}{|\operatorname{grad} f|}\right).$$

Hint: Take an orthonormal frame $E_1, \ldots, E_n$, $E_{n+1} = \frac{\operatorname{grad} f}{|\operatorname{grad} f|} = \eta$ in a neighborhood of $p \in M$ in $\overline{M}$ and use the definition of divergence in Exercise 8, Chapter 3, to obtain

$$\begin{aligned} nH &= \text{trace } S_\eta = \sum_{i=1}^{n} \langle S_\eta(E_i), E_i\rangle \\ &= -\sum_{i=1}^{n} \langle \overline{\nabla}_{E_i}\eta, E_i\rangle - \langle \overline{\nabla}_\eta \eta, \eta\rangle = -\sum_{i=1}^{n+1} \langle \overline{\nabla}_{E_i}\eta, E_i\rangle \\ &= -\operatorname{div}_{\overline{M}} \eta = -\operatorname{div}\left(\frac{\operatorname{grad} f}{|\operatorname{grad} f|}\right). \end{aligned}$$

d) Observe that every embedded hypersurface $M^n \subset \overline{M}^{n+1}$ is locally the inverse image of a regular value. Conclude from (c) that the mean curvature H of such a hypersuperface is given by

$$H = -\frac{1}{n}\operatorname{div} N,$$

where N is an appropriate local extension of the unit normal vector field on $M^n \subset \overline{M}^{n+1}$.

12. (*Singularities of a Killing field*). Let X be a Killing vector field on a Riemannian manifold M. Let $N = \{p \in M; X(p) = 0\}$. Prove that:

    a) If $p \in N$, and $V \subset M$ is a normal neighborhood of p, with $q \in N \cap V$, then the radial geodesic segment γ joining p to q is contained in N. Conclude that $\gamma \cap V \subset N$.

    b) If $p \in N$, there exists a neighborhood $V \subset M$ of p such that $V \cap N$ is a submanifold of M (this implies that every connected component of N is a submanifold of M).

    Hint: Proceed by induction, using (a). If p is isolated, nothing has to be done. In the contrary case, let $V \subset M$ be a normal neighborhood of p such that there exists $q_1 \in V \cap N$ and consider the radial geodesic γ_1 joining p to q_1. If $V \cap N = \gamma_1$, by (a), the proof is complete. Otherwise, let $q_2 \in V \cap N - \{\gamma_1\}$ and let γ_2 be the radial geodesic joining p to q_2. Let $Q \subset T_pM$ be the subspace generated by the vectors $\exp_p^{-1}(q_1)$ and $\exp_p^{-1}(q_2)$ and let $N_2 = \exp_p(Q \cap \exp_p^{-1}(V))$. Show that for all $t \in \mathbf{R}$, the restriction of the differential $(dX_t)_p$ of the flow $X_t \colon M \to M$, to Q, is the identity; conclude now that $N_2 \subset V \cap N$. Proceed in this way until the dimension of T_pM is exhausted.

    c) The codimension, as a submanifold of M, of a connected component N_k of N is even. Assume the following fact: if a sphere has a non-vanishing differentiable vector field on it then its dimension must be odd (for a proof, see Armstrong, [Ar], p. 198).

    Hint: Let $E_p = (T_pN_k)^{\perp}$ and let $V \subset M$ be a normal neighborhood of p. Set $N_k^{\perp} = \exp_p(E_p \cap \exp_p^{-1}(V))$. Since, for all t, $(dX_t)_p \colon E_p \to E_p$, we have that X is tangent to $N_k^{\perp}$. On the other hand, $X \neq 0$ is tangent to the geodesic spheres of $N_k^{\perp}$ with center p. From the theorem mentioned above, the dimension of such a sphere is odd. Hence $\dim N_k^{\perp} = \dim E_p$ is even.

CHAPTER 7

COMPLETE MANIFOLDS; HOPF-RINOW AND HADAMARD THEOREMS

1. Introduction

So far, we have essentially studied local properties of Riemannian manifolds. However, one of the most interesting aspects of differential geometry is the interplay that exists between the local properties and the global properties of a Riemannian manifold. By a local property, we mean a property that depends on the behavior of the manifold in the neighborhood of a point, and by a global property, we mean one which depends on the behavior of the manifold taken as a whole.

In this chapter, we begin the study of the relations between local and global properties. First, we define the natural "habitat" of global properties, namely, a complete Riemannian manifold M, as a manifold in which the geodesics are defined for all values of their parameter (Cf. Def. 2.2). Formally, this means that for any $p \in M$, $\exp_p$ is defined on all of T_pM; intuitively, this means that the manifold does not have any holes or boundaries.

What turns out to be very useful on complete manifolds is the fact that (Theorem of Hopf and Rinow) given any two points of such a manifold there exists a minimizing geodesic joining these two points. We are going to prove this statement in Theorem 2.8, together with other facts which imply, for instance, that a compact manifold is complete and that a closed submanifold of a complete manifold is a complete manifold.

As an application of the Theorem of Hopf and Rinow, we prove the theorem of Hadamard which states that there is a homeomorphism of a complete simply connected manifold of dimension n, whose sectional curvature satisfies $K \leq 0$, onto $\mathbf{R}^n$. This is an example of the relation between local and global properties, in which a local condition ($K \leq 0$) together with weak global restrictions (complete and simply connected) imply a strong global restriction (homeomorphic to $\mathbf{R}^n$).

From now on, except when explicitly mentioned otherwise, all manifolds will be supposed connected.

2. Complete manifolds; Hopf-Rinow Theorem

When we want to study global properties of a differentiable manifold M, we need to make sure that M is not a proper open submanifold of a manifold M'. The usual condition for guaranteeing this non-extendibility is compactness. In certain cases, however, we would like to use a weaker condition; a natural definition would be the following.

2.1 DEFINITION. A Riemannian manifold M is said to be *extendible* if there exists a Riemannian manifold M' such that M is isometric to a proper open subset of M'. In the opposite case, M is called *non-extendible*.

It happens that the class of non-extendible Riemannian manifolds is too large. A convenient subfamily of this class is given in the next definition.

2.2 DEFINITION. A Riemannian manifold M is (geodesically) *complete* if for all $p \in M$, the exponential map, $\exp_p$, is defined for all $v \in T_pM$, i.e., if any geodesic $\gamma(t)$ starting from p is defined for all values of the parameter $t \in \mathbf{R}$.

2.3 PROPOSITION. *If M is complete then M is non-extendible.*

Proof. Suppose that $M \subset M'$ is isometric to a proper open subset of a Riemannian manifold M'. Because M' is connected, the boundary ∂M of M in M' is non-empty. Let $p \in \partial M$ and let $U' \subset M'$ be a normal neighborhood of p in M'. Let $q \in U' \cap M$ and let $\tilde{\gamma}(t)$ be a geodesic in M' with $\tilde{\gamma}(0) = p$, $\tilde{\gamma}(1) = q$. Then $\gamma(t) = \tilde{\gamma}(1-t)$, $|t| < \delta$, is a geodesic in M with $\gamma(0) = q$. This geodesic is not defined for some $t \leq 1$, which contradicts the fact that M is complete. □

It is possible to show, by an example, that the converse is not true (cf. Exercise 4) and consequently that the class of non-extendible manifolds is actually larger than the the class of complete manifolds.

At this stage it is convenient to introduce a distance function on a Riemannian manifold (not necessarily complete) M as follows. Given two points $p, q \in M$, consider all the piecewise differentiable curves joining p to q. Since M is connected, such curves exist (cover a continuous curve joining p to q by a finite number of coordinate neighborhoods and replace each "piece" contained in a coordinate neighborhood by a differentiable curve).

2.4 DEFINITION. The *distance* $d(p, q)$ is defined by $d(p, q) =$ infimum of the lengths of all curves $f_{p,q}$, where $f_{p,q}$ is a piecewise differentiable curve joining p to q.

2.5 PROPOSITION. *With the distance d, M is a metric space, that is:*

1) $d(p, r) \leq d(p, q) + d(q, r)$,
2) $d(p, q) = d(q, p)$,
3) $d(p, q) \geq 0$, *and* $d(p, q) = 0 \Leftrightarrow p = q$.

Proof. (1), (2) and two of the assertions of (3) are immediate consequences of the definition of the infimum. It remains to show that if $d(p, q) = 0$ then $p = q$. Suppose to the contrary, and take a normal ball $B_r(p)$ that does not contain q. Since $d(p, q) = 0$, there exists a curve c joining p to q of length less than r. But the segment of c contained in $B_r(p)$ certainly has length greater than or equal to r, by Proposition 3.6 of Chapter 3, and that is a contradiction. □

Observe that if there exists a minimizing geodesic γ joining p to q (which is not always true) then $d(p, q) =$ length of γ.

2.6 PROPOSITION. *The topology induced by d on M coincides with the original topology on M.*

Proof. From the remark above, it follows that if r is sufficiently small, the normal ball $B_r(p)$ coincides with the metric ball of radius r, centered at p. Hence, metric balls contain normal balls, and conversely. □

2.7 COROLLARY. *If* $p_o \in M$, *the function* $f: M \to \mathbf{R}$ *given by* $f(p) = d(p, p_o)$ *is continuous.*

The fact which makes the concept of completeness relevant is the following theorem.

2.8 THEOREM. *(Hopf and Rinow [HR]). Let M be a Riemannian manifold and let* $p \in M$. *The following assertions are equivalent:*

a) $\exp_p$ *is defined on all of* $T_p(M)$.
b) *The closed and bounded sets of M are compact.*
c) *M is complete as a metric space.*
d) *M is geodesically complete.*
e) *Assume M is non-compact. There exists a sequence of compact subsets* $K_n \subset M$, $K_n \subset \operatorname{int}(K_{n+1})$ *and* $\bigcup_n K_n = M$, *such that if* $q_n \notin K_n$ *then* $d(p, q_n) \to \infty$.

In addition, any of the statements above implies that

f) *For any* $q \in M$ *there exists a geodesic* γ *joining p to q with* $\ell(\gamma) = d(p, q)$.

Proof. a) $\Rightarrow$ f). Let $d(p, q) = r$, and let $B_\delta(p)$ be a normal ball at p, with $S_\delta(p) = S$ the boundary of $B_\delta(p)$. Let x_o be a point where the continuous function $d(q,x)$, $x \in S$, attains a minimum. Then $x_o = \exp_p \delta v$, where $v \in T_pM$ and $|v| = 1$. Let γ be a geodesic given by $\gamma(s) = \exp_p sv$ (See Fig. 1). We are going to show that $\gamma(r) = q$.

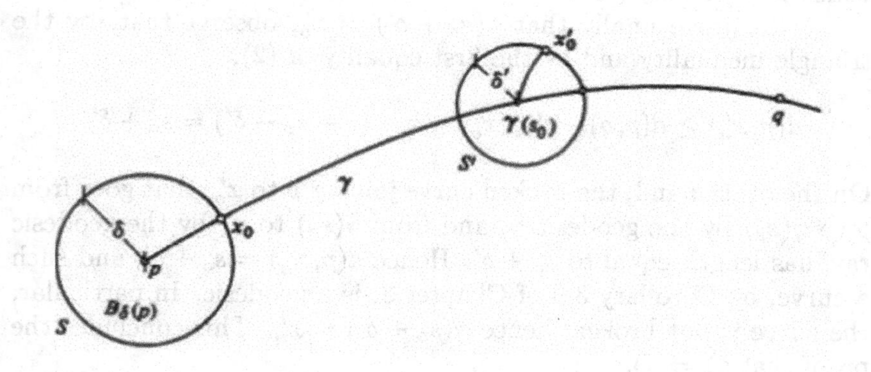

Figure 1

To prove this fact, consider the equation

$$d(\gamma(s), q) = r - s \tag{1}$$

and let $A = \{s \in [0, r]; (1) \text{ is valid}\}$. A is not empty, since (1) is true for $s = 0$. In addition, A is closed in $[0, r]$. Let $s_o \in A$. We are going to show that if $s_o < r$, then (1) is valid for $s_o + \delta'$, where $\delta' > 0$ is sufficiently small. This implies that sup $A = r$; since A is closed then $r \in A$, which shows that $\gamma(r) = q$.

In order to prove that (1) is true for $s_o + \delta'$, let $B_{\delta'}(\gamma(s_o))$ be a normal ball at $\gamma(s_o)$, with $S' = \partial B_{\delta'}(\gamma(s_o))$ its boundary and let x'_o be a point where $d(x, q)$, $x \in S'$ has a minimum. It suffices to show that $x'_o = \gamma(s_o + \delta')$. Indeed, if $x'_o = \gamma(s_o + \delta')$, since

$$d(\gamma(s_o), q) = \delta' + \min d(x, q) = \delta' + d(x'_o, q)$$

and

$$d(\gamma(s_o), q) = r - s_o,$$

we have

$$r - s_o = \delta' + d(x'_o, q) = \delta' + d(\gamma(s_o + \delta'), q), \tag{2}$$

or that

$$d(\gamma(s_o + \delta'), q) = r - (s_o + \delta'),$$

which is (1) for $s_o + \delta'$.

To prove finally that $\gamma(s_o + \delta') = x'_o$, observe that, by the triangle inequality and by the first equality of (2),

$$d(p, x'_o) \geq d(p, q) - d(q, x'_o) = r - (r - s_o - \delta') = s_o + \delta'.$$

On the other hand, the broken curve joining p to x'_o, that goes from p to $\gamma(s_o)$ by the geodesic γ, and from $\gamma(s_o)$ to x'_o by the geodesic ray, has length equal to $s_o + \delta'$. Hence $d(p, x'_o) = s_o + \delta'$, and such a curve, by Corollary 3.9 of Chapter 3, is a geodesic. In particular, the curve is not broken, hence $\gamma(s_o + \delta') = x'_o$. This concludes the proof that a) $\Rightarrow$ f).

a) $\Rightarrow$ b). Let $A \subset M$ be closed and bounded. Since A is bounded, $A \subset B$, where B is a ball with center p in the metric d. By (f), there exists a ball $B_r(0) \subset T_pM$, such that $B \subset \exp_p \overline{B_r(0)}$. Being the continuous image of a compact set, $\exp_p \overline{B_r(0)}$ is compact. Hence, A is a closed set contained in a compact set, and is therefore compact.

b) $\Rightarrow$ c). It suffices to observe that a subset $\{p_n\}$ formed by a Cauchy sequence is bounded, therefore, has compact closure by (b). Thus $\{p_n\}$ contains a convergent subsequence and, being Cauchy, converges.

c) $\Rightarrow$ d). Suppose that M is not geodesically complete. Then some normalized geodesic γ of M is defined for $s < s_o$ and is not defined for s_o. Let $\{s_n\}$ be a convergent sequence, converging to s_o with $s_n < s_o$. Given $\varepsilon > 0$, there exists an index n_o such that if $n, m > n_o$ then $|s_n - s_m| < \varepsilon$. It follows that

$$d(\gamma(s_n), \gamma(s_m)) \leq |s_n - s_m| < \varepsilon,$$

and hence the sequence $\{\gamma(s_n)\}$ is a Cauchy sequence in M. Since M is complete in the metric d, $\{\gamma(s_n)\} \to p_o \in M$.

Let (W, δ) be a totally normal neighborhood of p_o. Choose n_1 such that if $n, m > n_1$, then $|s_m - s_n| < \delta$ and $\gamma(s_n)$, $\gamma(s_m)$ belong to W. Then, there exists a unique geodesic g whose length is less than δ joining $\gamma(s_n)$ to $\gamma(s_m)$. It is clear that g coincides with γ, wherever γ is defined. Since $\exp_{\gamma(s_n)}$ is a diffeomorphism on $B_\delta(0)$ and $\exp_{\gamma(s_n)} (B_\delta(0)) \supset W$, g extends γ beyond s_0.
d) $\Rightarrow$ a). Obvious.
b) $\Leftrightarrow$ e). General topology. □

2.9 COROLLARY. *If M is compact then M is complete.*

2.10 COROLLARY. *A closed submanifold of a complete Riemannian manifold is complete in the induced metric; in particular, the closed submanifolds of Euclidean space are complete.*

3. The Theorem of Hadamard

As an application of the theorem of Hopf-Rinow, we are going to prove the following global fact.

3.1 THEOREM. *(Hadamard). Let M be a complete Riemannian manifold, simply connected, with sectional curvature $K(p, \sigma) \leq 0$, for all $p \in M$ and for all $\sigma \subset T_p(M)$. Then M is diffeomorphic to $\mathbf{R}^n$, $n = \dim M$; more precisely $\exp_p \colon T_pM \to M$ is a diffeomorphism.*

Before starting the proof, we need a few lemmas. The following lemma shows that the exponential map of a manifold with non-positive curvature is a local diffeomorphism.

3.2 LEMMA. *Let M be a complete Riemannian manifold with $K(p, \sigma) \leq 0$, for all $p \in M$ and for all $\sigma \subset T_pM$. Then for all $p \in M$, the conjugate locus $C(p) = \phi$; in particular the exponential map $\exp_p \colon T_pM \to M$ is a local diffeomorphism.*

Proof. Let J be a non-trivial (that is, not identically zero) Jacobi field along a geodesic $\gamma\colon [0,\infty) \to M$, where $\gamma(0) = p$ and $J(0) = 0$. Then from the hypothesis on the curvature and from the Jacobi equation

$$\begin{aligned}\langle J,J\rangle'' &= 2\langle J',J'\rangle + 2\langle J'',J\rangle\\ &= 2\langle J',J'\rangle - 2\langle R(\gamma',J)\gamma',J\rangle\\ &= 2|J'|^2 - 2K(\gamma',J)\,|\gamma'\wedge J|^2 \geq 0.\end{aligned}$$

Therefore $\langle J,J\rangle'(t_2) \geq \langle J,J\rangle'(t_1)$ whenever $t_2 > t_1$. Since $J'(0) \neq 0$ and $\langle J,J\rangle'(0) = 0$, it follows, in addition, that for t a sufficiently small positive number

$$\langle J,J\rangle(t) > \langle J,J\rangle(0).$$

It follows that for all $t > 0$, $\langle J,J\rangle(t) > 0$, and $\gamma(t)$ is not conjugate to $\gamma(0)$ along γ. □

The crucial point in the proof of the Hadamard theorem is given in the lemma below which is of independent interest.

3.3 LEMMA. *Let M be a complete Riemannian manifold and let $f\colon M \to N$ be a local diffeomorphism onto a Riemannian manifold N which has the following property: for all $p \in M$ and for all $v \in T_pM$, we have $|df_p(v)| \geq |v|$. Then f is a covering map.*

Proof. By a general property of covering spaces (Cf. M. do Carmo, [dC 2], p. 383), it suffices to show that f has the path lifting property for curves in N, that is, given a differentiable curve $c\colon [0,1] \to N$ and a point $q \in M$ with $f(q) = c(0)$, there exists a curve $\bar{c}\colon [0,1] \to M$ with $\bar{c}(0) = q$ and $f \circ \bar{c} = c$.

To prove what is required, observe that, since f is a local diffeomorphism at q, there exists an $\varepsilon > 0$ such that it is possible to define $\bar{c}\colon [0,\varepsilon] \to M$ with $\bar{c}(0) = q$ and $f \circ \bar{c} = c$; that is, c can be lifted to a small interval starting from q. Because f is a local diffeomorphism over all of M, the set of values $A \subset [0,1]$, such that c can be lifted on A starting from q, is an open interval on the right; that is, $A = [0,t_o)$. If we can show that $t_o \in A$, we shall have A open and closed in $[0,1]$, therefore $A = [0,1]$ and c can be lifted on the entire interval.

To show that $t_o \in A$, let $\{t_n\}$, $n = 1, \ldots$, be an increasing sequence in A with $\lim t_n = t_o$. Then the sequence $\{\bar{c}(t_n)\}$ is contained in a compact set $K \subset M$. Indeed, if this were not the case, since M is complete, the distance from $\bar{c}(t_n)$ to $\bar{c}(0)$ would be arbitrarily large. However, by hypothesis,

$$\ell_{0,t_n}(c) = \int_o^{t_n} \left|\frac{dc}{dt}\right| dt = \int_0^{t_n} \left|df_{\bar{c}(t)}\left(\frac{d\bar{c}}{dt}\right)\right| dt$$
$$\geq \int_0^{t_n} \left|\frac{d\bar{c}}{dt}\right| dt \geq d(\bar{c}(t_n), \bar{c}(0)),$$

implying that the length of c between 0 and t_o is arbitrarily large, which is absurd, and proves the assertion.

Since $\{\bar{c}(t_n)\} \subset K$, $n = 1, \ldots$, there exists an accumulation point $r \in M$ of $\{\bar{c}(t_n)\}$. Let V be a neighborhood of r such that $f \mid V$ is a diffeomorphism. Then $c(t_o) \in f(V)$ and, by continuity, there exists an interval $I \subset [0,1]$, $t_o \in I$, such that $c(I) \subset f(V)$. Choose an index n such that $\bar{c}(t_n) \in V$ and consider the lifting g of c on I passing through r. The liftings g and $\bar{c}$ coincide on $[0, t_n) \cap I$, because $f \mid V$ is bijective. Therefore, g is an extension of $\bar{c}$ to I, hence $\bar{c}$ is defined at t_o and $t_o \in A$.

Proof of the Hadamard theorem. Since M is complete, $\exp_p: T_pM \to M$ is defined for all $p \in M$ and is surjective. By Lemma 3.2, $\exp_p$ is a local diffeomorphism. This allows us to introduce a Riemannian metric on T_pM in such a way that $\exp_p$ is a local isometry. Such a metric is complete, because the geodesics of T_pM passing through the origin are straight lines (Cf. Theorem 2.8, (a) $\Rightarrow$ (d)). From Lemma 3.3, $\exp_p$ is a covering map. Since M is simply connected, $\exp_p$ is a diffeomorphism. □

3.4 Remark. The above proof gives a little more than what is stated. Call a point p of a complete Riemannian manifold M a *pole* if it has the property that it has no conjugate points. Any point of a complete manifold M with non-positive sectional curvature is a pole of M. However, poles can exist in non-compact manifolds which have positive sectional curvature (See Exercise 13). What we have just proved is the following general fact. *If a complete simply connected Riemannian manifold M has a pole, then M is diffeomorphic to* $\mathbf{R}^n$, $n = \dim M$.

EXERCISES

1. If M, N are Riemannian manifolds such that the inclusion $i: M \subset N$ is an isometric imersion, show by an example that the strict inequality $d_M > d_N$ can occur.
2. Let $\widetilde{M}$ be a covering space of a Riemannian manifold M. Show that it is possible to give $\widetilde{M}$ a Riemannian structure such that the covering map $\pi: \widetilde{M} \to M$ is a local isometry (this metric is called the *covering metric*). Show that $\widetilde{M}$ is complete in the covering metric if and only if M is complete.
3. Let $f: M_1 \to M_2$ be a local diffeomorphism of a manifold M_1 onto a Riemannian manifold M_2. Introduce on M_1 a Riemannian metric such that f is a local isometry. Show by an example that if M_2 is complete, M_1 need not be complete.
4. Consider the universal covering

$$\pi: M \to \mathbf{R}^2 - \{(0,0)\}$$

of the Euclidean plane minus the origin. Introduce the covering metric on M (Cf. Exercise 2). Show that M is not complete and not extendible, and that the Hopf-Rinow theorem is not true for M (this shows that the definition of non-extendibility, though natural, is not a satisfactory one).
Hint: The single difficult point is to prove that M is non-extendible. Suppose to the contrary, that is, $M \subset M'$, is an isometry and $M \neq M'$. Let $p' \in M'$ be a point in the boundary of M and let $W' \subset M'$ be a convex neighborhood of p'.
If we prove that $W' - \{p'\} \subset M$, we obtain a contradiction, since then $\pi(W' - \{p'\}) = U$ is a neighborhood of $(0,0) \in \mathbf{R}^2$ and considering the closed circle in U with center at $(0,0)$ we can lift the closed circle into M, which is impossible.
To prove that $W' - \{p'\} \subset M$, observe first that through any point $p \in M$ passes a *unique* geodesic in M that cannot be extended for all values of $t \in \mathbf{R}$. Let $x \in W' \cap M$ and join x to p' by a geodesic $\tilde{\gamma}$ of M'. $\tilde{\gamma}$ coincides initially with a geodesic

of M and therefore is the unique geodesic passing through x which cannot be extended for all values of t. From uniqueness and from the fact that p' is a point of the boundary it follows that all the points of $\tilde{\gamma} \cap W$, except p', belong to M, because $\tilde{\gamma}$ approaches the boundary of M arbitrarily close. Finally, if $z \in W'$ and $z \notin \tilde{\gamma}$, the geodesic joining z to x is, from the uniqueness above, entirely in M, hence $z \in M$.

5. A *divergent curve* in a non-compact Riemannian manifold M is a differentiable mapping $\alpha : [0,\infty) \to M$ such that for any compact set $K \subset M$ there exists $t_o \in (0,\infty)$ with $\alpha(t) \notin K$ for all $t > t_o$ (that is, α "escapes" every compact set in M). Define the length of a divergent curve by

$$\lim_{t\to\infty} \int_0^t |\alpha'(t)|\, dt.$$

Prove that M is complete if and only if the length of any divergent curve is unbounded.

6. A geodesic $\gamma : [0,\infty) \to M$ in a Riemannian manifold M is called a *ray starting from* $\gamma(0)$ if it minimizes the distance between $\gamma(0)$ and $\gamma(s)$, for any $s \in (0,\infty)$. Assume that M is complete, non- compact, and let $p \in M$. Show that M contains a ray starting from p.

7. Let M and $\overline{M}$ be non-compact Riemannian manifolds and let $f : M \to \overline{M}$ be a diffeomorphism. Assume that $\overline{M}$ is complete and that there exists a constant $c > 0$ such that

$$|v| \geq c|\, df_p(v)|,$$

for all $p \in M$ and all $v \in T_p(M)$. Prove that M is complete.

8. Let M be a complete Riemannian manifold, $\overline{M}$ a connected Riemannian manifold, and $f : M \to \overline{M}$ a differentiable mapping that is locally an isometry. Assume that any two points of $\overline{M}$ can be joined by a unique geodesic of $\overline{M}$. Prove that f is injective and surjective (and, therefore, f is a global isometry).

9. Consider the upper half-plane

$$\mathbf{R}^2_+ = \{(x,y) \in \mathbf{R}^2; y > 0\}$$

with the Riemannian metric given by

$$g_{11} = 1, \quad g_{12} = 0, \quad g_{22} = \frac{1}{y}.$$

Show that the length of the vertical segment

$$x = 0, \quad \varepsilon \leq y \leq 1, \quad \text{with } \varepsilon > 0,$$

tends to 2 as $\varepsilon \to 0$. Conclude from this that such a metric is not complete. (Observe, nevertheless, that when $y \to 0$ the length of vectors, in this metric, becomes arbitrarily large.)

10. Prove that the upper half-plane $\mathbf{R}^2_+$ with the Lobatchevski metric:

$$g_{11} = g_{22} = \frac{1}{y^2}, \quad g_{12} = 0,$$

is complete.

11. Let M be a complete Riemannian manifold, and let X be a differentiable vector field on M. Suppose that there exists a constant $c > 0$ such that $|X(p)| > c$, for all $p \in M$. Prove that the trajectories of X, that is, the curves $\varphi(t)$ in M with $\varphi'(t) = X(\varphi(t))$, are defined for all values of t.

12. A Riemannian manifold is said to be *homogeneous* if given $p, q \in M$ there exists an isometry of M which takes p into q. Prove that any homogeneous manifold is complete.

13. Show that the point $p = (0, 0, 0)$ of the paraboloid

$$S = \{(x, y, z) \in \mathbf{R}^3; z = x^2 + y^2\}$$

is a pole of S and, nevertheless, the curvature of S is positive.

CHAPTER 8

SPACES OF CONSTANT CURVATURE

1. Introduction

Among the Riemannian manifolds, those with constant sectional curvature are the most simple. They are related to classical non-Euclidean geometry, which is historically the first example of a geometric structure different from the euclidean.

An important property of the spaces with constant curvature is that they have a sufficiently large number of local isometries (Cf. Corollary 2.2). This means that in these spaces it is always possible to "displace" isometrically two small triangles situated in different positions and verify whether they can be superimposed (in which case we say they are equal or congruent). This property of "free mobility of small triangles" is fundamental in the constructions of elementary geometry and was considered as a postulate to be satisfied by the non-Euclidean geometries.

It is not difficult to verify that when we multiply a Riemannian metric by a positive constant c, then its sectional curvature is multiplied by $1/c$. Therefore, up to a similarity, we can suppose that the value of the constant sectional curvature of a Riemannian manifold is $1, 0$, or -1. In what follows, this hypothesis will be assumed without further comment.

So far, we have encountered two examples of Riemannian manifolds with constant sectional curvature K, namely, the Euclidean space $\mathbf{R}^n$ with $K \equiv 0$ and the unit sphere $S^n \subset \mathbf{R}^{n+1}$ with $K \equiv 1$. In this chapter, we shall introduce a Riemannian manifold, the hyperbolic space H^n of dimension n, which has sectional curvature $K \equiv -1$. The manifolds $\mathbf{R}^n$, S^n and H^n are complete and simply connected (Cf. Section 3). The main theorem of this chapter is that these are essentially the only complete, simply connected Riemannian manifolds, with constant sectional curvature. This allows us to reduce the problem of finding all the complete manifolds of constant curvature to the problem of determining certain subgroups of the group of isometries of $\mathbf{R}^n$, S^n and H^n (Cf. Section 4).

To prove the fact mentioned above, we introduce in Section 2 a slightly more general theorem on the determination of the metric by means of the curvature. Strictly speaking, this theorem does not refer to spaces of constant curvature, but this is a natural place to present it.

We frequently use the classical expression *spaces of constant curvature* to designate the Riemannian manifolds of constant sectional curvature.

In the last section of this chapter, we describe the isometries of the hyperbolic space H^n and identify certain important hypersurfaces of H^n, namely, the horospheres and the hyperspheres. In Exercise 6, we indicate how to calculate the mean and sectional curvature of such hypersuperfaces.

2. Theorem of Cartan on the determination of the metric by means of the curvature

Let M and $\tilde{M}$ be two Riemannian manifolds of dimension n and let $p \in M$ and $\tilde{p} \in \tilde{M}$. Choose a linear isometry $i: T_p(M) \to T_{\tilde{p}}(\tilde{M})$. Let $V \subset M$ be a normal neighborhood of p such that $\exp_{\tilde{p}}$ is defined at $i \circ \exp_p^{-1}(V)$. Define a mapping $f: V \to \tilde{M}$ by

$$f(q) = \exp_{\tilde{p}} \circ i \circ \exp_p^{-1}(q), \qquad q \in V.$$

For all $q \in V$ there exists a unique normalized geodesic $\gamma: [0,t] \to M$ with $\gamma(0) = p$, $\gamma(t) = q$. Denote by P_t the parallel transport along γ from $\gamma(0)$ to $\gamma(t)$. Define $\phi_t: T_q(M) \to T_{f(q)}(\tilde{M})$ by

$$\phi_t(v) = \tilde{P}_t \circ i \circ P_t^{-1}(v), \qquad v \in T_q(M),$$

where $\tilde{P}_t$ is the parallel transport along the normalized geodesic $\tilde{\gamma}: [0,t] \to \tilde{M}$ given by $\tilde{\gamma}(0) = \tilde{p}$, $\tilde{\gamma}'(0) = i(\gamma'(0))$. Finally, denote by R and $\tilde{R}$ the curvatures of M and $\tilde{M}$, respectively.

2.1 THEOREM. *(E. Cartan [Ca], p. 238). With the notation above, if for all $q \in V$ and all $x, y, u, v \in T_q(M)$ we have*

$$\langle R(x,y)u,v\rangle = \left\langle \tilde{R}(\phi_t(x),\phi_t(y))\phi_t(u),\phi_t(v)\right\rangle,$$

then $f\colon V \to f(V) \subset \tilde{M}$ is a local isometry and $df_p = i$.

Proof. Let $q \in V$ and let $\gamma\colon [0,\ell] \to M$ be a normalized geodesic with $\gamma(0) = p$, $\gamma(\ell) = q$. Let $v \in T_q(M)$ and let J be a Jacobi field along γ given by $J(0) = 0$ and $J(\ell) = v$. Let $e_1, \ldots, e_n = \gamma'(0)$ be an orthonormal basis of $T_p(M)$ and let $e_i(t)$, $i = 1, \ldots, n$, be the parallel transport of e_i along γ. Writing

$$J(t) = \sum_i y_i(t)e_i(t),$$

and using the Jacobi equation, we conclude that

$$y_j'' + \sum_i \langle R(e_n,e_i)e_n,e_j\rangle\, y_i = 0, \qquad j = 1,\ldots,n.$$

Now, let $\tilde{\gamma}\colon [0,\ell] \to \tilde{M}$ be a normalized geodesic given by $\tilde{\gamma}(0) = \tilde{p}$, $\tilde{\gamma}'(0) = i(\gamma'(0))$. Let $\tilde{J}(t)$ be the field along $\tilde{\gamma}$ given by

$$\tilde{J}(t) = \phi_t(J(t)), \qquad t \in [0,\ell].$$

Let $\tilde{e}_j(t) = \phi_t(e_j(t))$. Then, from the linearity of ϕ_t,

$$\tilde{J}(t) = \sum_i y_i(t)\tilde{e}_i(t).$$

Since, by hypothesis,

$$\langle R(e_n,e_i)e_n,e_j\rangle = \left\langle \tilde{R}(\tilde{e}_n,\tilde{e}_i)\tilde{e}_n,\tilde{e}_j\right\rangle,$$

we have that

$$y_j'' + \sum_i \left\langle \tilde{R}(\tilde{e}_n,\tilde{e}_i)\tilde{e}_n,\tilde{e}_j\right\rangle y_i = 0, \qquad j = 1,\ldots,n.$$

It follows that $\tilde{J}$ is a Jacobi field along $\tilde{\gamma}$ with $\tilde{J}(0) = 0$. Since parallel transport is an isometry, $|\tilde{J}(\ell)| = |J(\ell)|$. If we show that

$$\tilde{J}(\ell) = df_q(v) = df_q(J(\ell)),$$

we have completed the proof.

Since $\tilde{J}(t) = \phi_t(J(t))$, we have that $\tilde{J}'(0) = i(J'(0))$. On the other hand, since $J(t)$ and $\tilde{J}(t)$ are Jacobi fields which vanish at $t = 0$, we have from Corollary 2.5 of Chapter 5

$$J(t) = (d\exp_p)_{t\gamma'(0)}(tJ'(0)),$$

$$\tilde{J}(t) = (d\exp_{\tilde{p}})_{t\tilde{\gamma}'(0)}(t\tilde{J}'(0)).$$

Therefore,

$$\begin{aligned}\tilde{J}(\ell) &= (d\exp_{\tilde{p}})_{\ell\tilde{\gamma}'(0)}\ell i(J'(0)) \\ &= (d\exp_{\tilde{p}})_{\ell\tilde{\gamma}'(0)} \circ i \circ ((d\exp_p)_{\ell\gamma'(0)})^{-1}(J(\ell)) = df_q(J(\ell)),\end{aligned}$$

which proves what was asserted. □

Observe that the same proof shows that if $\exp_p$ and $\exp_{\tilde{p}}$ are diffeomorphisms, then, under the conditions of Theorem 2.1, f is defined on all of M and is an isometry.

The theorem implies that the metric is, in a certain sense, determined locally by the curvature. An equivalent assertion was made by Riemann ([Ri], p.289). As far as we know, the first proof of the local theorem was presented by E. Cartan ([Ca], p. 238). A global version of the theorem above, which will not be presented here, was given by W. Ambrose in 1956, [Am 1]. A simple proof of the theorem of Ambrose can be found in Cheeger and Ebin [CE].

For our purposes, we are only interested in the two corollaries below. In particular, Corollary 2.3 implies that a space of constant curvature is rich in isometries.

2.2 COROLLARY. *Let M and $\tilde{M}$ be spaces with the same constant curvature and the same dimension n. Let $p \in M$ and $\tilde{p} \in \tilde{M}$. Choose arbitrary orthonormal bases $\{e_j\} \in T_p(M)$ and $\{\tilde{e}_j\} \in T_{\tilde{p}}(\tilde{M})$, $j = 1, \dots, n$. Then there exist a neighborhood $V \subset M$ of p, a neighborhood $\tilde{V} \subset \tilde{M}$ of $\tilde{p}$, and an isometry $f: V \to \tilde{V}$ such that $df_p(e_j) = \tilde{e}_j$.*

Proof. Choose the isometry i of the theorem in such a way that $i(e_j) = \tilde{e}_j$. The condition on the curvature is immediately verified, and the conclusion follows from the theorem.

2.3 COROLLARY. *Let M be a space of constant curvature and let p and q be any two points of M. Let $\{e_j\}$ and $\{f_j\}$ be arbitrary orthonormal bases of $T_p(M)$ and $T_q(M)$, respectively. Then there exist neighborhoods U of p and V of q, and an isometry g: $U \to V$ such that $dg_p(e_j) = f_j$.*

Related to the problem above, is the problem of deciding if a diffeomorphism f: $M \to \tilde{M}$ which preserves the curvature in the sense that

$$\langle R(X,Y)Z,T\rangle_p = \left\langle \tilde{R}\left(df_p(X), df_p(Y)\right) df_p(Z), df_p(T)\right\rangle_{f(p)},$$

for all $p \in M$ and all $X,Y,Z,T \in T_p(M)$, is an isometry. In dimension two this is a kind of converse to the Theorem Egregium of Gauss, and is false, even in the compact case, as shown in the example of Fig. 1. (A non-trivial example for the non-compact case can be found in M. do Carmo [dC 2], p. 237.) For $n = \dim M = \dim \tilde{M} \geq 4$, the problem surprisingly admits an essentially affirmative solution (for further details, see Kulkarni, [Ku 1] and [Ku 2]). For $n = 3$, the problem was treated by Yau [Ya].

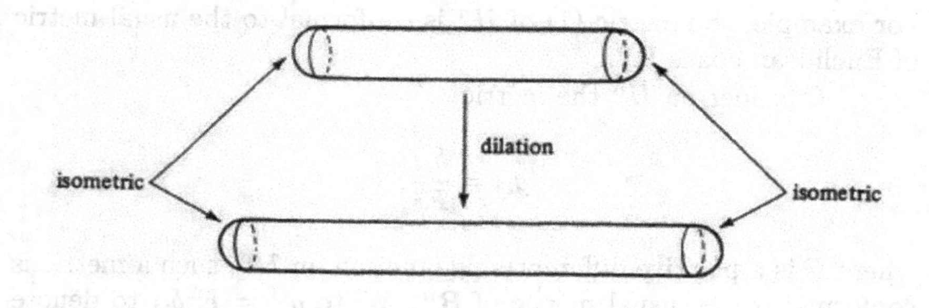

Figure 1. *A diffeomorphism which preserves curvature but is not an isometry.*

3. Hyperbolic space

We give now an example of a space of constant curvature -1.

Consider the half-space of $\mathbf{R}^n$ given by

$$H^n = \{(x_1, \ldots, x_n) \in \mathbf{R}^n; x_n > 0\}$$

and introduce on H^n the metric

$$g_{ij}(x_1, \ldots, x_n) = \frac{\delta_{ij}}{x_n^2}. \tag{1}$$

It is clear that H^n is simply connected. We are going to show that H^n with the metric (1) has constant sectional curvature equal to -1, and that H^n is complete. H^n is called the *hyperbolic space* of dimension n.

We start with the calculation of the curvature of H^n. A good part of the calculation can be carried out in the following more general situation: two metrics $\langle\ ,\ \rangle$ and $\langle\langle\ ,\ \rangle\rangle$ on a differentiable M are *conformal* if there exists a positive differentiable function $f: M \to \mathbf{R}$, such that for all $p \in M$ and all $u, v \in T_p(M)$

$$\langle u, v\rangle_p = f(p)\, \langle\langle u, v\rangle\rangle_p\,.$$

For example, the metric (1) of H^n is conformal to the usual metric of Euclidean space $\mathbf{R}^n$.

Consider on H^n the metric

$$g_{ij} = \frac{\delta_{ij}}{F^2},$$

where F is a positive differentiable function on H^n; such a metric is conformal to the usual metric of $\mathbf{R}^n$. Write $g^{ij} = F^2\delta_{ij}$ to denote the inverse matrix of g_{ij}, and put $\log F = f$. Under these conditions, denoting $\frac{\partial f}{\partial x_j} = f_j$, we have

$$\frac{\partial g_{ik}}{\partial x_j} = -\delta_{ik}\frac{2}{F^3}F_j = -2\frac{\delta_{ik}}{F^2}f_j.$$

To calculate the Christoffel symbols, observe that

$$\begin{aligned}\Gamma_{ij}^{k} &= \frac{1}{2}\sum_{m}\left\{\frac{\partial}{\partial x_i}g_{jm} + \frac{\partial}{\partial x_j}g_{mi} - \frac{\partial}{\partial x_m}g_{ij}\right\}g^{mk} \\ &= \frac{1}{2}\left\{\frac{\partial}{\partial x_i}g_{jk} + \frac{\partial}{\partial x_j}g_{ki} - \frac{\partial}{\partial x_k}g_{ij}\right\}F^2 \\ &= -\delta_{jk}f_i - \delta_{ki}f_j + \delta_{ij}f_k;\end{aligned}$$

therefore we can conclude that if all three indices are distinct, $\Gamma_{ij}^{k} = 0$, while if two indices are equal, we have

$$\Gamma_{ij}^{i} = -f_j, \quad \Gamma_{ii}^{j} = f_j, \quad \Gamma_{ij}^{j} = -f_i, \quad \Gamma_{ii}^{i} = -f_i.$$

To calculate the coefficients of the curvature, observe that

$$\begin{aligned}R_{ijij} &= \sum_{\ell} R_{iji}^{\ell}g_{\ell j} = R_{iji}^{j}g_{jj} = R_{iji}^{j}\frac{1}{F^2} \\ &= \frac{1}{F^2}\left\{\sum_{\ell}\Gamma_{ii}^{\ell}\Gamma_{j\ell}^{j} - \sum_{\ell}\Gamma_{ji}^{\ell}\Gamma_{i\ell}^{j} + \frac{\partial}{\partial x_j}\Gamma_{ii}^{j} - \frac{\partial}{\partial x_i}\Gamma_{ji}^{j}\right\}.\end{aligned}$$

Since $\frac{\partial}{\partial x_j}\Gamma_{ii}^{j} = f_{jj}$ and $\frac{\partial}{\partial x_i}\Gamma_{ji}^{j} = -f_{ii}$, we obtain

$$\begin{aligned}F^2R_{ijij} &= -\sum_{\substack{\ell \\ \ell\neq i,\ell\neq j}} f_\ell f_\ell + f_i^2 - f_j^2 - f_i^2 + f_j^2 + f_{jj} + f_{ii} \\ &= -\sum_{\ell} f_\ell^2 + f_i^2 + f_j^2 + f_{ii} + f_{jj}.\end{aligned}$$

In addition, $R_{ijk\ell} = 0$ if all four indices are distinct, and if any three indices are distinct, we have:

$$R_{ijk}^{i} = -f_kf_j - f_{kj}, \quad R_{ijk}^{j} = f_if_k + f_{ki}, \quad R_{ijk}^{k} = 0. \tag{2}$$

Finally, the sectional curvature with respect to the plane generated by $\frac{\partial}{\partial x_i}, \frac{\partial}{\partial x_j}$ is (observing that $\frac{\partial}{\partial x_i}, \frac{\partial}{\partial x_j}$ are orthogonal)

$$\begin{aligned}K_{ij} &= \frac{R_{ijij}}{g_{ii}g_{jj}} = R_{ijij}F^4 \\ &= (-\sum_{\ell} f_\ell^2 + f_i^2 + f_j^2 + f_{ii} + f_{jj})F^2.\end{aligned}$$

Now we are going to specialize to the case $F^2 = x_n^2$, which implies that $f = \log x_n$. In this case, if $i \neq n$ and $j \neq n$, we have

$$K_{ij} = (-\frac{1}{x_n^2})x_n^2 = -1;$$

if $i = n$, $j \neq n$, we have

$$K_{nj} = (-f_n^2 + f_n^2 + f_{nn})F^2 = -\frac{1}{x_n^2}x_n^2 = -1;$$

finally, if $i \neq j$, $j = n$, we have again $K_{in} = -1$. Using the expressions in (2) and Corollary 3.5 of Chapter 4, we conclude that the sectional curvature of H^n is a constant equal to -1.

In order to prove that H^n is complete, we use the following fact.

3.1 PROPOSITION. *The straight lines perpendicular to the hyperplane $x_n = 0$, and the circles of H^n whose planes are perpendicular to the hyperplane $x_n = 0$ and whose centers are in this hyperplane are the geodesics of H^n.*

Proof. Observe that an isometry of $\mathbf{R}^n$ which involves solely the variables $x_1, \ldots, x_{n-1}$ does not change the metric g_{ij}, and is, therefore, an isometry of H^n. Therefore it suffices to consider lines and circles in the x_1x_n-plane. The theorem follows now from Example 3.10 of Chapter 3. □

It is easy to verify, from the existence and uniqueness theorem for geodesics, that all the geodesics of H^n are of the type described in Proposition 3.1. This implies that all the geodesics of H^n are contained in planes perpendicular to the hyperplane $x_n = 0$. Since such planes are clearly isometric to the hyperbolic plane, the fact that H^n is complete is a consequence of the completeness of the hyperbolic plane (Cf. Exercise 10, in Chap. 7).

Another model of the hyperbolic space is given in Exercise 3 of this chapter.

4. Space forms

Now we are able to prove the main theorem of this chapter. As always, M^n denotes a manifold of dimension n.

4.1 THEOREM. *Let M^n be a complete Riemannian manifold with constant sectional curvature K. Then the universal covering $\tilde{M}$ of M, with the covering metric, is isometric to:*

a) H^n, *if* $K = -1$,

b) $\mathbf{R}^n$, *if* $K = 0$,

c) S^n, *if* $K = 1$.

In the proof, we make use of the following lemma which is interesting in itself.

4.2 LEMMA. *Let $f_i: M \to N$, $i = 1, 2$, be two local isometries of the (connected) Riemannian manifold M to the Riemannian manifold N. Suppose that there exists a point $p \in M$ such that $f_1(p) = f_2(p)$ and $(df_1)_p = (df_2)_p$. Then $f_1 = f_2$.*

Proof of the lemma. Let V be a normal neighborhood of p such that the restrictions $f_1|V$ and $f_2|V$ are diffeomorphisms. Let $\varphi = f_1^{-1} \circ f_2: V \to V$. Then $\varphi(p) = p$ and $d\varphi_p$ is the identity. If $q \in V$, there exists a unique $v \in T_pM$ with $\exp_p(v) = q$. It follows that $\varphi(q) = q$, hence $f_1 = f_2$ on V. Since M is connected, any point $r \in M$ can be joined to p by a path $\alpha: [0, 1] \to M$, $\alpha(0) = p$, $\alpha(1) = r$. Let

$$A = \{t \in [0, 1]; f_1(\alpha(t)) = f_2(\alpha(t)) \text{ and } (df_1)_{\alpha(t)} = (df_2)_{\alpha(t)}\}.$$

From the preceding, $\sup A$ is positive. If $\sup A = t_o \neq 1$, we repeat the argument above for the point $\alpha(t_o)$, obtaining a contradiction. Therefore $\sup A = 1$, hence $f_1(r) = f_2(r)$, for all $r \in M$. □

Proof of Theorem 4.1. $\tilde{M}$ is a simply connected, complete Riemannian manifold, with constant sectional curvature K. Let us consider, in the first place, the cases (a) and (b) and denote, for convenience, H^n as well as $\mathbf{R}^n$ by Δ. Fix points $p \in \Delta$, $\tilde{p} \in \tilde{M}$ and a linear isometry $i: T_p(\Delta) \to T_{\tilde{p}}(\tilde{M})$. Consider the map:

$$f = \exp_{\tilde{p}} \circ i \circ \exp_p^{-1}: \Delta \to \tilde{M}.$$

Since Δ and $\tilde{M}$ are complete with non-positive sectional curvature, f is well-defined. From the Theorem of Cartan, f is a local isometry. From Lemma 3.3 of Chap. 7, f is a diffeomorphism and this proves (a) and (b).

For the case (c), fix, once more, points $p \in S^n$, $\tilde{p} \in \tilde{M}$ and a linear isometry $i: T_p(S^n) \to T_{\tilde{p}}(\tilde{M})$. Let $q \in S^n$ be the antipodal point of p and define

$$f = \exp_{\tilde{p}} \circ i \circ \exp_p^{-1}: S^n - \{q\} \to \tilde{M}.$$

From the Theorem of Cartan, f is a local isometry. Now choose a point $p' \in S^n$; $p' \neq p$, $p' \neq q$. Set $\tilde{p}' = f(p')$, $i' = df_{p'}$ and define

$$f' = \exp_{\tilde{p}'} \circ i' \circ \exp_{p'}^{-1}: S^n - \{q'\} \to \tilde{M},$$

where q' is the antipodal point of p'.

Observe that $S^n - (\{q\} \cup \{q'\}) = W$ is connected, $p' \in W$, and

$$f(p') = \tilde{p}' = f'(p'), \quad df_{p'} = i' = df'_{p'}.$$

It follows from Lemma 4.2, that $f = f'$ on W. As a consequence, we can define a map $g: S^n \to \tilde{M}$ by

$$g(r) = \begin{cases} f(r), & \text{if } r \in S^n - \{q\} \\ f'(r), & \text{if } r \in S^n - \{q'\}. \end{cases}$$

It is clear that g is a local isometry, therefore a local diffeomorphism. By the compactness of S^n, g is a covering map, and since $\tilde{M}$ is simply connected, g is a diffeomorphism (see M. do Carmo [dC 2] §5.6). Therefore g is an isometry. □

The complete manifolds with constant sectional curvature are called *space forms*. The last theorem reduces the determination of all the space forms to a problem in group theory, as we show in what follows. We require some facts on covering spaces and group actions.

We say that a group G *acts* (on the left) on a set M if there exists a map of $G \times M$ onto M, denoted by

$$G \times M \ni (g, x) \to gx \in M$$

such that

$$ex = x, \quad (g_1 g_2)x = g_1(g_2 x)$$

where $e =$ the identity of G, $x \in M$ and $g_1, g_2 \in G$. We say that G acts *freely* (i.e., without fixed points) on M if $gx = x$ implies $g = e$. The *orbit* of a point $x \in M$ is the set

$$Gx = \{gx; g \in G\}.$$

The action of G is said to be *transitive* if $Gx = M$. The set of all orbits is denoted by M/G; there exists a natural projection $\pi: M \to M/G$ given by $\pi(x) = Gx$. When M has some additional structure (topological, differentiable, etc.), it is convenient to consider G as a group of isomorphisms (homeomorphisms, diffeomorphisms, etc.) of the structure under consideration.

If M is a topological space, we say that the group G (of homeomorphisms of M) *acts in a totally discontinuous manner* if every $x \in M$ has a neighborhood U such that $g(U) \cap U = \phi$, for all $g \in G$, $g \neq e$. In this case, the projection $\pi: M \to M/G$ (where M/G has the quotient topology) is a regular covering map and G is the group of covering transformations (See Massey [Ma], Prop. 8.2, p. 165).

Suppose now that M is a Riemannian manifold and let Γ be a subgroup of the group of isometries of M which acts in a totally discontinuous manner. We know that M/Γ has a differentiable structure in which $\pi: M \to M/\Gamma$ is a local diffeomorphism (See Chap. 0, Example 4.8). In addition, we can put a Riemannian metric on M/Γ in such a way that π is a local isometry. Indeed, given $p \in M/\Gamma$, choose $\tilde{p} \in \pi^{-1}(p)$; for every pair $u, v \in T_p(M/\Gamma)$, define

$$\langle u, v \rangle = \langle d\pi^{-1}(u), d\pi^{-1}(v) \rangle_{\tilde{p}}.$$

Since the covering π is regular, Γ is transitive on $\pi^{-1}(p)$ (See Massey [Ma], Lemma 8.1, p. 164). Therefore, given any $\tilde{q} \in \pi^{-1}(p)$, there exists $\gamma \in \Gamma$ with $\gamma(\tilde{p}) = \tilde{q}$, and the definition above does not depend on the choice of $\tilde{p} \in \pi^{-1}(p)$. It is easy to verify that with such a metric $\pi: M \to M/\Gamma$ is a local isometry; such a metric is called the *metric on M/Γ induced by the covering π*. Observe that M/Γ is complete if and only if M is complete and that M/Γ has constant curvature if and only if M has constant curvature. Taking $M = S^n$ or $\mathbf{R}^n$ or H^n, we conclude that M/Γ is a complete manifold of constant curvature 1 (if $M = S^n$), 0 (if $M = \mathbf{R}^n$) or -1 (if $M = H^n$). We are going to show that in this manner we obtain all such manifolds.

4.3 PROPOSITION. *Let M be a complete Riemannian manifold with constant sectional curvature K $(1, 0, -1)$. Then M is isometric to $\tilde{M}/\Gamma$, where $\tilde{M}$ is S^n (if $K = 1$), $\mathbf{R}^n$ (if $K = 0$) or H^n (if $K = -1$), Γ is a subgroup of the group of isometries of $\tilde{M}$ which* *acts in a totally discontinuous manner on $\tilde{M}$, and the metric on $\tilde{M}/\Gamma$ is induced from the covering $\pi: \tilde{M} \to \tilde{M}/\Gamma$.*

Proof. Consider the universal covering $p: \tilde{M} \to M$, and provide $\tilde{M}$ with the covering metric, that is, the metric such that p is a local isometry. Let Γ be the group of covering transformations of the covering p. Then Γ is a subgroup of the group of isometries of $\tilde{M}$ and acts on $\tilde{M}$ in a totally discontinuous manner. Therefore it is possible to introduce on $\tilde{M}/\Gamma$ the Riemannian metric induced by $\pi: \tilde{M} \to \tilde{M}/\Gamma$. Since the covering p is regular, given $\tilde{x}, \tilde{y} \in \tilde{M}$ then $p(\tilde{x}) = p(\tilde{y})$ if and only if $\Gamma\tilde{x} = \Gamma\tilde{y}$ which is equivalent to $\pi(\tilde{x}) = \pi(\tilde{y})$. The equivalence classes given by p and π on $\tilde{M}$ are, therefore, the same, and this induces a bijection $\xi: M \to \tilde{M}/\Gamma$ such that $\pi = \xi \circ p$. Since π and p are local isometries, ξ is a local isometry as well, and being a bijection, is an isometry of M onto $\tilde{M}/\Gamma$. □

The last proposition reduces the problem of finding all of the space forms to the problem of determining all the subgroups of the group of isometries that act in a totally discontinuous manner on each of the simply connected models (S^n, H^n and $\mathbf{R}^n$).

The determination of such subgroups is a difficult problem. The spherical problem ($\tilde{M} = S^n$) was solved during the sixties, and the interested reader will find an exposition in Wolf, J., "Spaces of Constant Curvature", McGraw-Hill, 1967 (Chapters 4–7). This book also contains a comprehensive study of what is known about related problems (Cf. the 2nd edition, published by Publish or Perish). Here we mention only two interesting facts.

4.4 PROPOSITION. *Let M^n be a complete Riemannian manifold of even dimension $n = 2m$, with constant sectional curvature $K = 1$. Then M^n is isometric to the sphere S^n or the real projective space P^n of the same dimension.*

Proof. The orthogonal group $0(2m+1)$ is the (transitive) group of isometries of S^n. Let Γ be a subgroup that acts in a totally discontinuous manner on S^n. If $\gamma \in \Gamma$ has determinant $+1$, there exists an eigenvalue of γ equal to 1. Then γ has a fixed point, which implies that $\gamma = e$. If $\tilde{\gamma} \in \Gamma$ has determinant -1, then $\tilde{\gamma}^2$ has determinant 1, therefore $\tilde{\gamma}^2 = e$. As a consequence, the eigenvalues of $\tilde{\gamma}$ are 1 or -1. If 1 is an eigenvalue of $\tilde{\gamma}$, then $\tilde{\gamma} = e$, which contradicts the fact that $\det \tilde{\gamma} = -1$. As a result $\tilde{\gamma} = -e$, hence,

$\Gamma = \{e, -e\}$. Using now Proposition 4.3 we obtain that M^n is either S^n, if $\Gamma = \{e\}$, or P^n, if $\Gamma = \{e, -e\}$. □

4.5 PROPOSITION. *Every compact orientable surface of genus $p > 1$ can be provided with a metric of constant negative curvature.*

Proof. In the hyperbolic plane H^2 take a closed geodesic polygon P with $4p$ sides of equal lengths. By a well known process of identifying the sides, P can be identified with a topological surface M^2 (See Massey [Ma], Chap. 1). Let Γ be the subgroup of isometries of H^2 generated by the isometries that identify the sides of P. It is possible to show that the transforms of P by Γ "tile H^2 without any gaps" if and only if the sum α of the interior angles at the vertices of P is equal to 2π (for a simple proof see Roger Fenn, "What is the geometry of a surface?", American Math. Monthly, Feb. 1983); in this case, $M^2 = H^2/\Gamma$ has a metric of constant curvature equal to -1. We assert that it is possible to find a polygon P which satisfies the condition $\alpha = 2\pi$, which will conclude the proof of the proposition.

To prove what is required, observe that by the Gauss-Bonnet Theorem (see M. do Carmo [dC 2], p. 274), we obtain that

$$\alpha = -A + 2\pi(2p - 1),$$

where A is the area of P in the hyperbolic metric. Therefore, on the one hand, if the polygon P is arbitrarily small, α is arbitrarily close to 2π $(2p - 1)$. On the other hand, we can enlarge the area A of polygon P in such a way that α is arbitrarily small. It follows that, by continuously deforming P, there exists a geodesic polygon such that $\alpha = 2\pi$. □

The spaces of constant curvature have had an important role in the historical development of Riemannian Geometry, due to their relationship with non-euclidean Geometry. A non-euclidean geometry is a complete Riemannian manifold M together with a transitive group of isometries G (the non-euclidean motions) which satisfy the *Axiom of free mobility: Let $p, \tilde{p} \in M$, γ_1, γ_2 be geodesics of M which start at p and form an angle α at p, and let $\tilde{\gamma}_1, \tilde{\gamma}_2$ be geodesics of $\tilde{M}$ with origin at $\tilde{p}$ forming an angle α at $\tilde{p}$. Then there exists $g \in G$, with $g(p) = \tilde{p}$, $g(\gamma_1) = \tilde{\gamma}_1$, $g(\gamma_2) = \tilde{\gamma}_2$.* This corresponds to the condition "side-angle-side" for congruence of triangles in Euclidean geometry and clearly implies that M has constant sectional curvature. It follows that the spaces of non-euclidean Geometry are included among the space forms, hence the importance given to determination of the space forms. The cases of S^n, P^n and H^n are called *spherical, elliptic* and *hyperbolic* geometry, respectively.

5. Isometries of the hyperbolic space; Theorem of Liouville

The isometries of hyperbolic space are closely related to the conformal transformations of $\mathbf{R}^n$. A map $f: U \subset \mathbf{R}^n \to \mathbf{R}^n$ of an open set $U \subset \mathbf{R}^n$ is called *conformal* if the (non-oriented) angles of intersecting curves are preserved, that is, if the (non-oriented) angle of any two vectors v_1 and v_2 at $p \in U$ is equal to the angle formed by $df_p(v_1)$ with $df_p(v_2)$. The principal objective of this section is to show that the isometries of the upper half-space $H^n \subset \mathbf{R}^n$ with the metric $\frac{\delta_{ij}}{x_n^2}$ (cf. Section 2) are the restrictions to H^n of conformal transformations of $\mathbf{R}^n$ that map H^n onto H^n. To prove this theorem, we have to describe the conformal transformations of $\mathbf{R}^n$.

For the case of $\mathbf{R}^2$, which we identify with the complex plane $\mathbf{C}$, it is well-known that the conformal transformations are holomorphic or anti-holomorphic functions (anti-holomorphic means that the complex conjugate function is holomorphic) with non-zero derivatives. In this regard, the Riemann mapping theorem guarantees that given two simply connected proper open sets in the plane, there exists a conformal mapping taking one into the other.

For the case of $\mathbf{R}^n$, $n > 2$, the situation is radically different, and the fact that f is a conformal transformation imposes strong restrictions on f. This is contained in the Theorem of Liouville below, for which we require some considerations.

First, observe that a necessary and sufficient condition so that $f: U \subset \mathbf{R}^n \to \mathbf{R}^n$ is conformal is that for all $p \in U$ and for all pairs of vectors v_1, v_2 at p it is true that:

$$(3) \qquad \langle df_p(v_1), df_p(v_2)\rangle = \lambda^2(p)\, \langle v_1, v_2\rangle\,, \quad \lambda^2 \neq 0.$$

Indeed, if (3) is satisfied, $|df_p(v)|^2 = \lambda^2 |v|^2$, for every vector v at p, hence

$$\cos \sphericalangle(df_p(v_1), df_p(v_2)) = \cos \sphericalangle(v_1, v_2), \tag{4}$$

that is, the (non-oriented) angles are preserved. Conversely, (4) implies that df_p takes a triangle with vertex at p into a similar triangle with vertex at $f(p)$. Therefore $|df_p(v)|^2 = \lambda^2 |v|^2$, for every v at p; applying this last relation to the sum of vectors $u + v$ we obtain (3).

The positive function $\lambda: U \to \mathbf{R}$ defined in (3) will be called the *coefficient of conformality* of f.

5.1 Example.. It is clear that an isometry of $\mathbf{R}^n$ (orthogonal linear transformation followed by a translation) is a conformal transformation of $\mathbf{R}^n$ with coefficient of conformality $\lambda \equiv 1$. The linear transformation $f(p) = \lambda I(p)$, $p \in \mathbf{R}^n$, where I is the identity matrix and $\lambda = \text{const.} > 0$, is evidently a conformal transformation with coefficient of conformality λ; f is called a *dilatation*. As a last example, we are going to show that the *inversion with respect to the unit sphere centered at* $p_o \in \mathbf{R}^n$, defined by

$$f(p) = \frac{p - p_o}{|p - p_o|^2} + p_o, \quad p \in \mathbf{R}^n - \{p_o\}, \tag{5}$$

is a conformal transformation.

Geometrically, f takes $p \in \mathbf{R}^n - \{p_o\}$ into a point $f(p)$ which is on the line through p and p_o, at a distance $|f(p) - p_o| = \frac{1}{|p - p_o|}$ from p_o. Therefore f keeps fixed the sphere of radius 1 around p_o, and permutes the interior region with the exterior of such a sphere; in particular, $f^2 = \text{identity}$.

To see that f is conformal, observe that, if v is a vector at p,

$$df_p(v) = \frac{v\,|p - p_o|^2 - 2\,\langle v, p - p_o\rangle\,(p - p_o)}{|p - p_o|^4},$$

therefore

$$|df_p(v)|^2 = \frac{\langle v,v\rangle}{|p-p_o|^4} + \frac{(4\langle v,p-p_o\rangle^2 - 4\langle v,p-p_o\rangle^2)\,|p-p_o|^2}{|p-p_o|^8}$$
$$= \frac{\langle v,v\rangle}{|p-p_o|^4},$$

that is, the inversion (5) is a conformal transformation with coefficient of conformality $\lambda = \frac{1}{|p-p_o|^2}$.

The Theorem of Liouville asserts the surprising fact that every conformal transformation f of an open set $U \subset \mathbf{R}^n$, $n \geq 3$, extends to a composition of isometries, dilatations and inversions. The theorem is, in reality, a little more precise and states that f is at most, composed of one isometry, one dilatation and one inversion.

The proof that we present below is contained, aside from notation and details, in the book of Dubrovin, Novikov and Fomenko, *Modern Geometry, Methods and Applications, Part I*, Springer-Verlag, New York, Berlin, 1984, pp. 138–141; no part of this proof will be used in the remainder of the book, and the reader eager for applications could omit it in a first reading.

5.2 THEOREM. *(Liouville). Let $f: U \to \mathbf{R}^n$, $n \geq 3$, be a conformal transformation of an open set $U \subset \mathbf{R}^n$. Then f is the restriction to U of a composition of isometries, dilatations or inversions, at most one of each.*

Proof. Let $a_1 = (1,0,\dots,0),\dots,a_n = (0,0,\dots,0,1)$ be the canonical basis of $\mathbf{R}^n$ and $(x_1,\dots,x_n)$ the cartesian coordinates of $\mathbf{R}^n$ relative to this basis. Let $e_1,\dots,e_n$ be parallel differentiable vector fields on U, such that at each point of U, $\langle e_i, e_j\rangle = \delta_{ij}$. If λ is the coefficient of conformality of f, we can write

$$\langle df(e_i), df(e_k)\rangle = \lambda^2\delta_{ik}, \quad i,k = 1,\dots,n. \tag{6}$$

Let d^2f be the second differential of f; that is, $d^2f = \mathbf{R}^n \times \mathbf{R}^n \to \mathbf{R}^n$ is a symmetric bilinear map with values in $\mathbf{R}^n$ and such that, in the canonical basis, $d^2f(a_i,a_j) = \frac{\partial^2 f}{\partial x_i \partial x_j}$. Taking the indices i,j,k distinct and differentiating (6), we obtain:

$$\left\langle d^2f(e_i,e_j), df(e_k)\right\rangle + \left\langle df(e_i), d^2f(e_k,e_j)\right\rangle = 0,$$

$$\langle d^2f(e_j,e_k), df(e_i)\rangle + \langle df(e_j), d^2f(e_i,e_k)\rangle = 0,$$
$$\langle d^2f(e_k,e_i), df(e_j)\rangle + \langle df(e_k), d^2f(e_j,e_i)\rangle = 0.$$

Summing the first two equations above and subtracting the third, we have

$$\langle d^2f(e_k,e_j), df(e_i)\rangle = 0, \quad \text{if } i,j,k \text{ are distinct.}$$

Fixing k,j and letting i vary in the remaining $(n-2)$ indices, we conclude that $d^2f(e_k,e_j)$ belongs to the plane generated by $df(e_j)$ and $df(e_k)$. Therefore,

$$d^2f(e_k,e_j) = \mu df(e_k) + \nu df(e_j),$$

in which, since $\langle df(e_k), df(e_k)\rangle = \langle df(e_j), df(e_j)\rangle = \lambda^2$,

$$\mu = \frac{\langle d^2f(e_k,e_j), df(e_k)\rangle}{\lambda^2} = \frac{\lambda d\lambda(e_j)}{\lambda^2} = \frac{d\lambda(e_j)}{\lambda},$$
$$\nu = \frac{d\lambda(e_k)}{\lambda},$$

that is,

$$d^2f(e_k,e_j) = \frac{1}{\lambda}(df(e_k)d\lambda(e_j) + df(e_j)d\lambda(e_k)). \tag{7}$$

It will be convenient, in what follows, to put $\rho = \frac{1}{\lambda}$. We are going to calculate the second differential $d^2(\rho f)$. Because $d(\rho f) = d\rho f + \rho df$, we obtain, using (7),

(8)
$$\begin{aligned} d^2(\rho f)(e_k,e_j) &= d^2\rho(e_k,e_j)f + \rho d^2f(e_k,e_j) + d\rho(e_k)df(e_j) \\ &+ d\rho(e_j)df(e_k) = d^2\rho(e_k,e_j)f + \frac{1}{\lambda}d^2f(e_k,e_j) \\ &- \frac{1}{\lambda^2}\{d\lambda(e_k)df(e_j) + d\lambda(e_j)df(e_k)\} = d^2\rho(e_k,e_j)f. \end{aligned}$$

We claim that $d^2\rho(e_k,e_j) = 0$, for $k \neq j$. To see this, calculate the third differential $d^3(\rho f)$, that is, the symmetric trilinear mapping, $d^3(\rho f)\colon \mathbf{R}^n \times \mathbf{R}^n \times \mathbf{R}^n \to \mathbf{R}^n$ with values in $\mathbf{R}^n$ and such that, in the canonical basis, $d^3(\rho f)(a_i,a_j,a_k) = \frac{\partial^3(\rho f)}{\partial x_i \partial x_j \partial x_k}$. Using (8), we obtain

$$d^3(\rho f)(e_k,e_j,e_i) = d^3\rho(e_k,e_j,e_i)f + d^2\rho(e_k,e_j)df(e_i).$$

In the expression above, the first member and the first part of the second member are symmetric in the three indices i, j, k. Therefore, the same thing happens with the second part of the second member. We conclude that

$$d^2\rho(e_k, e_j)df(e_i) = d^2\rho(e_k, e_i)df(e_j).$$

Since $df(e_i)$ and $df(e_j)$ are linearly independent and i, j, k are distinct but arbitrary indices, we obtain that $d^2\rho(e_k, e_j) = 0$, for all $j \neq k$, as we have asserted.

Now observe that, fixing $p \in U$, we can choose the vector fields $e_1, \ldots, e_n$ in such a way that they form an orthonormal basis previously given at p. Therefore the relation $d^2\rho(e_k, e_j) = 0$ is valid at p for every orthonormal basis at p and, because p is arbitrary, the same thing happens on U. Since $d^2\rho$ is a symmetric bilinear form and

$$0 = d^2\rho(\frac{e_j + e_k}{\sqrt{2}}, \frac{e_j - e_k}{\sqrt{2}}) = \frac{1}{2}\left\{d^2\rho(e_j, e_j) - d^2\rho(e_k, e_k)\right\},$$

we conclude that $d^2\rho(e_j, e_j) = d^2\rho(e_k, e_k)$, for all $j \neq k$.

In summary, for every $p \in U$ and for any orthonormal basis $e_1, \ldots, e_n$ at p, we have that $d^2\rho(e_j, e_k) = \sigma\delta_{jk}$. Taking, in particular, the canonical basis, we have

$$\frac{\partial^2\rho}{\partial x_i \partial x_j} = \sigma\delta_{ij}. \tag{9}$$

Calculating the derivative of both of the members of (9), we conclude that $(i \neq j)$

$$\frac{\partial\sigma}{\partial x_i} = \frac{\partial^3\rho}{\partial x_i \partial x_j \partial x_j} = \frac{\partial^3\rho}{\partial x_j \partial x_i \partial x_j} = 0,$$

that is, $\sigma = \text{const.}$

First we are going to consider the case $\sigma = \text{const.} \neq 0$ and show that Eq. (9) implies that

$$\rho = \frac{\sigma}{2}\sum x_i^2 + \sigma\sum b_i x_i + c, \quad b_i \text{ and } c \text{ constants.} \tag{10}$$

Indeed, from $\frac{\partial^2\rho}{\partial x_i^2} = \sigma$ we conclude that

$$\frac{\partial\rho}{\partial x_i} = \sigma x_i + \sigma b_i,$$

where b_i is a function that does not depend on x_i; since $\frac{\partial^2 \rho}{\partial x_i \partial x_j} = 0$, b_i also does not depend on x_j, $j \neq i$. Therefore b_i is a constant, and

$$\rho = \frac{1}{2}\sigma x_i^2 + \sigma b_i x_i + \varphi_i,$$

where φ_i does not depend on x_i. Because $\frac{\partial \rho}{\partial x_j} = \frac{\partial \varphi_i}{\partial x_j}$, $j \neq i$, we conclude that

$$\varphi_i = \frac{1}{2}\sigma x_j^2 + \sigma b_j x_j + \varphi_{ij},$$

where φ_{ij} does not depend on x_i and on x_j. Proceeding inductively, we obtain (10), as claimed.

Therefore, if $\sigma \neq 0$, we can write (10) in the form

$$\frac{1}{\lambda} = \rho = a_1 |p - p_o|^2 + k_1, \quad a_1 = \frac{\sigma}{2} \quad \text{and } k_1 = \text{const.},\ p_o \in \mathbf{R}^n.$$

The proof will be complete, for the case $\sigma \neq 0$, if we show that $k_1 = 0$. Because, considering the inversion $g \colon U \to \mathbf{R}^n$:

$$g(p) = \frac{p - p_o}{|p - p_o|^2} + p_o,$$

and taking the composition $h = g \circ f^{-1}$, we have that h is a conformal transformation whose coefficient of conformality is

$$a_1 |p - p_o|^2 \frac{1}{|p - p_o|^2} = a_1.$$

Therefore, h is an isometry followed by a dilatation, hence $f = h^{-1} \circ g$ is an inversion followed by a dilatation, followed by an isometry.

Now we are going to prove that $k_1 = 0$. Observe that applying to f^{-1} the above argument, we obtain

$$\lambda = a_2 |f(p) - q_o|^2 + k_2, \quad a_2 \text{ and } k_2 \text{ const.},$$

hence

$$(a_1 |p - p_o|^2 + k_1)(a_2 |f(p) - q_o|^2 + k_2) = 1. \tag{11}$$

Equation (11) shows that (the intersection with U of) a sphere of center p_o is mapped by f into a sphere of center q_o. Since f preserves angles, the radii of the first sphere are mapped into radii of the second. Let $p(s)$, $0 \leq s \leq s_o$, be a segment of a radius of the first sphere contained in U, where s is the arc length, and let $f \circ p(s)$ be its image. The length of the image segment is given by

$$\int_0^{s_o} \left| df\left(\frac{dp}{ds}\right) \right| ds = \int_0^{s_o} \frac{ds}{a_1 |p(s) - p_o|^2 + k_1} = |f(p(s_o)) - f(p(0)|.$$

If $k_1 \neq 0$, $|f(p(s_o)) - f(p(0))|$ is a transcendental function of $|p(s_o) - p_o|$. On the other hand, Eq. (11) implies that such a function is algebraic. This contradiction shows that $k_1 = 0$.

It remains to consider the case $\sigma = 0$. In this situation,

$$\rho = \frac{1}{\lambda} = \sum a_i x_i + c_1 = A_1(x) + c_1, \quad c_1 = \text{const.},$$

where we write, for convenience, $\sum a_i x_i = A_1(x)$, $x = (x_1, \ldots, x_n)$. In a similar manner as before, applying the initial argument to f^{-1}, we obtain

$$(11') \qquad (A_1(x) + c_1)(A_2(f(x)) + c_2) = 1.$$

Equation (11′) shows that (the intersection with U of) a hyperplane parallel to $A_1 = 0$ is taken by f into a hyperplane parallel to $A_2 = 0$. Because f preserves angles, a line perpendicular to the hyperplane $A_1 = 0$ is taken by f into a line (perpendicular to the hyperplane $A_2 = 0$). Considering a segment $p(s)$, $0 \leq s \leq s_o$, of such a line, parametrized by arc length s, we obtain, in an analogous manner as before, that

$$|f(p(s_o)) - f(p(0))| = \int_0^{s_o} \frac{ds}{A_1(p(s)) + c_1}.$$

The expression above contradicts (11′), except if the linear expression $A_1(p(s))$ were zero.

We conclude that if $\sigma = 0$, $\lambda = \text{const}$. In this case, the lengths of tangent vectors are multiplied by a constant λ and, as is easy to verify, f is an isometry, followed by a dilatation. This concludes the case $\sigma = 0$ and the proof of Liouville's Theorem. □

Now we can describe the isometries of hyperbolic space using the model of the upper half-space $H^n = \{(x_1, \ldots, x_n); x_n > 0\}$ with the metric $g_{ij} = \frac{\delta_{ij}}{x_n^2}$.

5.3 THEOREM. *The isometries of H^n are the restrictions to $H^n \subset \mathbf{R}^n$ of the conformal transformations of $\mathbf{R}^n$ that take H^n onto itself.*

Proof. Suppose first that $n \geq 3$. Let $f\colon H^n \to H^n$ be an isometry in the metric g_{ij}. Then, by Liouville's Theorem, f extends to a conformal map of $\mathbf{R}^n$ with the metric δ_{ij}. Since H^n is complete in the metric g_{ij}, f maps H^n onto H^n.

Conversely, let $f\colon H^n \subset \mathbf{R}^n \to H^n$ be a conformal transformation of H^n onto H^n and let $e_1, \ldots, e_n$ be an orthonormal basis, in the metric g_{ij}, at a point $p \in H^n$. Since $g_{ij} = \frac{\delta_{ij}}{x_n^2}$ and f is conformal, there exists a $\lambda^2 > 0$ such that $\langle df_p(e_i), df_p(e_j)\rangle = \lambda^2 \delta_{ij}$, where $\langle\ ,\ \rangle$ is the inner product in the metric g_{ij}. Therefore, the basis $\left\{\frac{df_p(e_i)}{\lambda}\right\}$ is orthonormal at $f(p)$, and by Corollary 2.3 of Cartan's Theorem, there exists an isometry g of H^n taking p to $f(p)$, with $dg(e_i) = \frac{df(e_i)}{\lambda}$. From what was proven in the first part, g is conformal. Hence $h = g^{-1} \circ f$ is the restriction to H^n of a conformal mapping of $\mathbf{R}^n$ which takes H^n onto H^n, leaving p fixed and satisfying $dh_p = \lambda I$.

The proof will be concluded if we show that $h =$ identity. In other words, we must show that if a conformal map h of $\mathbf{R}^n$ takes H^n onto H^n , leaving a point $p \in H^n$ fixed and satisfying $dh_p =$ multiple of the identity, then h is the identity.

To see this, let P be a hyperplane passing through p. From Liouville's theorem, $h(P)$ is a hyperplane or a sphere passing through p. Because dh_p is a multiple of the identity, P and $h(P)$ are tangent at P. Because h takes the boundary ∂H^n of H^n into itself, and h is conformal, the angle of P with ∂H^n is the same as the angle of $h(P)$ with ∂H^n.

We claim that $h(P) = P$. To see this, consider a straight line r_1 passing through p and perpendicular to ∂H^n, and let $q_1 = r_1 \cap \partial H^n$. Since the image $h(r_1)$ of r_1 is a circle or a line, and makes the same angle with ∂H^n that r_1 does, it is clear that $h(r_1) = r_1$. Hence $h(q_1) = q_1$. This shows that if P is perpendicular to ∂H^n, then $h(P) = P$. If P is not perpendicular to ∂H^n, let r be a line contained in P. Again, $h(r)$ is a circle or a line. In order that $h(r)$

be a circle, making the same angle α with ∂H^n that r makes with ∂H^n, it is necessary that $q_1 \in h(r)$ (See Fig. 2), which contradicts the fact that $h(q_1) = q_1$. Hence $h(r)$ is a line, $h(r) = r$, and because r is an arbitrary line in P, $h(P) = P$, as we have claimed.

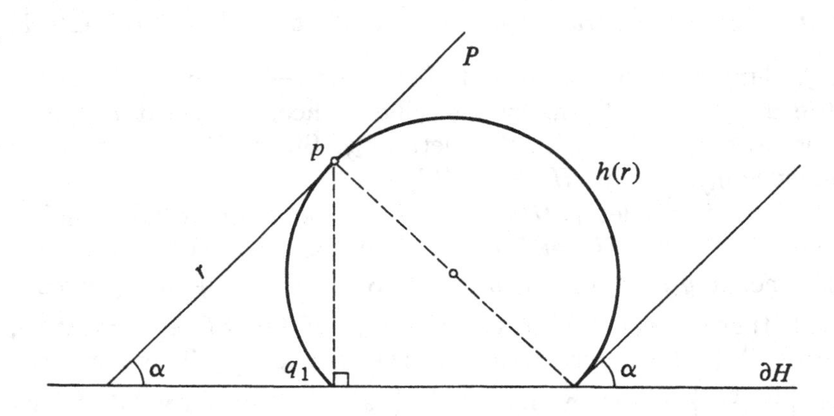

Figure 2

It follows that h cannot be an inversion (since the image of some plane will be a sphere) or an isometry distinct from the identity. From Liouville's Theorem, h is a dilatation. Because h takes H^n onto H^n, h is the identity.

For $n = 2$, the argument above does not apply. However, a simple calculation (cf. Exercise 4 of Chap. 1) shows that the conformal transformations of the form:

$$(12) \quad f(z) = \frac{az+b}{cz+d}, \quad z \in H^2 \subset \mathbf{C}, \quad a, b, c, d \in \mathbf{R}, \quad ad - bc \neq 0$$

(observe that f maps H^2 onto H^2) are isometries of H^2 with the metric g_{ij}. Moreover it is not difficult to show that for a fixed point $p_o \in H^2$ and unit vector v_o at p_o, there exists a transformation f of the form (12) which takes an arbitrary point $p \in H^2$ and a unit vector v at p into p_o and v_o, respectively (it suffices to observe that p and v are determined by three parameters which is the number of parameters of f). Since there exists a unique isometry of H^2 which takes (p, v) into (p_o, v_o), we conclude that all isometries of H^2 are of the form (12). This completes the proof of the case $n = 2$ and of Theorem 5.3. □

To conclude this Section, we are going to identify some important hypersurfaces of the hyperbolic space H^n. It is not difficult to verify that the intersection with H^n of hyperplanes of $\mathbf{R}^n$, orthogonal to ∂H^n and the intersection with H^n of spheres of $\mathbf{R}^n$, with center on ∂H^n, are totally geodesic submanifolds of H^n (see Exercise 2). Now we are going to determine the intersection with H^n of planes and spheres of $\mathbf{R}^n$ in any position.

For this, and for many other purposes, it is convenient to consider another model of the hyperbolic space, the so-called *ball model*. Let $B^n \subset \mathbf{R}^n$ be the open ball of radius 2 and center at the origin,

$$B^n = \{p \in \mathbf{R}^n; |p| < 2\}, \quad p = (x_1, \ldots, x_n),$$

and introduce on B^n the metric

$$h_{ij}(p) = \frac{\delta_{ij}}{(1 - \frac{1}{4}|p|^2)^2}.$$

We are going to show that B^n is isometric to H^n. Indeed, we shall show that the map $f\colon B^n \to H^n$ given by

$$f(p) = 4\frac{p - p_o}{|p - p_o|^2} - (0, \ldots, 1), \quad \text{where } p_o = (0, \ldots, -2), \tag{13}$$

is an isometry. In fact, if v is a vector at p and $\langle\ ,\ \rangle$ denotes the inner product in the Euclidean metric,

$$\langle df_p(v), df_p(v)\rangle = \frac{16\langle v, v\rangle}{|p - p_o|^4}.$$

On the other hand, letting $f(p) = (f_1(p), \ldots, f_n(p))$, we obtain

$$f_n(p) = \frac{4(x_n + 2)}{|p - p_o|^2} - 1 = \frac{4 - |p|^2}{|p - p_o|^2}.$$

Therefore,

$$\frac{\langle df_p(v), df_p(v)\rangle}{(f_n(p))^2} = \frac{16|p - p_o|^4 \langle v, v\rangle}{(4 - |p|^2)^2 |p - p_o|^4} = \frac{\langle v, v\rangle}{(1 - \frac{1}{4}|p|^2)^2}.$$

Because f is injective, we conclude that f is an isometry of B^n into H^n. Observe that f takes $\partial B^n - \{p_o\}$ into ∂H^n.

Notice that, by Liouville's Theorem, a mapping $g\colon B^n \to B^n$ is an isometry of B^n in the metric h_{ij} if and only if g is the restriction to B^n of a conformal transformation of $\mathbf{R}^n$ that takes B^n onto B^n.

Now let $S \subset H^n$ be an $(n-1)$-sphere of Euclidean space completely contained in the upper-half space H^n. We claim that S is a geodesic sphere of H^n in the metric g_{ij}. To see this, let $f^{-1}(S)$ be the image of S by the isometry $f^{-1}\colon H^n \to B^n$ given in (13). It is easy to verify that $f^{-1}(S)$ is a Euclidean sphere contained in B^n. It is possible to map the (Euclidean) center of S onto the origin of B^n by an isometry of B^n. Since the metric h_{ij} of B^n is symmetric with respect to the origin, the sphere so obtained is a geodesic sphere in B^n, and the same thing happens with its isometric image $S \subset H^n$.

Consider next an $(n-1)$-dimensional Euclidean sphere S tangent to ∂H^n at p and such that $S - \{p\} \subset H^n$. By an inversion of $\mathbf{R}^n$ at p (which is an isometry of H^n), S is mapped in a hyperplane P parallel to ∂H^n. Because the induced metric on P from H^n is a multiple of the Euclidean metric, P is a manifold of constant curvature zero and the same thing happens with its isometric image $S - \{p\}$. Such submanifolds are called *horospheres* of H^n.

Consider finally a Euclidean sphere S which cuts ∂H^n at an angle α, and denote its intersection $S \cap H^n$ with H^n, by $\sum$. Through an inversion of $\mathbf{R}^n$ at a point of $S \cap \partial H^n$, $\sum$ is mapped isometrically into the intersection, with H^n, of a hyperplane P which cuts ∂H^n at the same angle α. Consider the hyperplane Q which is orthogonal to ∂H^n and contains $P \cap \partial H^n$. We are going to prove that P is a hypersurface equidistant from the totally geodesic hypersurface Q. For this, let γ_r be a geodesic, represented in H^n by a semi-circle of radius r, with center 0 in $P \cap \partial H^n$ and in the plane perpendicular to $P \cap \partial H^n$. Since there exists a homothety with center 0 (hyperbolic isometry) taking the circle of radius r into a circle of any radius, the length of γ_r between the points of intersection of γ_r with P and Q does not depend on r. We conclude that P, or its isometric image $\sum$, is obtained by taking geodesics perpendicular to the totally geodesic hypersurface Q and marking on it a fixed distance. Such hypersurfaces are called *equidistant surfaces* (or *hyperspheres*).

In Exercise 6, we shall show that the hypersurfaces described above are characterized by the fact that they have, at each point, all of their principal curvatures equal, that is, they are umbilic, and we shall calculate their mean and sectional curvature.

EXERCISES

1. Consider, on a neighborhood in $\mathbf{R}^n$, $n > 2$ the metric

$$g_{ij} = \frac{\delta_{ij}}{F^2}$$

where $F \neq 0$ is a function of $(x_1, \ldots, x_n) \in \mathbf{R}^n$. Denote by $F_i = \frac{\partial F}{\partial x_i}$, $F_{ij} = \frac{\partial^2 F}{\partial x_i \partial x_j}$, etc.

a) Show that a necessary and sufficient condition for the metric to have constant curvature K is

$$(*) \qquad \begin{cases} F_{ij} = 0, & i \neq j, \\ F(F_{jj} + F_{ii}) = K + \sum_{i=1}^{n} (F_i)^2. \end{cases}$$

b) Use (*) to prove that the metric g_{ij} has constant curvature K if and only if

$$F = G_1(x_1) + G_2(x_2) + \cdots + G_n(x_n),$$

where

$$G_i(x_i) = ax_i^2 + b_i x_i + c_i$$

and

$$\sum_{i=1}^{n} (4c_i a - b_i^2) = K.$$

c) Put $a = K/4$, $b_i = 0$, $c_i = 1/n$ and obtain the formula of Riemann (See [Ri])

$$(**) \qquad g_{ij} = \frac{\delta_{ij}}{(1 + \frac{K}{4} \sum x_i^2)^2}$$

for a metric g_{ij} of constant curvature K. If $K < 0$ the metric g_{ij} is defined in a ball of radius $\sqrt{\frac{4}{-K}}$.

d) If $K > 0$, the metric (**) is defined on all of $\mathbf{R}^n$. Show that such a metric on $\mathbf{R}^n$ is not complete.

2. Show that if M^k is a closed, totally geodesic submanifold of H^n, $k \leq n$, then M^k is isometric to H^k. Determine all the totally geodesic submanifolds of H^n.

3. (*Another model of the hyperbolic space*). Consider on $\mathbf{R}^{n+1}$ the quadratic form

$$Q(x_o, x_1, \dots, x_n) = -(x_o)^2 + \sum_{i=1}^{n} (x_i)^2, \qquad (x_o, \dots, x_n) \in \mathbf{R}^{n+1}.$$

With the pseudo-Riemannian metric (,) induced by Q (Cf. Exercise 9, Chap. 2), $\mathbf{R}^{n+1}$ will be denoted by L^{n+1} (The Lorentzian space). Denote by H^n_k, $k = -\frac{1}{r^2}$, the connected component corresponding to $x_o > 0$ of the regular surface of $\mathbf{R}^{n+1}$ given by $Q(x) = -r^2$, $r > 0$. (Geometrically $Q(x) = -r^2$ is a hyperboloid of two sheets and H^n_k is the sheet contained in the half-space $x_o > 0$.)

a) Show that for all $x \in H^n_k$, the vector $\eta = \frac{x}{r}$ is normal to the tangent space $T_x(H^n_k)$.

b) Prove that $(\eta, \eta) = -1$, and that it is possible to choose a basis $b_o, \dots, b_n$ of L^{n+1} with $b_o = \eta$, $(b_i, b_j) = \delta_{ij}$, $(b_i, b_o) = 0$, $i, j = 1, \dots, n$. (Use the fact that the index of a quadratic form does not depend on the basis chosen to represent it.) Conclude that the metric induced by L^{n+1} on H^n_k is Riemannian.

c) Use the pseudo-Riemannian connection $\overline{\nabla}$ of L^{n+1} (Cf. Exercise 9, Chap. 2) to show that $S_\eta = (-\frac{1}{r})I$, where I is the identity map. Conclude that $B(X, Y) = \frac{\eta}{r}(X, Y)$, and use the Gauss formula to show that the sectional curvature of H^n_k is constant and equal $k = -\frac{1}{r^2}$.

d) Let $0(1, n+1)$ be the subgroup of linear transformations of $\mathbf{R}^{n+1}$ which preserve the metric (,). Show that the elements of $0(1, n+1)$ with det > 0 are isometries of

H_k^n and that given $X, Y \in H_k^n$ and orthonormal bases $\{v_i\} \in T_X(H_k^n)$ and $\{w_i\} \in T_Y(H_k^n)$, $i = 1, \ldots, n$, the restriction to H_k^n of the "linear" transformation which takes

$$\frac{X}{r} \to \frac{Y}{r} \qquad \text{and} \qquad v_i \to w_i$$

is an isometry of H_k^n. Conclude then that H_k^n has constant curvature (which we already know from (c)) and that H_k^n is complete.

e) Show that H_{-1}^n is isometric to the hyperbolic space H^n.

f) Show that the symmetries of H_k^n with respect to the plane P which passes through the origin of $\mathbf{R}^{n+1}$ and contains the x_o-axis are isometries of H_k^n. Conclude that all of the geodesics of H_n^k which pass through $(r, 0, \ldots, 0)$ are obtained as intersections $H_k^n \cap P$.

4. Identify $\mathbf{R}^4$ with $\mathbf{C}^2$ by letting (x_1, x_2, x_3, x_4) correspond to $(x_1 + ix_2, x_3 + ix_4)$. Let

$$S^3 = \left\{(z_1, z_2) \in \mathbf{C}^2; |z_1|^2 + |z_2|^2 = 1\right\},$$

and let $h: S^3 \to S^3$ be given by

$$h(z_1, z_2) = (e^{\frac{2\pi i}{q}} z_1, e^{\frac{2\pi i r}{q}} z_2), \quad (z_1, z_2) \in S^3,$$

where q and r are relatively prime integers, $q > 2$.

a) Show that $G = \{\text{id}, h, \ldots, h^{q-1}\}$ is a group of isometries of the sphere S^3, with the usual metric, which operates in a totally discontinuous manner. The manifold S^3/G is called a *lens space*.

b) Consider S^3/G with the metric induced by the projection $p: S^3 \to S^3/G$. Show that all the geodesics of S^3/G are closed but can have different lengths.

5. (*Connections of conformal metrics*) Let M be a differentiable manifold. Two Riemannian metrics g and $\bar{g}$ on M are *conformal* if there exists a positive function $\mu: M \to \mathbf{R}$ such that $\bar{g}(X, Y) = \mu g(X, Y)$, for all $X, Y \in \mathcal{X}(M)$. Let ∇ and $\overline{\nabla}$ be the Riemannian connections of g and $\bar{g}$, respectively. Prove that

$$\overline{\nabla}_X Y = \nabla_X Y + S(X, Y),$$

where $S(X,Y) = \frac{1}{2\mu}\{(X\mu)Y + (Y\mu)X - g(X,Y)\operatorname{grad}\mu\}$ and $\operatorname{grad}\mu$ is calculated in the metric g, that is,

$$X(\mu) = g(X, \operatorname{grad}\mu).$$

Hint: Since $\overline{\nabla}$ is obviously symmetric, it suffices to show that $\overline{\nabla}$ is compatible with $\bar{g}$, that is, that

$$X(\bar{g}(Y,Z)) = \bar{g}(\overline{\nabla}_X Y, Z) + \bar{g}(Y, \overline{\nabla}_X Z).$$

But the first member of the equality above is

$$X(\mu g(Y,Z)) = X(\mu)g(Y,Z) + \mu g(\nabla_X Y, Z) + \mu g(Y, \nabla_X Z),$$

and the second is

$$\mu g(\nabla_X Y, Z) + \mu g(Y, \nabla_X Z)$$
$$+\mu\{g(S(X,Y),Z) + g(Y, S(X,Z))\}.$$

Therefore, it is enough to prove that

$$X(\mu)g(Y,Z) = \mu\{g(S(X,Y),Z) + g(Y,S(X,Z))\},$$

which follows from a direct calculation.

6. (*Umbilic hypersurfaces of the hyperbolic space*). Let (M^{n+1}, g) be a manifold with a Riemannian metric g and let ∇ be its Riemannian connection. We say an immersion $x\colon N^n \to M^{n+1}$ is (totally) *umbilic* if for all $p \in N$, the second fundamental form B of x at p satisfies

$$\langle B(X,Y), \eta\rangle(p) = \lambda(p)\langle X,Y\rangle, \qquad \lambda(p) \in \mathbf{R},$$

for all $X, Y \in \mathcal{X}(N)$ and for a given unit field η normal to $x(N)$; here we are using $\langle\ ,\ \rangle$ to denote the metric g on M and the metric induced by x on N.

a) Show that if M^{n+1} has constant sectional curvature, λ does not depend on p.

Hint: Let $T, X, Y \in \mathcal{X}(N)$. The given condition implies that

$$-\langle\nabla_X\eta, Y\rangle = \lambda\langle X,Y\rangle \qquad \text{and} \qquad -\langle\nabla_T\eta, Y\rangle = \lambda\langle T,Y\rangle.$$

Differentiate the first equation with respect to T and the second with respect to X, obtaining, for all Y,

$$\langle \nabla_T \nabla_X \eta - \nabla_X \nabla_T \eta, Y \rangle$$
$$= -\langle T(\lambda)X - X(\lambda)T + \nabla_{[X,T]}\eta, Y \rangle.$$

Use the fact that M has constant sectional curvature to conclude that $T(\lambda)X - X(\lambda)T = 0$. Because T and X can be chosen linearly independently, this implies that $X(\lambda) = 0$, for all $X \in \mathcal{X}(N)$; therefore $\lambda = \text{const}$.

b) Use Exercise 5 to show that if we change the metric g to a metric $\bar{g} = \mu g$, conformal to g, the immersion $x\colon N^n \to (M^{n+1}, \bar{g})$ continues being umbilic, that is, if (using the notation of Exercise 5) $\langle \nabla_X \eta, Y \rangle_g = -\lambda \langle X, Y \rangle_g$, then

$$\left\langle \overline{\nabla}_X\left(\frac{\eta}{\sqrt{\mu}}\right), Y \right\rangle_{\bar{g}} = \frac{-2\lambda\mu + \eta(\mu)}{2\mu\sqrt{\mu}} \langle X, Y \rangle_{\bar{g}}.$$

c) Take $M^{n+1} = \mathbf{R}^{n+1}$ with the euclidean metric. Show that if $x\colon N^n \to \mathbf{R}^{n+1}$ is umbilic, then $x(N)$ is contained in an n-plane or an n-sphere in $\mathbf{R}^{n+1}$.

Hint: From (a), $\lambda = \text{constant}$. If $\lambda = 0$, $\langle \nabla_X \eta, Y \rangle = 0$ for all $X, Y \in \mathcal{X}(N)$ and all $\eta \in \mathcal{X}(N)^{\perp}$. It follows that $x(N)$ is contained in an affine n-plane in $\mathbf{R}^{n+1}$. If $\lambda \neq 0$, consider the map $y\colon N \to \mathbf{R}^{n+1}$ given by

$$y(p) = x(p) - \frac{\eta(p)}{\lambda}, \qquad p \in N.$$

Let $T, Y \in \mathcal{X}(N)$. Observe that

$$\langle \nabla_T y, Y \rangle = \langle T, Y \rangle - \frac{1}{\lambda} \langle \nabla_T \eta, Y \rangle = 0.$$

It follows that $y(N)$ reduces to a point, call it x_o, and that x satisfies

$$|x(p) - x_o|^2 = 1/\lambda^2,$$

that is, $x(N)$ is contained in a sphere of center x_o and radius $1/\lambda$.

d) Use (b) and (c) to establish that the umbilic hypersurfaces of the hyperbolic space, in the upper half-space model H^{n+1}, are the intersections with H^{n+1} of n-planes or n-spheres of $\mathbf{R}^{n+1}$. Therefore, the umbilic hypersurfaces of the hyperbolic space are the geodesic spheres, the horospheres and the hyperspheres. Conclude that such hypersurfaces have constant sectional curvature.

e) Calculate the mean curvature and the sectional curvature of the umbilic hypersurfaces of the hyperbolic space.

Hint: Consider the model of H^n as the upper half-space. Let $\sum = S \cap H^n$ be the intersection of H^n with a Euclidean $(n-1)$-sphere $S \subset \mathbf{R}^n$ of radius 1 and center in H^n. Since $\sum$ is umbilic, all of the directions are principal, and it is enough to calculate the curvature of the curves of intersection of $\sum$ with the $x_1 x_n$-plane. Use the expression obtained in part (b) of this exercise to establish that the mean curvature of $\sum$ (in the metric of H^n) is equal to 1 if S is tangent to ∂H^n, is equal to $\cos\alpha$ if S makes an angle α with ∂H^n, and is equal to the "height" of the Euclidean center of S relative to ∂H^n, if $S \subset H^n$. To calculate the sectional curvature, use the Gauss formula.

7. Define a "stereographic projection" $f\colon H^n_{-1} \to D^n$ from the model of the hyperbolic space H^n_{-1} of curvature -1 given in Exercise 3 onto the open ball

$$D^n = \{(x_o, \ldots, x_n); x_o = 0, \sum_{\alpha=1}^{n} x_\alpha^2 < 1\}$$

in the following way: If $p \in H^n_{-1} \subset L^{n+1}$, join p to $p_o = (-1, 0, \ldots, 0)$ by a line r; $f(p)$ is the intersection of r with D^n (See Fig. 3). Let $p = (x_o, \ldots, x_n)$ and $f(p) = (0, u_1, \ldots, u_n)$.

a) Prove that:

$$x_\alpha = \frac{2u_\alpha}{1 - \sum_\alpha u_\alpha^2}, \quad \alpha = 1, \ldots, n,$$

$$x_o = \frac{2}{1 - \sum_\alpha u_\alpha^2} - 1.$$

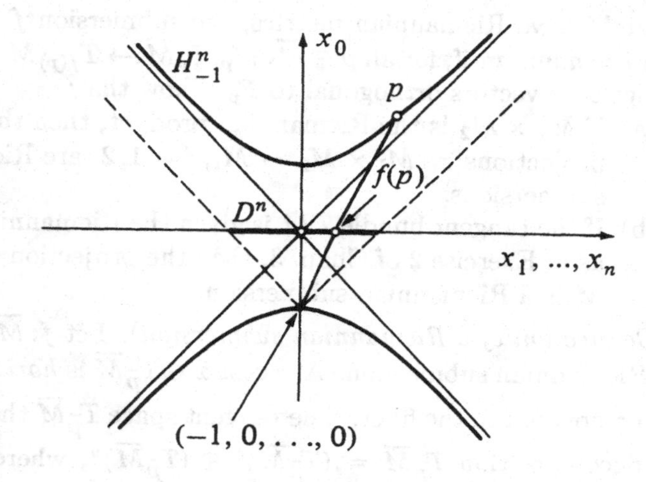

Figure 3

b) Show that:

$$-(dx_o)^2 + (dx_1)^2 + \cdots + (dx_n)^2 = \frac{4\{(du_1)^2 + \cdots + (du_n)^2\}}{(1 - \sum_\alpha u_\alpha^2)^2}.$$

Conclude that $f^{-1}: D^n \to H^n_{-1}$ induces on D the metric $g_{ij} = \frac{4\delta_{ij}}{(1-\sum_\alpha u_\alpha^2)^2}$. Therefore, D^n with the metric g_{ij} has constant curvature -1 (cf. Exercise 1(c)).

c) Show that the images by f of the non-empty intersections of affine hyperplanes P of L^{n+1} with H^n_{-1} are intersections with D^n of spheres (or planes, when P passes through p_o) contained in the hyperplane $x_o = 0$. Conclude that the umbilic hypersurfaces of H^n_{-1} (Cf. Exercise 6) are of the form $P \cap H^n_{-1}$.

8. (*Riemannian submersions*). A differentiable mapping $f: \overline{M}^{n+k} \to M^n$ is called *submersion* if f is surjective, and for all $\overline{p} \in \overline{M}$, $df_{\overline{p}}: T_{\overline{p}}\overline{M} \to T_{f(p)}M$ has rank n. In this case, for all $p \in M$, the *fiber* $f^{-1}(p) = F_p$ is a submanifold of $\overline{M}$ and a tangent vector of $\overline{M}$, tangent to some F_p, $p \in M$, is called a *vertical vector* of the submersion. If, in addition, $\overline{M}$

and M have Riemannian metrics, the submersion f is said to be *Riemannian* if, for all $p \in \overline{M}$, $df_p: T_p\overline{M} \to T_{f(p)}M$ preserves lengths of vectors orthogonal to F_p. Show that:

a) If $M_1 \times M_2$ is the Riemannian product, then the natural projections $\pi_i: M_1 \times M_2 \to M_i$, $i = 1, 2$, are Riemannian submersions.

b) If the tangent bundle TM is given the Riemannian metric as in Exercise 2 of Chap. 3, then the projection $\pi: TM \to M$ is a Riemannian submersion.

9. (*Connection of a Riemannian submersion*). Let $f: \overline{M} \to M$ be a Riemannian submersion. A vector $\overline{x} \in T_{\overline{p}}\overline{M}$ is *horizontal* if it is orthogonal to the fiber. The tangent space $T_{\overline{p}}\overline{M}$ then admits a decomposition $T_{\overline{p}}\overline{M} = (T_{\overline{p}}\overline{M})^h \oplus (T_{\overline{p}}\overline{M})^v$, where $(T_{\overline{p}}\overline{M})^h$ and $(T_{\overline{p}}\overline{M})^v$ denote the subspaces of horizontal and vertical vectors, respectively. If $X \in \mathcal{X}(M)$, the *horizontal lift* $\overline{X}$ of X is the horizontal field defined by $df_{\overline{p}}(\overline{X}(\overline{p})) = X(f(p))$.

a) Show that $\overline{X}$ is differentiable.

b) Let ∇ and $\overline{\nabla}$ be the Riemannian connections of M and $\overline{M}$ respectively. Show that

$$\overline{\nabla}_{\overline{X}}\overline{Y} = \overline{(\nabla_X Y)} + \frac{1}{2}[\overline{X}, \overline{Y}]^v, \qquad X, Y \in \mathcal{X}(M),$$

where Z^v is the vertical component of Z.

c) $[\overline{X}, \overline{Y}]^v(\overline{p})$ depends only on $\overline{X}(\overline{p})$ and $\overline{Y}(\overline{p})$.

Hint for (b): Let $X, Y, Z \in \mathcal{X}(M)$. Let $T \in \mathcal{X}(\overline{M})$ be a vertical field. Observe that:

$$\left\langle \overline{X}, T \right\rangle = \left\langle \overline{Y}, T \right\rangle = \left\langle \overline{Z}, T \right\rangle = 0, \quad \overline{X}\left\langle \overline{Y}, \overline{Z} \right\rangle = X\langle Y, Z\rangle,$$

$$df[\overline{X}, T] = 0, \quad [X, Y] = [df\overline{X}, df\overline{Y}] = df[\overline{X}, \overline{Y}] \quad \text{and}$$

$$T\left\langle \overline{X}, \overline{Y} \right\rangle = 0.$$

Conclude that $\left\langle [\overline{X}, \overline{Y}], \overline{Z} \right\rangle = \langle [X, Y], Z\rangle$, $\left\langle [\overline{X}, T], \overline{Y} \right\rangle = 0$ and use the formula for the Riemannian connection as a function of the metric to obtain

$$\left\langle \overline{\nabla}_{\overline{X}}\overline{Y}, \overline{Z} \right\rangle = \langle \nabla_X Y, Z\rangle, \quad 2\left\langle \overline{\nabla}_{\overline{X}}\overline{Y}, T \right\rangle = \left\langle T, [\overline{X}, \overline{Y}] \right\rangle,$$

which implies (b).

Hint for (c): Use the fact that

$$\left\langle [\overline{X},\overline{Y}],T\right\rangle = \left\langle \overline{\nabla}_{\overline{X}}\overline{Y} - \overline{\nabla}_{\overline{Y}}\overline{X},T\right\rangle.$$

10. (*Curvature of a Riemannian submersion*). Let $f:\overline{M}\to M$ be a Riemannian submersion. Let $X,Y,Z,W\in\mathcal{X}(M)$, $\overline{X},\overline{Y},\overline{Z},\overline{W}$ be their horizontal lifts, and let R and $\overline{R}$ be the curvature tensors of M and $\overline{M}$ respectively. Prove that:

(a) $$\left\langle \overline{R}(\overline{X},\overline{Y})\overline{Z},\overline{W}\right\rangle = \langle R(X,Y)Z,W\rangle - \frac{1}{4}\left\langle [\overline{X},\overline{Z}]^v,[\overline{Y},\overline{W}]^v\right\rangle$$

$$+\frac{1}{4}\left\langle [\overline{Y},\overline{Z}]^v,[\overline{X},\overline{W}]^v\right\rangle - \frac{1}{2}\left\langle [\overline{Z},\overline{W}]^v,[\overline{X},\overline{Y}]^v\right\rangle.$$

b) $K(\sigma)=\overline{K}(\overline{\sigma})+\frac{3}{4}\left|[\overline{X},\overline{Y}]^v\right|^2\geq \overline{K}(\overline{\sigma})$,

where σ is the plane generated by the orthonormal vectors $X,Y\in\mathcal{X}(M)$ and $\overline{\sigma}$ is the plane generated by $\overline{X}$, $\overline{Y}$.

Hint for (a): We shall use the notation of Exercise 9. Observe that $\overline{X}\left\langle \overline{\nabla}_{\overline{Y}}\overline{Z},\overline{W}\right\rangle = X\langle\nabla_Y Z,W\rangle$. Therefore

$$\left\langle \overline{\nabla}_{\overline{X}}\overline{\nabla}_{\overline{Y}}\overline{Z},\overline{W}\right\rangle = \overline{X}\left\langle \overline{\nabla}_{\overline{Y}}\overline{Z},\overline{W}\right\rangle - \left\langle \overline{\nabla}_{\overline{Y}}\overline{Z},\overline{\nabla}_{\overline{X}}\overline{W}\right\rangle$$

$$= \langle \nabla_X\nabla_Y Z,W\rangle - \frac{1}{4}\left\langle [\overline{Y},\overline{Z}]^v,[\overline{X},\overline{W}]^v\right\rangle.$$

On the other hand, if $T\in\mathcal{X}(\overline{M})$ is vertical,

$$\left\langle \overline{\nabla}_T\overline{X},\overline{Y}\right\rangle = \left\langle \overline{\nabla}_{\overline{X}}T,\overline{Y}\right\rangle + \left\langle [T,\overline{X}],\overline{Y}\right\rangle = -\left\langle T,\overline{\nabla}_{\overline{X}}\overline{Y}\right\rangle.$$

Therefore,

$$\left\langle \overline{\nabla}_{[\overline{X},\overline{Y}]}\overline{Z},\overline{W}\right\rangle = \left\langle \overline{\nabla}_{[\overline{X},\overline{Y}]^h}\overline{Z},\overline{W}\right\rangle + \left\langle \overline{\nabla}_{[\overline{X},\overline{Y}]^v}\overline{Z},\overline{W}\right\rangle$$

$$= \left\langle \overline{\nabla}_{[X,Y]}Z,W\right\rangle - \frac{1}{2}\left\langle [\overline{X},\overline{Y}]^v,[\overline{Z},\overline{W}]^v\right\rangle.$$

Putting the above together, we obtain (a).

11. (*The complex projective space*). Let

$$\mathbf{C}^{n+1}-\{0\} = \{(z_o,\dots,z_n) = Z \neq 0, z_j = x_j+iy_j, j=0,\dots,n\}$$

be the set of all non-zero $(n+1)$-tuples of complex numbers z_j. Define an equivalence relation on $\mathbf{C}^{n+1}-\{0\}: Z = (z_o,\dots,z_n) \sim W = (w_o,\dots,w_n)$ if $z_j = \lambda w_j$, $\lambda \in \mathbf{C}$, $\lambda \neq 0$. The equivalence class of Z will be denoted by $[Z]$ (= the complex line passing through the origin and through Z). The set of such classes is called, by analogy with the real case, *the complex projective space* $P^n(\mathbf{C})$ of complex dimension n.

a) Show that $P^n(\mathbf{C})$ has a differentiable structure of a manifold of real dimension $2n$ and that $P^1(\mathbf{C})$ is diffeomorphic to S^2.

b) Let $(Z,W) = z_o\overline{w}_o + \cdots + z_n\overline{w}_n$ be the hermitian product on $\mathbf{C}^{n+1}$, where the bar denotes complex conjugation. Identify $\mathbf{C}^{n+1} \approx \mathbf{R}^{2n+2}$ by putting $z_j = x_j + iy_j = (x_j, y_j)$. Show that

$$S^{2n+1} = \{N \in \mathbf{C}^{n+1} \approx \mathbf{R}^{2n+2}; (N,N) = 1\}$$

is the unit sphere in $\mathbf{R}^{2n+2}$.

c) Show that the equivalence relation $\sim$ induces on S^{2n+1} the following equivalence relation: $Z \sim W$ if $e^{i\theta}Z = W$. Establish that there exists a differentable map (the Hopf fibering) $f\colon S^{2n+1} \to P^n(\mathbf{C})$ such that

$$\begin{aligned} &f^{-1}([Z]) \\ &= \{e^{i\theta}N \in S^{2n+1}, N \in [Z]\cap S^{2n+1}, 0 \leq \theta \leq 2\pi\} \\ &= [Z] \cap S^{2n+1}. \end{aligned}$$

d) Show that f is a submersion.

12. (*Curvature of the complex projective space*). Define a Riemannian metric on $\mathbf{C}^{n+1}-\{0\}$ in the following way: If $Z \in \mathbf{C}^{n+1}-\{0\}$ and $V, W \in T_Z(\mathbf{C}^{n+1}-\{0\})$,

$$\langle V, W\rangle_Z = \frac{\text{Real }(V,W)}{(Z,Z)}.$$

Observe that the metric $\langle\ ,\ \rangle$ restricted to $S^{2n+1} \subset \mathbf{C}^{n+1} - \{0\}$ coincides with the metric induced from $\mathbf{R}^{2n+2}$.

a) Show that, for all $0 \leq \theta \leq 2\pi$, $e^{i\theta}: S^{2n+1} \to S^{2n+1}$ is an isometry, and that, therefore, it is possible to define a Riemannian metric on $P^n(\mathbf{C})$ in such a way that the submersion f is Riemannian.

b) Show that, in this metric, the sectional curvature of $P^n(\mathbf{C})$ is given by

$$K(\sigma) = 1 + 3\cos^2\varphi,$$

where σ is generated by the orthonormal pair X, Y, $\cos\varphi = \left\langle \overline{X}, i\overline{Y} \right\rangle$, and $\overline{X}$, $\overline{Y}$ are the horizontal lifts of X and Y, respectively. In particular, $1 \leq K(\sigma) \leq 4$.

Hint for (b): Let Z be the position vector describing S^{2n+1}. Since $(\frac{d}{d\theta}e^{i\theta}Z)_{\theta=0} = iZ$, $iZ \in T_Z(S^{2n+1})$ and is vertical. Let $\overline{\nabla}$ be the Riemannian connection of $\mathbf{R}^{2n+2} \approx \mathbf{C}^{n+1}$ and $X, Y \in \mathcal{X}(P^n(\mathbf{C}))$. Take $\alpha: (-\varepsilon, \varepsilon) \to S^{2n+1}$ with $\alpha(0) = Z$, $\alpha'(0) = \overline{X}$. Then

$$\begin{aligned}(\overline{\nabla}_{\overline{X}} iZ)_Z &= \frac{d}{dt} iZ \circ \alpha(t))\Big|_{t=0} \\ &= \frac{d}{dt} i\alpha(t)\Big|_{t=0} = i\alpha'(0) = i\overline{X}\end{aligned}$$

Therefore,

$$\begin{aligned}\left\langle [\overline{X}, \overline{Y}], iZ \right\rangle &= \left\langle \overline{\nabla}_{\overline{X}}\overline{Y} - \overline{\nabla}_{\overline{Y}}\overline{X}, iZ \right\rangle \\ &= -\left\langle i\overline{X}, \overline{Y} \right\rangle + \left\langle i\overline{Y}, \overline{X} \right\rangle = 2\cos\varphi.\end{aligned}$$

Now use Exercise 10 (b).

13. Let $p \in M$ and let $\sigma: M \to M$ be an isometry such that $\sigma(p) = p$ and $d\sigma_p(v) = -v$, for all $v \in T_pM$. Let X be a parallel field along a geodesic γ in M with $\gamma(0) = p$. Show that $d\sigma_{\gamma(t)}X(\gamma(t)) = -X(\gamma(-t))$.
Hint: It is clear that $\sigma(\gamma(t)) = \gamma(-t)$. Prove that $d\sigma_{\gamma(t)}X(\gamma(t))$ is a parallel field along $\gamma(t)$. Note that for $t = 0$,

$d\sigma_{\gamma(t)}X(\gamma(t)) = -X(\gamma(-t))$, and use the uniqueness of parallel fields, with given initial conditions.

14. (*Geometric characterization of locally symmetric spaces*). Let M be a Riemannian manifold. A *local symmetry* at $p \in M$ is a map $\sigma\colon B_\varepsilon(p) \to B_\varepsilon(p)$ of a normal geodesic ball centered at p such that $\sigma(\gamma(t)) = \gamma(-t)$, where γ is a radial geodesic ($\gamma(0) = p$) of $B_\varepsilon(p)$. Prove that: *M is locally symmetric $\Leftrightarrow$ every local symmetry is an isometry.*
Hint $\Rightarrow$: Consider a geodesic frame $e_1, \dots, e_n$ in the ball $B_\varepsilon(p)$ (Cf. Exercise 7 of Chap. 3) and put $R_{ijk\ell} = R(e_i, e_j, e_k, e_\ell)$. Since $\nabla R = 0$, $R_{ijk\ell}$ is constant along the geodesics which start from p. Let $i\colon T_pM \to T_pM$ be a linear isometry given by $i(v) = -v$, $v \in T_pM$, and observe that $\sigma = \exp_p \circ i \circ \exp_p^{-1}$. Use Cartan's Theorem to establish that σ is an isometry.
Hint $\Leftarrow$: Let $p \in M$ and $Z \in T_p(M)$. Consider a geodesic $\gamma\colon (-\varepsilon, \varepsilon) \to M$ with $\gamma(0) = p$, $\gamma'(0) = Z$. Take an orthonormal basis $e_1, \dots, e_n$ in T_pM, and obtain, by parallel transport, a frame $e_1(t), \dots, e_n(t)$ along γ. Put $R_{ijk\ell}(t) = R(e_i(t), e_j(t), e_k(t), e_\ell(t))$. Then

$$\begin{aligned}(\nabla_Z R)(p) &= \frac{d}{dt} R_{ijk\ell}\Big|_{t=0} \\ &= \lim_{t\to 0} \frac{R_{ijk\ell}(t) - R_{ijk\ell}(-t)}{2t} = 0\end{aligned}$$

where, in the last equality, we use that σ is an isometry and we also use the last exercise. Since p and Z are arbitrary, $\nabla R = 0$.

CHAPTER 9

VARIATIONS OF ENERGY

1. Introduction

In Chapter 3, we defined geodesics as curves with zero acceleration and we saw that they are characterized by the fact that they locally minimize the arc length. In this chapter, we present a further characterization of a geodesic as a "solution of a variational problem". For this, we have to introduce certain ideas that are adaptations to Differential Geometry of concepts and techniques from the Calculus of Variations. No knowledge of Calculus of Variations will be assumed.

The fundamental point of the chapter is the calculation of the formula for the second variation of the energy of a geodesic, which will be presented in Section 2. In Section 3, we shall make two geometric applications of this formula. The first (Theorem of Bonnet-Myers), states that a complete manifold whose curvature is positive and does not approach zero is compact, and its diameter can be estimated in terms of the bounds of the curvature. The second is an extension, due to A. Weinstein, of a theorem of Synge which asserts the simple connectivity of a compact, orientable, manifold of even dimension whose curvature is positive.

Together with the theorem of Hadamard, the applications included in this chapter concern investigations which attempt to determine the influence of curvature on the topology of Riemannian manifolds. These results culminate with the Sphere Theorem (see Chap. 13) and have ramifications which extend to current research.

2. Formulas for the first and second variations of energy

We start by making precise the idea of "neighboring curves" to a given curve.

2.1 DEFINITION. Let $c\colon [0,a] \to M$ be a piecewise differentiable curve in a manifold M. A *variation* of c is a continuous mapping $f\colon (-\varepsilon, \varepsilon) \times [0,a] \to M$ such that:

a) $f(0,t) = c(t)$, $t \in [0,a]$,

b) there exists a subdivision of $[0,a]$ by points $0 = t_o < t_1 < \cdots < t_{k+1} = a$, such that the restriction of f to each $(-\varepsilon,\varepsilon) \times [t_i, t_{i+1}]$, $i = 0, 1, \ldots, k$, is differentiable.

A variation is said to be *proper* if

$$f(s,0) = c(0) \qquad \text{and} \qquad f(s,a) = c(a),$$

for all $s \in (-\varepsilon, \varepsilon)$. If f is differentiable, the variation is said to be *differentiable*.

For each $s \in (-\varepsilon, \varepsilon)$, the parametrized curve $f_s\colon [0,a] \to M$ given by $f_s(t) = f(s,t)$ is called a *curve in the variation*. In this way, a variation determines a family $f_s(t)$ of neighboring curves of $f_o(t) = c(t)$, and a variation is proper if and only if the curves of this family have the same initial point $c(0)$ and the same endpoint $c(a)$.

It is customary to call the parametrized differentiable curve given by $f_t(s) = f(s,t)$, t fixed, a *transversal curve of the variation.* The velocity vector of a transversal curve at $s = 0$, defined by $V(t) = \frac{\partial f}{\partial s}(0,t)$, is a (piecewise differentiable) vector field along $c(t)$ and is called the *variational field* of f (Fig. 1).

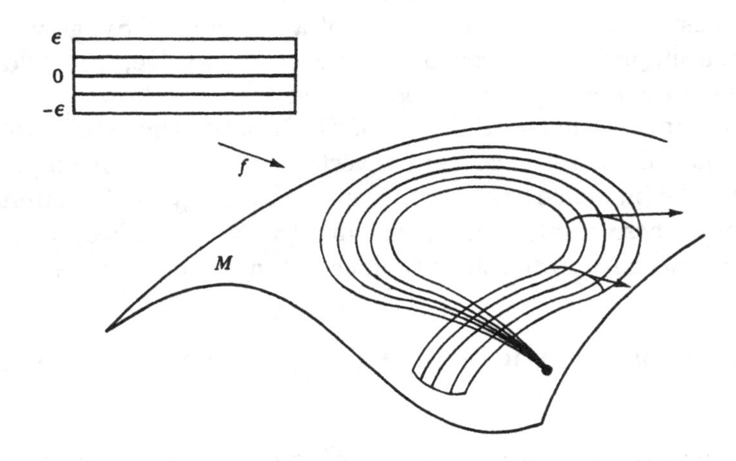

Figure 1

2.2 PROPOSITION. *Given a piecewise differentiable field $V(t)$, along the piecewise differentiable curve $c\colon [0,a] \to M$, there exists a variation $f\colon (-\varepsilon,\varepsilon) \times [0,a] \to M$, of c, such that $V(t)$ is the variational field of f; in addition, if $V(0) = V(a) = 0$, it is possible to choose f as a proper variation.*

Proof. Since $c([0,a]) \subset M$ is compact it is possible to find a $\delta > 0$ such that $\exp_{c(t)}$, $t \in [0,a]$, is well-defined for all $v \in T_{c(t)}M$, with $|v| < \delta$. Indeed, for each $c(t)$ consider a totally normal neighborhood W_t of $c(t)$ and the number $\delta_t > 0$ associated to this neighborhood (Theor. 3.7, Chap. 3). The union $\bigcup_t W_t$ covers the compact set $c([0,a])$ and, therefore, a finite number of the W_t's, say $W_1, \dots, W_n$, still cover $c([0,a])$. Taking $\delta = \min(\delta_1, \dots, \delta_n)$, where $\delta_i > 0$ is the number corresponding to the neighborhoods W_i, $i = 1, \dots, n$, we see that this δ satisfies the conditions of the assertion that was made.

Consider $N = \max_{t\in[0,a]} |V(t)|$, $\varepsilon < \frac{\delta}{N}$ and define $f(s,t) = \exp_{c(t)} sV(t)$, $s \in (-\varepsilon,\varepsilon)$, $t \in [0,a]$.

By the choice of ε, the map $f\colon (-\varepsilon,\varepsilon) \times [0,a] \to M$ is well-defined. In addition, since

$$\exp_{c(t)} sV(t) = \gamma(1, c(t), sV(t))$$

and the geodesic $\gamma(1, c(t), sV(t))$ depends differentiably on the initial conditions, the map f is piecewise differentiable. It is easy to verify that $f(0,t) = c(t)$.

Finally, the variational field of f is given by:

$$\frac{\partial f}{\partial s}(0,t) = \frac{d}{ds}(\exp_{c(t)} sV(t))\bigg|_{s=0} = (d\exp_{c(t)})_o V(t) = V(t),$$

and it is clear that, from the definition of f, if $V(0) = V(a) = 0$ then f is proper. □

To compare the arc length of c with the arc length of neighboring curves in a variation $f\colon (-\varepsilon,\varepsilon) \times [0,a] \to M$ of c, we define a function $L\colon (-\varepsilon,\varepsilon) \to \mathbf{R}$ by

$$L(s) = \int_0^a \left|\frac{\partial f}{\partial t}(s,t)\right| dt, \qquad s \in (-\varepsilon,\varepsilon),$$

that is, $L(s)$ is the length of the curve $f_s(t)$. It will be more convenient, however, to work with the *energy function* $E(s)$ given by

$$E(s) = \int_0^a \left|\frac{\partial f}{\partial t}(s,t)\right|^2 dt, \qquad s \in (-\varepsilon, \varepsilon).$$

For that, we need some general facts about the energy function.

Let $c\colon [0,a] \to M$ be a curve and let

$$L(c) = \int_0^a \left|\frac{dc}{dt}\right| dt \quad \text{and} \quad E(c) = \int_0^a \left|\frac{dc}{dt}\right|^2 dt.$$

Putting $f \equiv 1$ and $g = |\frac{dc}{dt}|$ in the Schwarz inequality:

$$\left(\int_0^a fg\, dt\right)^2 \leq \int_0^a f^2 dt \cdot \int_0^a g^2 dt,$$

we obtain

$$L(c)^2 \leq aE(c),$$

and equality occurs if and only if g is constant, that is, if and only if t is proportional to arc length.

The lemma below shows that the curves which minimize energy are automatically parametrized by a parameter proportional to arc length. This is one of the advantages of working with the energy function rather than the arc length function.

2.3 LEMMA. *Let $p, q \in M$ and let $\gamma\colon [0,a] \to M$ be a minimizing geodesic joining p to q. Then, for all curves $c\colon [0,a] \to M$ joining p to q,*

$$E(\gamma) \leq E(c)$$

with equality holding if and only if c is a minimizing geodesic.

Proof. From the considerations above, it follows that

$$aE(\gamma) = (L(\gamma))^2 \leq (L(c))^2 \leq aE(c),$$

which proves the first assertion. If the equality holds, we have $(L(c))^2 = aE(c)$, which implies that the parameter of c is proportional to arc length, and $L(\gamma) = L(c)$, which implies that c is clearly a minimizing geodesic (see Corollary 3.9 of Chapter 3). The converse is obvious. □

Now, we return to the energy function $E(s)$ defined by a variation. An initial piece of information on the behavior of E is given by the value of its first derivative.

2.4 PROPOSITION. *(Formula for the first variation of the energy of a curve). Let $c\colon [0,a] \to M$ be a piecewise differentiable curve and let $f\colon (-\varepsilon,\varepsilon) \times [0,a] \to M$ be a variation of c. If $E\colon (-\varepsilon,\varepsilon) \to \mathbf{R}$ is the energy of f then*

$$
\begin{aligned}
\frac{1}{2}E'(0) = &-\int_0^a \left\langle V(t), \frac{D}{dt}\frac{dc}{dt} \right\rangle dt \\
&- \sum_{i=1}^{k} \left\langle V(t_i), \frac{dc}{dt}(t_i^+) - \frac{dc}{dt}(t_i^-) \right\rangle \qquad (1) \\
&- \left\langle V(0), \frac{dc}{dt}(0) \right\rangle + \left\langle V(a), \frac{dc}{dt}(a) \right\rangle,
\end{aligned}
$$

where $V(t)$ is the variational field of f, and

$$
\frac{dc}{dt}(t_i^+) = \lim_{\substack{t\to t_i \\ t>t_i}} \frac{dc}{dt}, \quad \frac{dc}{dt}(t_i^-) = \lim_{\substack{t\to t_i \\ t<t_i}} \frac{dc}{dt}.
$$

Proof. By definition,

$$
E(s) = \int_0^a \left\langle \frac{\partial f}{\partial t}, \frac{\partial f}{\partial t} \right\rangle dt = \sum_{i=0}^{k} \int_{t_i}^{t_{i+1}} \left\langle \frac{\partial f}{\partial t}, \frac{\partial f}{\partial t} \right\rangle dt.
$$

Differentiating under the integral sign and using the symmetry of the Riemannian connection, we obtain

$$
\begin{aligned}
\frac{d}{ds}\int_{t_i}^{t_{i+1}} \left\langle \frac{\partial f}{\partial t}, \frac{\partial f}{\partial t} \right\rangle dt &= \int_{t_i}^{t_{i+1}} 2\left\langle \frac{D}{ds}\frac{\partial f}{\partial t}, \frac{\partial f}{\partial t} \right\rangle dt \\
&= 2\int_{t_i}^{t_{i+1}} \left\langle \frac{D}{dt}\frac{\partial f}{\partial s}, \frac{\partial f}{\partial t} \right\rangle dt \\
&= 2\int_{t_i}^{t_{i+1}} \frac{d}{dt}\left\langle \frac{\partial f}{\partial s}, \frac{\partial f}{\partial t} \right\rangle dt - 2\int_{t_i}^{t_{i+1}} \left\langle \frac{\partial f}{\partial s}, \frac{D}{dt}\frac{\partial f}{\partial t} \right\rangle dt \\
&= 2\left\langle \frac{\partial f}{\partial s}, \frac{\partial f}{\partial t} \right\rangle \Bigg|_{t_i}^{t_{i+1}} - 2\int_{t_i}^{t_{i+1}} \left\langle \frac{\partial f}{\partial s}, \frac{D}{dt}\frac{\partial f}{\partial t} \right\rangle dt.
\end{aligned}
$$

Therefore,

$$
\frac{1}{2}\frac{dE}{ds} = \sum_{i=0}^{k} \left\langle \frac{\partial f}{\partial s}, \frac{\partial f}{\partial t} \right\rangle \Bigg|_{t_i}^{t_{i+1}} - \int_0^a \left\langle \frac{\partial f}{\partial s}, \frac{D}{dt}\frac{\partial f}{\partial t} \right\rangle dt. \qquad (2)
$$

Putting $s = 0$ in (2), this yields (1). □

An immediate application of the formula for the first variation is the following characterization of geodesics.

2.5 PROPOSITION. *A piecewise differentiable curve $c\colon [0,a] \to M$ is a geodesic if and only if, for every proper variation f of c, we have $\frac{dE}{ds}(0) = 0$.*

Proof. If c is a geodesic, $\frac{D}{dt}\frac{dc}{dt} = 0$ and c is regular. Therefore if f is proper, $V(0) = V(a) = 0$, and all the terms of (1) are zero. Conversely, suppose that $\frac{dE}{ds}(0) = 0$, for all proper variations of c. Let $V(t) = g(t)\frac{D}{dt}\frac{dc}{dt}$, where $g\colon [0,a] \to \mathbf{R}$ is a piecewise differentiable function with $g(t) > 0$ if $t \neq t_i$ and $g(t_i) = 0$, $i = 0, 1, \ldots, k+1$. Construct a variation of c having $V(t)$ as variational field. Then, since

$$\frac{1}{2}\frac{dE}{ds}(0) = -\int_o^a g(t)\left\langle \frac{D}{dt}\frac{dc}{dt}, \frac{D}{dt}\frac{dc}{dt}\right\rangle dt = 0,$$

we have $\frac{D}{dt}\frac{dc}{dt} = 0$ on each interval (t_i, t_{i+1}), that is, c is a geodesic on each (t_i, t_{i+1}), $i = 0, 1, \ldots, k$.

To see what happens at the points t_i, consider another variational field $\overline{V}(t)$ such that $\overline{V}(0) = \overline{V}(a) = 0$ and if $t \neq 0, a$, $\overline{V}(t_i) = \frac{dc}{dt}(t_i^+) - \frac{dc}{dt}(t_i^-)$. Then, using the fact that c is a geodesic on (t_i, t_{i+1}), we obtain

$$\begin{aligned}
0 &= \frac{1}{2}\frac{dE}{ds}(0) \\
&= -\sum_{i=1}^{k}\left\langle \frac{dc}{dt}(t_i^+) - \frac{dc}{dt}(t_i^-), \frac{dc}{dt}(t_i^+) - \frac{dc}{dt}(t_i^-)\right\rangle \\
&= -\sum_{i=1}^{k}\left|\frac{dc}{dt}(t_i^+) - \frac{dc}{dt}(t_i^-)\right|^2,
\end{aligned}$$

or that c is of class C^1 at each t_i. Since $\frac{D}{dt}\frac{dc}{dt} = 0$ at t_i, c satisfies the equation for geodesics on $(0, a)$. By the uniqueness of the solutions to ordinary differential equations, $c \in C^\infty$ and, therefore, c is a geodesic. □

2.6 REMARK. Proposition 2.5 furnishes a characterization of geodesics as critical points of the energy for all proper variations.

It is in this sense that geodesics can be thought of as solutions to a variational problem. Observe that, in contrast to previous characterizations, such a characterization is not local but involves the behavior of the geodesic as a whole.

2.7 REMARK. There is a certain analogy between the first variation of a proper variation and the usual derivative of a function defined on a differentiable manifold. It is natural to think of the set $\Omega_{p,q}$ of piecewise differentiable curves c, joining two points p and q of M, as a manifold, a tangent vector at the point c being a piecewise differentiable vector field V along c which vanishes at the extremities of c. The energy E is then a differentiable function on such a manifold, and $\frac{dE}{ds}(0)$ is the derivative of E in the direction of V; geodesics joining p to q are critical points of the function E.

The difficulty with this point of view is that the tangent space of such a manifold at a point c (that is, the set of piecewise differentiable vector fields along c) does not have finite dimension. Thus the impossibility of finding local parametrizations of this manifold by means of open subsets of $\mathbf{R}^n$. This tangent space, nevertheless, is a vector space (infinite dimensional) and this suggests the possibility of having manifolds with infinitely many dimensions. Such manifolds can indeed be constructed but will not be treated further here. The interested reader can consult Palais, R., "Morse theory on Hilbert manifolds", Topology 2 (1963), 299–340, for more information.

Because $\frac{dE}{ds}(0)$ is zero for every proper variation of a geodesic, our next information on the energy of neighboring curves is given by $\frac{d^2E}{ds^2}(0)$, which we are going to calculate.

2.8 PROPOSITION. *(Formula for the second variation). Let $\gamma\colon [0,a] \to M$ be a geodesic and let $f\colon (-\varepsilon,\varepsilon) \times [0,a] \to M$ be a proper variation of γ. Let E be the energy function of the variation. Then*

$$
(3) \qquad \frac{1}{2}E''(0) = -\int_0^a \left\langle V(t), \frac{D^2V}{dt^2} + R(\frac{d\gamma}{dt}, V)\frac{d\gamma}{dt} \right\rangle dt
- \sum_{i=1}^{k} \left\langle V(t_i), \frac{DV}{dt}(t_i^+) - \frac{DV}{dt}(t_i^-) \right\rangle,
$$

where V is the variational field of f, R is the curvature of M and

$$\frac{DV}{dt}(t_i^+) = \lim_{\substack{t\to t_i\\ t>t_i}} \frac{DV}{dt}, \quad \frac{DV}{dt}(t_i^-) = \lim_{\substack{t\to t_i\\ t<t_i}} \frac{DV}{dt}.$$

Proof. Taking the derivative of (2), we obtain

$$\frac{1}{2}\frac{d^2E}{ds^2} = \sum_{i=0}^{k}\left\langle \frac{D}{ds}\frac{\partial f}{\partial s}, \frac{\partial f}{\partial t}\right\rangle\Big|_{t_i}^{t_{i+1}} + \sum_{i=0}^{k}\left\langle \frac{\partial f}{\partial s}, \frac{D}{ds}\frac{\partial f}{\partial t}\right\rangle\Big|_{t_i}^{t_{i+1}}$$

$$-\int_0^a \left\langle \frac{D}{ds}\frac{\partial f}{\partial s}, \frac{D}{dt}\frac{\partial f}{\partial t}\right\rangle dt - \int_0^a \left\langle \frac{\partial f}{\partial s}, \frac{D}{ds}\frac{D}{dt}\frac{\partial f}{\partial t}\right\rangle dt.$$

Putting $s = 0$ in the expression above, we obtain that the first and the third terms are zero, since f is proper and γ is a geodesic. In addition, because

$$\frac{D}{ds}\frac{D}{dt}\frac{\partial f}{\partial t} = \frac{D}{dt}\frac{D}{ds}\frac{\partial f}{\partial t} + R(\frac{\partial f}{\partial t}, \frac{\partial f}{\partial s})\frac{\partial f}{\partial t},$$

we have, at $s = 0$,

$$\frac{D}{ds}\frac{D}{dt}\frac{\partial f}{\partial t} = \frac{D^2}{dt^2}V + R(\frac{d\gamma}{dt}, V)\frac{d\gamma}{dt}.$$

Further use of the fact that the variation is proper yields

$$(4) \quad \sum_{i=0}^{k}\left\langle \frac{\partial f}{\partial s}, \frac{D}{ds}\frac{\partial f}{\partial t}\right\rangle\Big|_{t_i}^{t_{i+1}} = -\sum_{i=1}^{k}\left\langle V(t_i), \frac{DV}{dt}(t_i^+) - \frac{DV}{dt}(t_i^-)\right\rangle.$$

Putting all these facts together, we obtain (3). □

2.9 Remark. If the variation is not proper, the first term at the beginning of Proposition 2.8 need not be zero. Taking this with the corresponding terms at $i = 0$ and $i = k+1$ of the second member of (4), we obtain the following expression:

$$(5) \quad \frac{1}{2}E''(0) = -\int_0^a \left\langle V(t), \frac{D^2V}{dt^2} + R(\frac{d\gamma}{dt}, V)\frac{d\gamma}{dt}\right\rangle dt$$

$$-\sum_{i=1}^{k}\left\langle V(t_i),\frac{DV}{dt}(t_i^+)-\frac{DV}{dt}(t_i^-)\right\rangle-\left\langle\frac{D}{ds}\frac{\partial f}{\partial s},\frac{d\gamma}{dt}\right\rangle(0,0)$$

$$+\left\langle\frac{D}{ds}\frac{\partial f}{\partial s},\frac{d\gamma}{dt}\right\rangle(0,a)-\left\langle V(0),\frac{DV}{dt}(0)\right\rangle+\left\langle V(a),\frac{DV}{dt}(a)\right\rangle$$

2.10 REMARK. It is often convenient to write the expression (5) in the following manner. Since, on each interval where V is differentiable, we have

$$\frac{d}{dt}\left\langle V,\frac{DV}{dt}\right\rangle=\left\langle V,\frac{D^2V}{dt^2}\right\rangle+\left\langle\frac{DV}{dt},\frac{DV}{dt}\right\rangle,$$

we can write, taking a geodesic $\gamma\colon[0,a]\to M$, and a partition $0=t_o<t_1<\cdots<t_k<t_{k+1}=a$,

$$\int_0^a\left\langle V(t),\frac{D^2V}{dt^2}+R(\gamma',V)\gamma'\right\rangle dt$$

$$=\sum_{i=0}^{k}\Big\{\int_{t_i}^{t_{i+1}}\frac{d}{dt}\left\langle V,\frac{DV}{dt}\right\rangle dt$$

$$-\int_{t_i}^{t_{i+1}}\left(\left\langle\frac{DV}{dt},\frac{DV}{dt}\right\rangle-\langle R(\gamma',V)\gamma',V\rangle\right)dt\Big\}.$$

Therefore

$$\frac{1}{2}E''(0)=\sum_{i=0}^{k}\left\{\int_{t_i}^{t_{i+1}}(\langle V',V'\rangle-\langle R(\gamma',V)\gamma',V\rangle)dt\right\}$$

$$-\left\langle\frac{D}{ds}\frac{\partial f}{\partial s},\frac{d\gamma}{dt}\right\rangle(0,0)+\left\langle\frac{D}{ds}\frac{\partial f}{\partial s},\frac{d\gamma}{dt}\right\rangle(0,a)$$

$$=\int_0^a\{\langle V',V'\rangle-\langle R(\gamma',V)\gamma',V\rangle\}\,dt$$

$$-\left\langle\frac{D}{ds}\frac{\partial f}{\partial s},\gamma'\right\rangle(0,0)+\left\langle\frac{D}{ds}\frac{\partial f}{\partial s},\gamma'\right\rangle(0,a).$$

For reasons which will be clear later, it is convenient to write

$$\int_0^a\{\langle V',V'\rangle-\langle R(\gamma',V)\gamma',V\rangle\}\,dt=I_a(V,V).$$

Observe that, if the variation is proper, $I_a(V,V)=\frac{1}{2}E''(0)$ depends only on V. In the general case, we have

$$(6)\quad \frac{1}{2}E''(0)=I_a(V,V)-\left\langle \frac{D}{ds}\frac{\partial f}{\partial s},\gamma'\right\rangle(0,0)+\left\langle \frac{D}{ds}\frac{\partial f}{\partial s},\gamma'\right\rangle(0,a).$$

2.11 REMARK. It is possible to establish formulas for the first and second variations for the function $L(s)$ which represent the arc length of the curves in the variation. The expressions are entirely analogous to those obtained for the energy and, since we are not going to make use of them, they will not be treated here (See, however, [dC 2] paragraph 5.4).

2.12 REMARK. The analogy mentioned in Remark 2.7 can be carried further, considering $2I_a(V, V)$ as the hessian $d^2E(V, V)$ of the energy function $E:\Omega_{p,q}\to\mathbf{R}$ at the critical point $\gamma\in\Omega_{p,q}$ with respect to the "vector" V.

3. The theorems of Bonnet-Myers and of Synge-Weinstein

We now go into some applications of the formula for the second variation of the energy.

3.1 THEOREM. *(Bonnet* [Bo], *Myers* [My]*). Let M^n be a complete Riemannian manifold. Suppose that the Ricci curvature of M satisfies*

$$\text{Ric}_p(v)\geq\frac{1}{r^2}>0,$$

for all $p\in M$ and for all $v\in T_pM, |v|=1$. Then M is compact and the diameter $\text{diam}(M)\leq\pi r$.

Proof. Let p and q be any two points in M. Since M is complete, there exists a minimizing geodesic γ: $[0, 1]\to M$ joining p to q. It is enough to show that the length $\ell(\gamma)\leq\pi r$, because then M is bounded and complete, therefore compact; in addition, since $d(p, q)\leq\pi r$, for any $p,q\in M$, it follows that $\text{diam}(M)\leq\pi r$, as asserted.

Suppose, to the contrary, that $\ell(\gamma)=\ell>\pi r$. Let us consider parallel fields $e_1(t),\ldots,e_{n-1}(t)$ along γ which are orthonormal, for each $t \in [0,1]$, and belong to the orthogonal complement of $\gamma'(t)$. Let $e_n(t) = \frac{\gamma'(t)}{\ell}$ and let V_j be a vector field along γ given by

$$V_j(t) = (\sin \pi t) e_j(t), \qquad j = 1, \ldots, n-1.$$

It is clear that $V_j(0) = V_j(1) = 0$, therefore V_j generates a proper variation of γ, whose energy we denote by E_j.

Using the formula for the second variation of energy and the fact that e_j is parallel, we obtain

$$\begin{aligned} \frac{1}{2} E_j''(0) &= -\int_0^1 \langle V_j, V_j'' + R(\gamma', V_j)\gamma' \rangle \, dt \\ &= \int_0^1 \sin^2 \pi t (\pi^2 - \ell^2 K(e_n(t), e_j(t))) dt, \end{aligned}$$

where $K(e_n(t), e_j(t))$ is the sectional curvature at $\gamma(t)$ with respect to the plane generated by $e_n(t)$, $e_j(t)$. Summing on j and using the definition of the Ricci curvature, we get

$$\begin{aligned} &\frac{1}{2} \sum_{j=1}^{n-1} E_j''(0) \\ &= \int_0^1 \left\{ \sin^2 \pi t ((n-1)\pi^2 - (n-1)\ell^2 \operatorname{Ric}_{\gamma(t)}(e_n(t))) \right\} dt. \end{aligned}$$

Since $\operatorname{Ric}_{\gamma(t)}(e_n(t)) \geq \frac{1}{r^2}$ and $\ell > \pi r$, we have

$$(n-1)\ell^2 \operatorname{Ric}_{\gamma(t)}(e_n(t)) > (n-1)\pi^2,$$

hence

$$\frac{1}{2} \sum_{j=1}^{n-1} E_j''(0) < \int_0^1 \sin^2 \pi t ((n-1)\pi^2 - (n-1)\pi^2) dt = 0.$$

As a result there exists an index j such that $E_j''(0) < 0$, which, by Lemma 2.3, contradicts the fact that γ is minimizing. Therefore $\ell \leq \pi r$. □

In the corollaries that follow, we use some facts on the fundamental group and on covering spaces (see Massey [Ma], Chapters 2 and 5).

3.2 COROLLARY. *Let M be a complete Riemannian manifold with* $\operatorname{Ric}_p(v) \geq \delta > 0$, *for all $p \in M$ and all $v \in T_p(M)$. Then, the universal cover of M is compact. In particular, the fundamental group $\pi_1(M)$ is finite.*

Because, introducing on the universal cover $\pi: \tilde{M} \to M$ the covering metric (that is, the metric such that π is a local isometry), we conclude that $\tilde{M}$ is complete and that its Ricci curvature satisfies $\tilde{\operatorname{Ric}}_p \geq \delta > 0$. From the theorem, $\tilde{M}$ is compact. Hence the number of sheets of the covering is finite; since that is the number of elements in the fundamental group $\pi_1(M)$ of M, we conclude that $\pi_1(M)$ is finite.

3.3 COROLLARY. *Let M be a complete Riemannian manifold with sectional curvature $K \geq \frac{1}{r^2} > 0$. Then M is compact,* $\operatorname{diam}(M) \leq \pi r$ *and $\pi_1(M)$ is finite.*

3.4 REMARK. The hypothesis $K \geq \delta > 0$ cannot be relaxed to $K > 0$. Indeed, the paraboloid

$$\{(x, y, z) \in \mathbf{R}^3; z = x^2 + y^2\}$$

has curvature $K > 0$, but is complete and non-compact.

3.5 REMARK. In fact it is not necessary that K be bounded away from zero but only that K not approach zero too fast. In this respect, see E. Calabi, "On Ricci curvatures and geodesics", Duke Math. J. 34 (1967), 667–676 and R. Schneider, "Konvexe Flächen mit langsam abnehmender Krümmung", Archiv der Math. 23 (1972), 650–654.

3.6 REMARK. The estimate for the diameter given by Theorem 3.1 cannot be improved. Indeed, the unit sphere $S^n \subset \mathbf{R}^{n+1}$ has constant sectional curvature equal to 1 (therefore its Ricci curvature also is a constant equal to 1) and $\operatorname{diam}(S^n) = \pi$. A surprising theorem is that this example is unique in the following sense: *Let M^n be complete with* $\operatorname{Ric}_p(v) \geq 1/r^2$, *for all $p \in M$ and all $v \in T_pM$; if* $\operatorname{diam}(M) = \pi r$, *then M^n is isometric to the sphere S^n of curvature* $1/r^2$ (see S.Y. Cheng, "Eigenvalues comparison theorems and its geometric applications", Math. Z. 143 (1975), 289–297 and, for another proof, K. Shiohama, "A sphere theorem for manifolds of positive Ricci curvature", Trans. A.M.S. 275 (1983), 811–819).

Another application of the formula for the second variation is the theorem below, due essentially to A. Weinstein. A special case of the theorem (which appears as Corollary 3.10) had been proved earlier by Synge.

3.7 THEOREM. *(Weinstein [We 2] and Synge [Sy]). Let f be an isometry of a compact oriented Riemannian manifold M^n. Suppose that M has positive sectional curvature and that f preserves the orientation of M if n is even, and reverses it if n is odd. Then f has a fixed point, i.e., there exists $p \in M$ with $f(p) = p$.*

In the proof of Theorem 3.7 we need the following fact from linear algebra.

3.8 LEMMA. *Let A be an orthogonal linear transformation of $\mathbf{R}^{n-1}$ and suppose that $\det A = (-1)^n$. Then A leaves invariant some non-zero vector of $\mathbf{R}^{n-1}$.*

Proof of the Lemma. If n is even, $\det(A - \lambda I)$ is a real polynomial in λ of odd degree ($= n - 1$). Therefore, A has a real eigenvalue. Since A is orthogonal, such eigenvalues are of the form ± 1. Since the product of the complex eigenvalues of A is non-negative, and $\det A = 1$, at least one of the eigenvalues of A equals 1. This proves the lemma in this case.

If n is odd, $\det A = -1$. Because the product of the complex eigenvalues is non-negative, there is at least one pair of real eigenvalues, one of which is positive, hence equal to 1. □

Proof of Theorem 3.7. . Suppose, to the contrary, that $f(q) \neq q$ for all $q \in M$. Let $p \in M$ such that $d(p, f(p))$ attains a minimum. Since M is complete, there exists a normalized minimizing geodesic $\gamma: [0, \ell] \to M$, joining p to $f(p)$, that is, $\gamma(0) = p$, $\gamma(\ell) = f(p)$.

Let $\tilde{A} = P \circ df_p: T_p(M) \to T_p(M)$, where P is the parallel transport along γ from $f(p) = \gamma(\ell)$ to p. Then $\tilde{A}$ is an isometry. We are going to show that

$$(f \circ \gamma)'(0) = \gamma'(\ell).$$

In fact, consider the geodesic $f \circ \gamma$ which joins $f(p)$ to $f^2(p)$. Let $p' = \gamma(t')$, $t' \neq 0$, $t' \neq \ell$, and $f(p') = f \circ \gamma(t')$.

Since f is an isometry, $d(p, p') = d(f(p), f(p'))$ and, from the triangle inequality (see Fig. 2),

$$\begin{aligned} d(p', f(p')) &\leq d(p', f(p)) + d(f(p), f(p')) \\ &= d(p', f(p)) + d(p, p') \\ &= d(p, f(p)). \end{aligned}$$

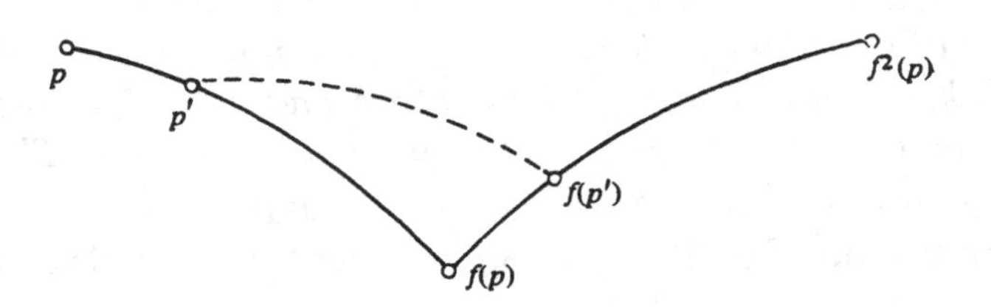

Figure 2

Because $d(p, f(p))$ is a minimum,

$$d(p', f(p')) = d(p', f(p)) + d(f(p), f(p')).$$

Therefore, the curve formed by γ and $f \circ \gamma$ is a geodesic, hence

$$(f \circ \gamma)'(0) = \gamma'(\ell),$$

as we have asserted.

It follows that $\tilde{A}$ leaves $\gamma'(0)$ fixed, since

$$\tilde{A}(\gamma'(0)) = P \circ df_p(\gamma'(0)) = P((f \circ \gamma)'(0)) = P(\gamma'(\ell)) = \gamma'(0).$$

Let A be the restriction of $\tilde{A}$ to the orthogonal complement of $\gamma'(0)$. Then A is orthogonal on $\mathbf{R}^{n-1}$ and, since P is an isometry which preserves orientation,

$$\det A = \det \tilde{A} = \det(P \circ df_p) = (-1)^n,$$

where in the last equality we use the hypothesis on f and the fact that P preserves orientation. From the lemma, A leaves a vector invariant. Let $e_1(t)$ be a unit parallel field along γ such that, for each t, $e_1(t)$ belongs to the orthogonal complement of $\gamma'(t)$ and $e_1(0)$ is invariant by A.

Let $\beta(s)$, $s \in (-\varepsilon, \varepsilon)$, be a geodesic such that $\beta(0) = p$ and $\beta'(0) = e_1(0)$. Because $P \circ df_p(e_1(0)) = e_1(0)$, we have

$$df_p(e_1(0)) = e_1(\ell),$$

that is, the geodesic $f \circ \beta$ is such that $f \circ \beta(0) = f(p)$ and $(f \circ \beta)'(0) = e_1(\ell)$.

Let h be a variation of γ given by

$$h(s,t) = \exp_{\gamma(t)}(se_1(t)), \quad s \in (-\varepsilon, \varepsilon), \quad t \in [0, \ell].$$

Observe that, since $h(s,0) = \beta(s)$, then (see Fig. 3)

$$h(s,\ell) = \exp_{f(p)}(se_1(\ell)) = (f \circ \beta)(s).$$

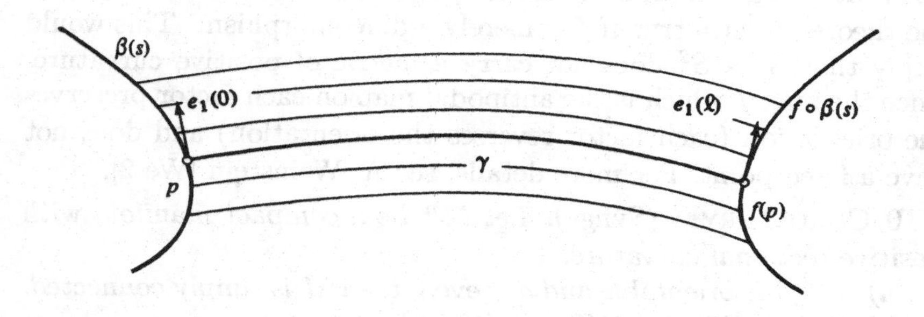

Figure 3

Therefore,

$$V(t) = \frac{\partial}{\partial s} \exp_{\gamma(t)}(se_1(t))\Big|_{s=0} = e_1(t),$$

hence $\frac{D^2 V}{dt^2} = 0$. Using the formula for the second variation (see Remark 2.9) and the fact that $\frac{\partial h}{\partial s}(0,t) = e_1(t)$, we obtain

$$\begin{aligned}\frac{1}{2}\frac{d^2E}{ds^2}(0) &= -\int_0^\ell \left\langle V(t), R(\frac{d\gamma}{dt}, V)\frac{d\gamma}{dt} \right\rangle dt \\ &\quad + \left\langle \frac{d\gamma}{dt}, \frac{D}{ds}\frac{\partial h}{\partial s} \right\rangle (0,\ell) - \left\langle \frac{d\gamma}{dt}, \frac{D}{ds}\frac{\partial h}{\partial s} \right\rangle (0,0) \\ &= -\int_0^\ell K(e_1(t), \frac{d\gamma}{dt})dt.\end{aligned}$$

Because the sectional curvature is positive,

$$\frac{1}{2}\frac{d^2E}{ds^2}(0) < 0,$$

and therefore $\frac{dE}{ds}$ is strictly decreasing on a neighborhood of zero. This shows that there exists a curve c in the variation, such that $(\ell(c))^2 \leq \ell E(c) < \ell E(\gamma) = \ell(\gamma)^2$. Since the curves in the variation join q to $f(q)$, we obtain a contradiction to the fact that $d(p, f(p))$ is a minimum. □

3.9 REMARK. The theorem remains true under the weaker hypothesis that f is a conformal diffeomorphism. It is not known whether the theorem is still true if f is merely a diffeomorphism. This would imply that $S^2 \times S^2$ does not carry a metric of positive curvature, since the map f which is the antipodal map on each factor preserves the orientation (each factor reverses the orientation) and does not have a fixed point. For more details, see A. Weinstein [We 2].

3.10 COROLLARY. *(Synge). Let M^n be a compact manifold with positive sectional curvature.*

a) If M^n is orientable and n is even, then M is simply connected.

b) If n is odd, then M^n is orientable.

Proof. a) Let $\pi: \widetilde{M} \to M$ be the universal covering of M. Introduce on $\widetilde{M}$ the covering metric, and orient $\widetilde{M}$ in such a way that π preserves the orientation. Because M is compact and has positive curvature, we must have $K \geq \delta > 0$. From the fact that π is a local isometry, the same curvature condition holds on $\widetilde{M}$. Since $\widetilde{M}$ is complete, by Corollary 3.3, $\widetilde{M}$ is compact. Let $k: \widetilde{M} \to \widetilde{M}$ be a covering transformation of $\widetilde{M}$, that is, $\pi \circ k = \pi$ (see Massey [Ma], p. 159). Then k is an isometry of $\widetilde{M}$ which, from the way that we oriented $\widetilde{M}$, preserves the orientation. Because n is even, we can use the theorem to conclude that k has a fixed point. But a covering transformation which has a fixed point is the identity. It follows that the group of covering transformations of $\widetilde{M}$ (which is isomorphic to the fundamental group of M; see Massey [Ma], p. 163) reduces to the identity. Therefore M is simply connected.

b) Suppose that M is not orientable, and consider the orientable double cover $\overline{M}$ of M (see Exercise 12 of Chap.0). We introduce on $\overline{M}$ the covering metric. Since $\overline{M}$ is the double cover of a compact manifold, $\overline{M}$ is compact. Let k be a covering transformation of $\overline{M}$, $k \neq \text{id}$. Because M is not orientable, k is an isometry which reverses the orientation of $\overline{M}$. Since n is odd, we can apply Theorem 3.7 which guarantees that k has a fixed point. Therefore $k = \text{id}.$, which is a contradiction. □

3.11 REMARK. The real projective space $P^2(\mathbf{R})$ of dimension two, which is compact, non-orientable and not simply connected is an example which shows the necessity of orientability in a) and odd dimension in b). On the other hand, $P^3(\mathbf{R})$ which is compact, orientable and not simply connected is an example which shows the necessity of even dimension in (a).

EXERCISES

1. Let M be a complete Riemannian manifold, and let $N \subset M$ be a closed submanifold of M. Let $p_o \in M$, $p_o \notin N$, and let $d(p_o, N)$ be the distance from p_o to N. Show that there exists a point $q_o \in N$ such that $d(p_o, q_o) = d(p_o, N)$ and that a minimizing geodesic which joins p_o to q_o is orthogonal to N at q_o.

2. Introduce a complete Riemannian metric on $\mathbf{R}^2$. Prove that

$$\lim_{r\to\infty} \left(\inf_{x^2+y^2 \geq r^2} K(x,y) \right) \leq 0,$$

where $(x, y) \in \mathbf{R}^2$ and $K(x, y)$ is the Gaussian curvature of the given metric at (x, y).

3. Prove the following generalization of the Theorem of Bonnet-Myers: Let M^n be a complete Riemannian manifold. Suppose that there exist constants $a > 0$ and $c \geq 0$ such that for all pairs of points in M^n and for all minimizing geodesics $\gamma(s)$, parametrized by arc length s, joining these points, we have

$$\text{Ric}(\gamma'(s)) \geq a + \frac{df}{ds}, \quad \text{along } \gamma,$$

where f is a function of s, satisfying $|f(s)| \leq c$ along γ. Then M^n is compact.
Calculate an estimate for the diameter of M^n, and observe that if $f \equiv 0$ and $c = 0$, we obtain the Theorem of Bonnet-Myers.
The theorem above has application to Relativity, see G.J. Galloway, "A generalization of Myers' Theorem and an application to relativistic cosmology", J. Diff. Geometry, 14 (1979), 105–116.

4. Let M^n be an orientable Riemannian manifold with positive curvature and even dimension. Let γ be a closed geodesic in M, that is, γ is an immersion of the circle S^1 in M that is geodesic at all of its points. Prove that γ is homotopic to a closed curve whose length is strictly less than that of γ.
Hint: The parallel transport along the closed curve γ leaves a vector orthogonal to γ invariant (this comes from the orientability of M and the fact that the dimension is even). Therefore there exists a vector field $V(t)$ parallel along the closed curve γ. Calculate $E''_V(0)$ and show that it is strictly negative. Therefore, close to γ, there exists a closed curve of length smaller than that of γ.

5. Let N_1 and N_2 be two closed disjoint submanifolds of a compact Riemannian manifold.
   a) Show that the distance between N_1 and N_2 is assumed by a geodesic γ perpendicular to both N_1 and N_2.
   b) Show that, for any orthogonal variation $h(t,s)$ of γ, with $h(0,s) \in N_1$ and $h(\ell,s) \in N_2$, we have the following expression for the formula for the second variation

$$\frac{1}{2}E''(0) = I_\ell(V,V) + \left\langle V(\ell), S^{(2)}_{\gamma'(\ell)}V(\ell)\right\rangle - \left\langle V(0), S^{(1)}_{\gamma'(0)}(V(0))\right\rangle$$

   where V is the variational vector and $S^{(i)}_{\gamma'}$ is the linear map associated to the second fundamental form of N_i in the direction of γ', $i = 1, 2$.

6. Let $\widetilde{M}$ be a complete simply connected Riemannian manifold, with curvature $K \leq 0$. Let $\gamma\colon (-\infty, \infty) \to \widetilde{M}$ be a normalized

geodesic and let $p \in \tilde{M}$ be a point which does not belong to γ. Let $d(s) = d(p, \gamma(s))$.

a) Consider the minimizing geodesic $\sigma_s\colon [0, d(s)] \to \tilde{M}$ joining p to $\gamma(s)$, that is, $\sigma_s(0) = p$, $\sigma_s(d(s)) = \gamma(s)$. Consider the variation $h(t,s) = \sigma_s(t)$, and show that:
   (i) $\frac{1}{2}E'(s) = \langle \gamma'(s), \sigma_s'(d(s))\rangle + \sigma_s'(d(s))\cdot d'(s)$,
   (ii) $\frac{1}{2}E''(s_0) > 0$, if $d'(s_0) = 0$.

b) Conclude from (i) that s_o is a critical point of d if and only if $\langle \gamma'(s_o), \sigma_s'(d(s_o))\rangle = 0$. Conclude from (ii) that d has a unique critical point, which is a minimum.

c) From (b), it follows that if $\tilde{M}$ is complete, simply connected and has curvature $K \leq 0$, then a point off the geodesic γ of $\tilde{M}$ can be connected by a unique perpendicular to γ. Show by examples that the condition on the curvature and the condition of simple connectivity are essential to the theorem.

CHAPTER 10

THE RAUCH COMPARISON THEOREM

1. Introduction

Let M be a Riemannian manifold. As we saw in Chapter 5, if $\gamma\colon [0,\ell] \to M$ is a normalized geodesic and J is a Jacobi field along γ with $J(0) = 0$, $|J'(0)| = 1$ and $\langle J'(0), \gamma'(0)\rangle = 0$, then

$$|J(t)| = t - \frac{K}{6}t^3 + R, \qquad \lim_{t\to 0} \frac{R}{t^3} = 0,$$

where K is the sectional curvature at $\gamma(0)$ with respect to the plane generated by $\gamma'(0)$ and $J'(0)$. Therefore, if t is small, the smaller K is, the larger $|J(t)|$ will be. Consider now another Riemannian manifold $\tilde{M}$, a geodesic $\tilde{\gamma}\colon [0,\ell] \to \tilde{M}$ and a Jacobi field $\tilde{J}$ along $\tilde{\gamma}$ satisfying: $\tilde{J}(0) = 0$, $|\tilde{J}'(0)| = 1$, $\left\langle \tilde{J}'(0), \tilde{\gamma}'(0)\right\rangle = 0$. Suppose that

$$\tilde{K}(\tilde{\gamma}'(0), \tilde{J}'(0)) \geq K(\gamma'(0), J'(0)).$$

It follows from the expression above that, for small t, $|\tilde{J}(t)| \leq |J(t)|$.

The Theorem of Rauch, which we intend to prove in this chapter, furnishes conditions so that the inequality above is valid without the restriction that t be small. More precisely, the theorem asserts (see the statement in Section 2) that if $\tilde{\gamma}\colon [0,\ell] \to \tilde{M}$ does not have conjugate points and

$$\tilde{K}(\tilde{\gamma}'(t), \tilde{J}(t)) \geq K(\gamma'(t), J(t)), \quad t \in (0,\ell),$$

then $|\tilde{J}(t)| \leq |J(t)|$.

Rauch's Theorem is one of the basic facts in Riemannian Geometry. Intuitively, it expresses the plausible fact that as the curvature grows, lengths shorten.

In dimension two, such a theorem is an easy consequence of the classical theorem of Sturm on ordinary differential equations. In fact, in this case, the Jacobi equations for $J = fe_2$, $\tilde{J} = \tilde{f}\tilde{e}_2$ ($e_2(t)$ and $\tilde{e}_2(t)$ are unit parallel vector fields along and normal to the geodesics $\gamma(t)$ and $\tilde{\gamma}(t)$, respectively) can be written:

$$f''(t) + K(t)f(t) = 0, \quad f(0) = 0, \quad t \in [0, \ell],$$
$$\tilde{f}''(t) + \tilde{K}(t)\tilde{f}(t) = 0, \quad \tilde{f}(0) = 0, \quad t \in [0, \ell].$$

Sturm's Theorem asserts that if $f'(0) = \tilde{f}'(0) > 0$, $\tilde{f}(t) \neq 0$ on $(0, \ell]$, and $\tilde{K}(t) \geq K(t)$, then $\tilde{f}(t) \leq f(t)$ (see Fig. 1 and Exercise 5 of this chapter).

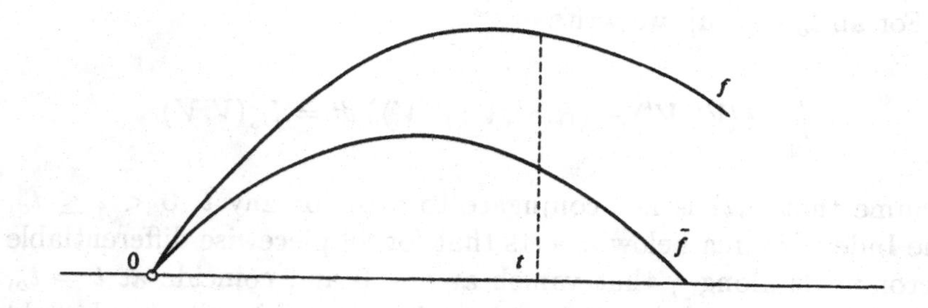

Figure 1

In dimension higher than two, the proof is much less simple, and a presentation of the theorem was made for the first time in 1951 by Rauch [R 1] (for another exposition, see Rauch [R 3]). The proof that we present in Section 2 is an elaboration due to various mathematicians in the fifties, among whom we should mention Ambrose and Singer.

In Section 3 we make an application of the ideas in the Rauch Theorem to the theory of isometric immersions. In Section 4, we introduce the notion of focal point, which generalizes the idea of conjugate point, and we extend the Theorem of Rauch to this situation. The last two sections are not used in the rest of the book and can be omitted on a first reading.

2. The Theorem of Rauch

In the proof of Rauch's Theorem, we need a certain number of facts that we now establish in the form of lemmas. The Index Lemma (Lemma 2.2) is a basic fact that has application to many other situations.

The following lemma is a particular case of Lemma 5.5 of Chapter 0.

2.1 LEMMA. *Let $h: [0,1] \to \mathbf{R}$ be a differentiable function with $h(0) = 0$. Then there exists a differentiable function $\phi: [0,1] \to \mathbf{R}$, with $\phi(0) = \frac{dh}{dt}(0)$, $h(t) = t\phi(t)$, $t \in [0,1]$.*

Let M be a Riemannian manifold and let $\gamma: [0,a] \to M$ be a geodesic of M. Let V be a piecewise differentiable vector field along γ. For all $t_o \in [0,a]$, we write

$$\int_0^{t_o} \{\langle V', V'\rangle - \langle R(\gamma', V)\gamma', V\rangle\}\, dt = I_{t_o}(V,V).$$

Assume that $\gamma(t)$ is not conjugate to $\gamma(0)$ for any t, $0 < t \leq t_o$. The Index Lemma below asserts that for all piecewise differentiable vector fields along γ that vanish at $t = 0$ and coincide at $t = t_o$, the minimum of the expression above is assumed by the Jacobi field that vanishes at $t = 0$ and takes on the same value at $t = t_o$. More precisely, we have the following fundamental lemma.

2.2 LEMMA. *(The Index Lemma). Let $\gamma: [0,a] \to M$ be a geodesic without conjugate points to $\gamma(0)$ in the interval $(0,a]$. Let J be a Jacobi field along γ, with $\langle J, \gamma'\rangle = 0$, and let V be a piecewise differentiable vector field along γ, with $\langle V, \gamma'\rangle = 0$. Suppose that $J(0) = V(0) = 0$ and that $J(t_o) = V(t_o)$, $t_o \in (0,a]$. Then*

$$I_{t_o}(J,J) \leq I_{t_o}(V,V)$$

and equality occurs if and only if $V = J$ on $[0,t_o]$.

Proof. The vector space $\mathcal{J}$ of Jacobi fields J along γ with $J(0) = 0$ and $\langle J, \gamma'\rangle = 0$ has dimension $n-1$, where $n = \dim M$. Let $\{J_1, \ldots, J_{n-1}\}$ be a basis for this space. Then $J = \sum_i \alpha_i J_i$, $i = 1, \ldots, n-1$, where the α_i are constants. Since there are no conjugate points in the interval $(0,a]$, for all $t \neq 0$, the vectors $J_1(t),\dots,J_{n-1}(t)$ form a basis of the orthogonal complement of $\gamma'(t)$ in $T_{\gamma(t)}(M)$. Therefore, for $t \neq 0$, we can write

$$V(t) = \sum_i f_i(t)J_i(t),$$

where f_i are piecewise differentiable functions on $(0,a]$. We are going to show that f_i can be extended continuously and differentiably to $t=0$, that is, f_i is piecewise differentiable on $[0,a]$.

For this, use Lemma 2.1 to write $J_i(t) = t A_i(t)$. Then $A_i(0) = J_i'(0)$, hence the $A_i(0)$ are linearly independent. Therefore, the $A_i(t)$ are linearly independent for all $t \in [0,a]$, and we can write $V(t) = \sum_i g_i(t)A_i(t)$ where the g_i are piecewise differentiable functions on $[0,a]$ and $g_i(0) = 0$. Applying Lemma 2.1 again, we have that $g_i(t) = th_i(t)$, where the $h_i(t)$ are piecewise differentiable on $[0,a]$. Since we have $f_i(t) = h_i(t)$, for $t \neq 0$, the claim is proved.

We are going to show that, on the interior of each subinterval where f_i is differentiable,

$$\text{(1)}\quad \langle V',V'\rangle - \langle R(\gamma',V)\gamma',V\rangle = \langle \sum_i f_i'J_i, \sum_j f_j'J_j\rangle + \frac{d}{dt}\langle \sum_i f_iJ_i, \sum_j f_jJ_j'\rangle.$$

Indeed, since

$$R(\gamma',V)\gamma' = R(\gamma',\sum_i f_iJ_i)\gamma' = \sum_i f_iR(\gamma',J_i)\gamma' = -\sum_i f_iJ_i'',$$

we have

$$\begin{aligned}
&\langle V',V'\rangle - \langle R(\gamma',V)\gamma',V\rangle \\
&\quad = \langle \sum_i f_i'J_i + \sum_i f_iJ_i', \sum_j f_j'J_j + \sum_j f_jJ_j'\rangle - \langle R(\gamma',V)\gamma',V\rangle \\
&\quad = \langle \sum_i f_i'J_i, \sum_j f_j'J_j\rangle + \langle \sum_i f_i'J_i, \sum_j f_jJ_j'\rangle \\
&\qquad + \langle \sum_i f_iJ_i', \sum_j f_j'J_j\rangle + \langle \sum_i f_iJ_i', \sum_j f_jJ_j'\rangle \\
&\qquad + \langle \sum_i f_iJ_i'', \sum_j f_jJ_j\rangle.
\end{aligned}$$

On the other hand,

$$\begin{aligned}\frac{d}{dt}\langle\sum_i f_iJ_i,\sum_j f_jJ_j'\rangle &= \langle\sum_i f_i'J_i+\sum_i f_iJ_i',\sum_j f_jJ_j'\rangle\\ &\quad+\langle\sum_i f_iJ_i,\sum_j f_j'J_j'+\sum_j f_jJ_j''\rangle\\ &= \langle\sum_i f_i'J_i,\sum_j f_jJ_j'\rangle+\langle\sum_i f_iJ_i',\sum_j f_jJ_j'\rangle\\ &\quad+\langle\sum_i f_iJ_i,\sum_j f_jJ_j''\rangle+\langle\sum_i f_iJ_i,\sum_j f_j'J_j'\rangle.\end{aligned}$$

Therefore, to prove (1), it suffices to show that

$$\langle\sum_i f_iJ_i',\sum_j f_j'J_j\rangle=\langle\sum_i f_iJ_i,\sum_j f_j'J_j'\rangle. \tag{2}$$

To prove (2), we write

$$h(t)=\langle J_i',J_j\rangle-\langle J_i,J_j'\rangle;$$

since $h(0)=0$ and

$$\begin{aligned}h'(t)&=\langle J_i'',J_j\rangle+\langle J_i',J_j'\rangle-\langle J_i',J_j'\rangle-\langle J_i,J_j''\rangle\\ &=-\langle R(\gamma',J_i)\gamma',J_j\rangle+\langle J_i,R(\gamma',J_j)\gamma'\rangle=0\end{aligned}$$

we conclude that $h(t)\equiv 0$. By distributivity, we then obtain (2), which concludes the proof of (1).

Applying (1) to V and J, we obtain sucessively:

$$I_{t_o}(V,V)=\langle\sum_i f_iJ_i,\sum_j f_jJ_j'\rangle(t_o)+\int_0^{t_o}\langle\sum_i f_i'J_i,\sum_j f_j'J_j\rangle dt,$$

$$I_{t_o}(J,J)=\langle\sum_i \alpha_iJ_i,\sum_j \alpha_jJ_j'\rangle(t_o).$$

Because $J(t_o)=V(t_o)$, we have that $\alpha_i=f_i(t_o)$, hence

$$I_{t_o}(V,V)=I_{t_o}(J,J)+\int_0^{t_o}\left|\sum_i f_i'J_i\right|^2 dt. \tag{3}$$

It follows from (3) that $I_{t_o}(V,V) \geq I_{t_o}(J,J)$, which proves the first part of the lemma. If $I_{t_o}(V,V) = I_{t_o}(J,J)$, then $\sum_i f_i' J_i = 0$. Because the J_i are linearly independent for $t \neq 0$, we conclude, by continuity, that $f_i' = 0$, for all i and for all $t \in [0, t_o]$. Therefore, $f_i = \text{const.}$, and since $f_i(t_o) = \alpha_i$, we have that $f_i(t) = \alpha_i$, that is, $V = J$, as claimed. □

We are now in a position to prove Rauch's Theorem. In what follows M^n will denote a manifold of dimension n.

2.3 THEOREM. *(Rauch). Let $\gamma\colon [0,a] \to M^n$ and $\tilde{\gamma}\colon [0,a] \to \widetilde{M}^{n+k}$, $k \geq 0$, be geodesics with the same velocity (i.e., $|\gamma'(t)| = |\tilde{\gamma}'(t)|$), and let J and $\tilde{J}$ be Jacobi fields along γ and $\tilde{\gamma}$, respectively, such that*

$$J(0) = \tilde{J}(0) = 0, \quad \langle J'(0), \gamma'(0)\rangle = \langle \tilde{J}'(0), \tilde{\gamma}'(0)\rangle,$$
$$|J'(0)| = |\tilde{J}'(0)|.$$

Assume that $\tilde{\gamma}$ does not have conjugate points on $(0,a]$ and that, for all t and all $x \in T_{\gamma(t)}(M)$, $\tilde{x} \in T_{\tilde{\gamma}(t)}(\widetilde{M})$, we have

$$\tilde{K}(\tilde{x}, \tilde{\gamma}'(t)) \geq K(x, \gamma'(t)),$$

where $K(x,y)$ denotes the sectional curvature with respect to the plane generated by x and y. Then

$$|\tilde{J}| \leq |J|.$$

In addition, if for some $t_o \in (0,a]$, we have $|\tilde{J}(t_o)| = |J(t_o)|$, then $\tilde{K}(\tilde{J}(t), \tilde{\gamma}'(t)) = K(J(t), \gamma'(t))$, for all $t \in [0, t_o]$.

Proof. Observe that, from Proposition 3.6 of Chapter 5, the condition $\langle J'(0), \gamma'(0)\rangle = \langle \tilde{J}'(0), \tilde{\gamma}'(0)\rangle$ is equivalent (with $J(0) = \tilde{J}(0) = 0$) to the condition $\langle J, \gamma'\rangle = \langle \tilde{J}, \tilde{\gamma}'\rangle$. In addition, since

$$\langle J, \gamma'\rangle\gamma' = \langle J'(0), \gamma'(0)\rangle t\gamma' + \langle J(0), \gamma'(0)\rangle\gamma',$$

the tangential components of J and $\tilde{J}$ have, by hypothesis, the same length. Therefore, we can suppose that

$$\langle J, \gamma'\rangle = 0 = \langle \tilde{J}, \tilde{\gamma}'\rangle.$$

If $|J'(0)| = |\tilde{J}'(0)| = 0$, then $|J| = |\tilde{J}| = 0$. In the contrary case, we put $|J(t)|^2 = v(t)$ and $|\tilde{J}(t)|^2 = \tilde{v}(t)$. Since $\tilde{\gamma}$ does not have any conjugate points on $(0, a]$, $\frac{v(t)}{\tilde{v}(t)}$ is well-defined for $t \in (0, a]$. From L'Hospital's rule,

$$\lim_{t\to 0} \frac{v(t)}{\tilde{v}(t)} = \lim_{t\to 0} \frac{v''(t)}{\tilde{v}''(t)} = \frac{|J'(0)|^2}{|\tilde{J}'(0)|^2} = 1.$$

Therefore, to prove that $|\tilde{J}| \leq |J|$, it is enough to prove that $\frac{d}{dt}\left(\frac{v(t)}{\tilde{v}(t)}\right) \geq 0$. This is equivalent to proving that $v'\tilde{v} \geq v\tilde{v}'$.

To prove what we want, fix $t_o \in (0, a]$. If $v(t_o) = 0$ we have that

$$v'(t_o) = 2\langle J'(t_o), J(t_o)\rangle = 0,$$

and the inequality is satisfied trivially. Suppose, therefore, that $v(t_o) \neq 0$. Put

$$U(t) = \frac{1}{\sqrt{v(t_o)}} J(t), \quad \tilde{U}(t) = \frac{1}{\sqrt{\tilde{v}(t_o)}} \tilde{J}(t),$$

and observe that

$$\begin{aligned}
\frac{v'(t_o)}{v(t_o)} &= \frac{2\langle J'(t_o), J(t_o)\rangle}{\langle J(t_o), J(t_o)\rangle} = 2\langle U'(t_o), U(t_o)\rangle = \langle U, U\rangle'(t_o) \\
&= \int_0^{t_o} \langle U, U\rangle'' dt = 2\int_0^{t_o} \{\langle U', U'\rangle - \langle U, R(\gamma', U)\gamma'\rangle\}\, dt \\
&= 2I_{t_o}(U, U).
\end{aligned}$$

Similarly,

$$\frac{\tilde{v}'(t_o)}{\tilde{v}(t_o)} = 2I_{t_o}(\tilde{U}, \tilde{U}).$$

By the arbitrariness of t_o, it suffices, therefore, to prove that $I_{t_o}(\tilde{U}, \tilde{U}) \leq I_{t_o}(U, U)$ to complete the proof of the inequality.

For this, let $\{e_1, \ldots, e_n\}$ and $\{\tilde{e}_1, \ldots, \tilde{e}_{n+k}\}$ be parallel orthonormal bases along γ and $\tilde{\gamma}$, respectively, such that:

$$e_1(t) = \gamma'(t)/|\gamma'(t)|, \quad e_2(t_o) = U(t_o),$$

$$\tilde{e}_1(t) = \tilde{\gamma}'(t)/|\tilde{\gamma}'(t)|, \quad \tilde{e}_2(t_o) = \tilde{U}(t_o).$$

To each vector field $V(t) = \sum_i g_i(t)e_i(t)$ along γ associate the field ϕV along $\tilde{\gamma}$ given by

$$(\phi V)(t) = \sum_{i=1}^{n} g_i(t)\tilde{e}_i(t).$$

The map ϕ defined above satisfies the following properties:

$$\langle \phi V_1, \phi V_2 \rangle = \langle V_1, V_2 \rangle \tag{4}$$

$$(\phi V)' = \phi(V'). \tag{5}$$

It follows, from the hypothesis on the curvature and the fact that the geodesics have the same velocity, that

$$I_{t_o}(\phi(U), \phi(U)) \leq I_{t_o}(U, U).$$

Observe now that $\tilde{U}$ and $\phi(U)$ are vector fields along $\tilde{\gamma}$ which satisfy the hypothesis of Lemma 2.2, and that $\tilde{U}$ is a Jacobi field. From Lemma 2.2,

$$I_{t_o}(\tilde{U}, \tilde{U}) \leq I_{t_o}(\phi(U), \phi(U)) \leq I_{t_o}(U, U),$$

which proves the inequality in the Theorem.

Suppose now that $|J(t_o)| = |\tilde{J}(t_o)|$, for some $t_o \in (0, a]$. For all $t \neq 0$, we have $I_t(\tilde{U}, \tilde{U}) \leq I_t(U, U)$, and, therefore, $v'\tilde{v}(t) \geq v\tilde{v}'(t)$. Since $|J(t_o)| = |\tilde{J}(t_o)|$, it follows that

$$v'\tilde{v}(t) = v\tilde{v}'(t), \quad t \in (0, t_o],$$

hence the inequalities above are equalities for $t \in (0, t_o]$, i.e.,

$$I_t(\phi(U), \phi(U)) = I_t(U, U), \quad t \in (0, t_o].$$

Since ϕ satisfies (4) and (5), we conclude from the equality above and from the hypothesis on the curvature that $K(J(t), \gamma'(t)) = \tilde{K}(\tilde{J}(t), \tilde{\gamma}'(t))$, for $t \in (0, t_o]$, hence, by continuity, for $t \in [0, t_o]$, as has been asserted. □

An immediate application of Rauch's Theorem allows us to obtain information on the location of conjugate points from bounds on the curvature.

2.4 PROPOSITION. *Suppose that the sectional curvature K of a Riemannian manifold M satisfies the inequality*

$$0 < L \leq K \leq H,$$

where H and L are constants. Let γ be geodesic in M. Then the distance d along γ between two consecutive conjugate points of γ satisfies

$$\frac{\pi}{\sqrt{H}} \leq d \leq \frac{\pi}{\sqrt{L}}.$$

Proof. To get the inequality $d \geq \pi/\sqrt{H}$, it suffices to compare the manifold M^n with the sphere $S^n(H)$ of curvature H. Let $\gamma\colon [0,\ell] \to M$ be a normalized geodesic in M with $\gamma(0) = p$, and let J be a Jacobi field along γ with $J(0) = 0$ and $\langle J, \gamma' \rangle = 0$. Choose a point $\tilde{p} \in S^n(H)$ and a normalized geodesic $\tilde{\gamma}\colon [0,\ell] \to S^n(H)$ with $\tilde{\gamma}(0) = \tilde{p}$. Let $\tilde{J}$ be a Jacobi field along $\tilde{\gamma}$ with $\tilde{J}(0) = 0$, $\langle \tilde{J}, \tilde{\gamma}' \rangle = 0$, $|\tilde{J}'(0)| = |J'(0)|$. Since $\tilde{\gamma}$ does not have conjugate points in the interval $(0, \pi/\sqrt{H})$, by Rauch's Theorem, $|J(t)| \geq |\tilde{J}(t)| > 0$, $t \in (0, \pi/\sqrt{H})$. Therefore, the distance d from p to its first conjugate point satisfies $d \geq \pi/\sqrt{H}$.

To obtain the inequality $d \leq \pi/\sqrt{L}$, we make an analogous comparison with the sphere $S^n(L)$ of dimension n and constant curvature L. If $d > \pi/\sqrt{L}$, Rauch's Theorem applies, hence $S^n(L)$ has all its conjugate points after $\pi/\sqrt{L}$, which is absurd. Therefore $d \leq \pi/\sqrt{L}$. $\square$

One of the typical applications of Rauch's Theorem consists in estimating the lengths of curves in a Riemannian manifold in which we can estimate the curvature. The proposition below is an example of this situation.

2.5 PROPOSITION. *Let M^n and $\tilde{M}^n$ be Riemannian manifolds and suppose that for all $p \in M$, $\tilde{p} \in \tilde{M}$, $\sigma \subset T_pM$, $\tilde{\sigma} \subset T_{\tilde{p}}\tilde{M}$, we have that $\tilde{K}_{\tilde{p}}(\tilde{\sigma}) \geq K_p(\sigma)$. Let $p \in M$, $\tilde{p} \in \tilde{M}$ and fix a linear isometry $i\colon T_pM \to T_{\tilde{p}}\tilde{M}$. Let $r > 0$ be such that the restriction $\exp_p |B_r(0)$ is a diffeomorphism and $\exp_{\tilde{p}} |\tilde{B}_r(0)$ is non-singular. Let $c\colon [0,a] \to \exp_p(B_r(0)) \subset M$ be a differentiable curve and define a curve $\tilde{c}\colon [0,a] \to \exp_{\tilde{p}}(\tilde{B}_r(0)) \subset \tilde{M}$ by*

$$\tilde{c}(s) = \exp_{\tilde{p}} \circ i \circ \exp_p^{-1}(c(s)), \quad s \in [0,a].$$

Then $\ell(c) \geq \ell(\tilde{c})$.

Proof. Consider the curve $\bar{c}(s) = \exp_p^{-1} c(s)$ in T_pM. For s fixed, consider the radial geodesic $\gamma_s(t) = \exp_p t\bar{c}(s)$. The mapping

$$f(t,s) = \gamma_s(t), \quad 0 \leq s \leq a, \quad 0 \leq t \leq 1,$$

is a parametrized surface (see Fig. 2). Therefore, for all s, $\frac{\partial f}{\partial s}(t,s) = J_s(t)$ is a Jacobi field along γ_s, with $J_s(0) = 0$, $J_s(1) = \frac{\partial f}{\partial s}(1,s) = c'(s)$. In addition,

$$\frac{DJ_s}{dt}(0) = \frac{D}{dt}\left\{(d\exp_p)_{t\bar{c}(s)}(t\bar{c}'(s))\right\}\Big|_{t=o} = \bar{c}'(s).$$

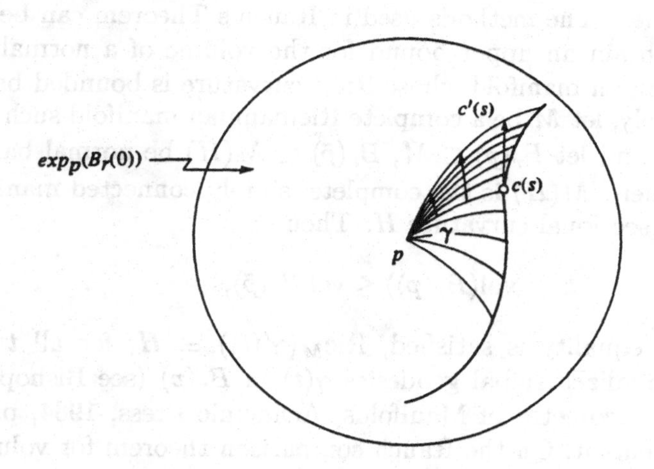

Figure 2

Consider now a parametrized surface $\tilde{f}(t,s)$ in $\tilde{M}$ given by

$$\tilde{f}(t,s) = \exp_{\tilde{p}} ti(\bar{c}(s)) = \tilde{\gamma}_s(t),$$

and observe that $\tilde{\gamma}_s$ is a geodesic. Then the Jacobi field $\tilde{J}_s(t) = \frac{\partial \tilde{f}}{\partial s}(t,s)$ satisfies

$$\tilde{J}_s(0) = 0, \quad \tilde{J}_s(1) = \tilde{c}'(s), \quad \frac{D\tilde{J}_s}{dt}(0) = i\bar{c}'(s).$$

Because i is an isometry,

$$|J_s(0)| = |\tilde{J}_s(0)| = 0, \quad |J_s'(0)| = |\tilde{J}_s'(0)|$$

and

$$\begin{aligned}\langle \tilde{J}_s'(0), \tilde{\gamma}_s'(0) \rangle &= \langle i\tilde{c}'(s), i\gamma_s'(0) \rangle \\ &= \langle \tilde{c}'(s), \gamma_s'(0) \rangle = \langle J_s'(0), \gamma_s'(0) \rangle.\end{aligned}$$

We can, therefore, apply Rauch's Theorem and conclude that

$$|\tilde{c}'(s)| = |\tilde{J}_s(1)| \le |J_s(1)| = |c'(s)|,$$

hence, by integration, $\ell(c) \ge \ell(\tilde{c})$. □

2.6 REMARK. The methods used in Rauch's Theorem can be employed to obtain an upper bound for the volume of a normal ball in a Riemannian manifold whose Ricci curvature is bounded below. More precisely, let M be a complete Riemannian manifold such that $\mathrm{Ric}_M \ge H$, and let $B_r(p) \subset M$, $B_r(\tilde{p}) \subset \widetilde{M}(H)$ be normal balls of radius r, where $\widetilde{M}(H)$ is the complete, simply connected manifold, of constant sectional curvature H. Then

$$\mathrm{vol}(B_r(p)) \le \mathrm{vol}\, B_r(\tilde{p}),$$

and, if the equality is satisfied, $\mathrm{Ric}_M(\gamma'(t)) \equiv H$, for all $t < r$ and all normalized radial geodesics $\gamma(t)$ in $B_r(p)$ (see Bishop and Crittenden, Geometry of Manifolds, Academic Press, 1964, p. 256 and K. Tenenblat, On the Rauch comparison theorem for volumes, Bol. Soc. Bras. Mat. 4 (1973), 31-39).

It is curious to observe that, in contrast to Rauch's Theorem, the theorem above does not extend to the case $\mathrm{Ric}_M \le L$; a counterexample is described in the article cited above, by K. Tenenblat.

2.7 REMARK. The relation between the Ricci curvature and the volume can be extended to the following global theorem: With the same notation as in Remark 2.6, suppose that $\mathrm{Ric}_M \ge H$ and denote the corresponding volumes by $V_r = \mathrm{vol}\, B_r(p)$ and $\tilde{V}_r = \mathrm{vol}\, B_r(\tilde{p})$, where r is now an arbitrary positive real number. Then for all $R \ge r > 0$, we have

$$V_r/\tilde{V}_r \ge V_R/\tilde{V}_R.$$

In addition, the equality occurs for $R \leq \operatorname{diam}(M)$ if and only if $B_R(p) \subset M$ is isometric to the ball $B_R(\tilde{p}) \subset \tilde{M}(H)$. For a proof, see J. Eschenburg "Comparison theorems and hypersurfaces", Manuscripta Math. 59 (1987), 295–323. This reference contains an exposition of comparison theorems from a point of view different than that presented here; in particular, the theorem above is used to establish the uniqueness theorem mentioned in Remark 3.6 of Chapter 9.

3. Applications of the Index Lemma to immersions

In this Section, we shall present an application of the Index Lemma to the theory of isometric immersions. We prove the following theorem, which generalizes previous theorems of Tompkins, O'Neill, and Chern-Kuiper (see Corollaries 3.4, 3.5 and Remarks 3.6 and 3.7).

3.1 THEOREM. *(J.D. Moore, [Mo]). Let $\overline{M}$ be a complete simply connected, Riemannian manifold, whose sectional curvature satisfies*

$$\overline{K} \leq b \leq 0.$$

Let M be a compact Riemannian manifold whose sectional curvature satisfies $K - \overline{K} \leq -b$. If $\dim \overline{M} < 2 \dim M$, there does not exist an isometric immersion $f\colon M \to \overline{M}$.

Proof. We suppose that such an immersion exists, and obtain a contradiction. Choose a point $\overline{p} \in \overline{M}$, $\overline{p} \notin f(M)$, and let $q \in M$ be such that

$$d(f(q),\overline{p}) \geq d(f(p),\overline{p}), \quad \text{for all} \quad p \in M.$$

Let $U \subset M$ be a neighborhood of q in which f is an embedding, and identify U with $f(U)$. Let $\gamma\colon [0,\ell] \to \overline{M}$ be the normalized minimizing geodesic in $\overline{M}$, with $\gamma(0) = \overline{p}$, $\gamma(\ell) = q$. Using the formula for the first variation, we see that γ is perpendicular to U at q.

For all $v \in T_qM$, $|v| = 1$, consider a curve $c(s)$ in U with $c(0) = q$, $c'(0) = v$, $s \in (-\varepsilon,\varepsilon)$. Let $\tilde{c}(s) = \exp_{\overline{p}}^{-1}(c(s))$. Then, the parametrized surface

$$f(t,s) = \exp_{\overline{p}} \frac{t}{\ell}\tilde{c}(s), \qquad t \in [0,\ell], s \in (-\varepsilon,\varepsilon),$$

generates a Jacobi field $V(t) = \frac{\partial f}{\partial s}(t,0)$ with $V(0) = 0$, $V(\ell) = v$ (Fig. 3). Observe that the radial geodesics $\gamma_s: t \to f(t,s)$ in this variation are shorter than γ.

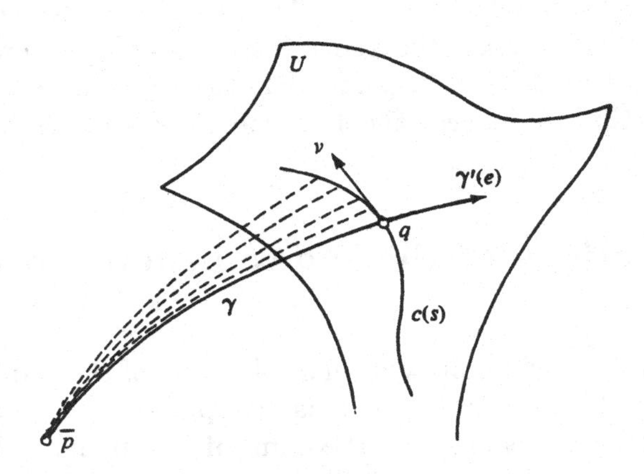

Figure 3

We need the following lemma.

3.2 LEMMA. *Let $E(s)$ be the energy of the curve $t \to f(t,s)$. Then*

$$\frac{1}{2}E''(0) = I_\ell(V,V) + \langle S_{\gamma'(\ell)}V(\ell), V(\ell)\rangle,$$

where $S_{\gamma'(\ell)}$ is the linear operator associated to the second fundamental form of the immersion f at the point q with respect to the normal $\gamma'(\ell)$.

Proof of the Lemma. We know that (see Remark 2.10, Chap. 9)

$$\begin{aligned}\frac{1}{2}E''(0) &= I_\ell(V,V) + \langle \frac{D}{ds}\frac{\partial f}{\partial s}, \frac{\partial f}{\partial t}\rangle(\ell,0)\\ &= I_\ell(V,V) + \langle(\overline{\nabla}_V\overline{V})(q), \overline{N}(q)\rangle,\end{aligned}$$

where $\overline{\nabla}$ is the covariant derivative of $\overline{M}$, and $\overline{V}$, $\overline{N}$ are local extensions, in a neighborhood of q, of the vectors $V(\ell)$ and $\gamma'(\ell)$, respectively, in such a way that $\overline{V}(c(s)) = \frac{\partial f}{\partial s}(\ell,s)$. It is possible to choose such extensions in such a way that they satisfy the condition $\langle\overline{V},\overline{N}\rangle = 0$. Therefore,

$$\langle\overline{\nabla}_V\overline{V},\overline{N}\rangle(q) = -\langle\overline{\nabla}_V\overline{N},\overline{V}\rangle(q) = \langle V(\ell), S_{\gamma'(\ell)}V(\ell)\rangle,$$

hence the assertion. □

Consider now a complete Riemannian manifold $\tilde{M}(b)$, with constant curvature b, whose dimension equals $\dim \overline{M} = n$. Let $\tilde{p} \in \tilde{M}(b)$ be an arbitrary point and let $\tilde{\gamma}: [0,\ell] \to \tilde{M}(b)$ be a normalized geodesic with $\tilde{\gamma}(0) = \tilde{p}$. As usual, choose parallel orthonormal bases $\{e_1(t), \ldots, e_n(t)\}$ of $T_{\gamma(t)}(\overline{M})$ and $\{\tilde{e}_1(t), \ldots, \tilde{e}_n(t)\}$ of $T_{\tilde{\gamma}(t)}(\tilde{M}(b))$, such that $e_1(0) = \gamma'(0)$ and $\tilde{e}_1(0) = \tilde{\gamma}'(0)$, and define a map ϕ which takes a vector field $W = \sum_{i=1}^n a_i e_i$ along γ into the vector field $\phi(W) = \sum_{i=1}^n a_i \tilde{e}_i$ along $\tilde{\gamma}$. It is clear that $\langle \phi(W_1), \phi(W_2) \rangle = \langle W_1, W_2 \rangle$ and $(\phi(W))' = \phi(W')$. Put $\phi(V) = \tilde{V}$. Since $\overline{K} \leq \tilde{K} \equiv b$, we conclude that

$$I_\ell(V,V) \geq I_\ell(\tilde{V},\tilde{V}).$$

Observe that $\tilde{V}(0) = 0$ and put $\tilde{V}(\ell) = \tilde{v}$.

We are going to estimate $I_\ell(\tilde{V},\tilde{V})$ in order to obtain an estimate for $I_\ell(V,V)$.

To estimate $I_\ell(\tilde{V},\tilde{V})$, we have to obtain an expression for the Jacobi field $\tilde{J}$ along $\tilde{\gamma}$, with $\tilde{J}(0) = 0$, $\tilde{J}(\ell) = \tilde{v}$. It is a simple exercise (Cf. Exercise 4, Chap. 5) to verify that $\tilde{J}$ is given by

$$\tilde{J}(t) = \frac{\sinh(t\sqrt{-b})}{\sinh(\ell\sqrt{-b})}\tilde{w}(t) \quad (\text{if} \quad b < 0),$$

$$\tilde{J}(t) = \frac{t}{\ell}\tilde{w}(t) \quad (\text{if} \quad b = 0),$$

where $\tilde{w}(t)$ is the parallel transport along $\tilde{\gamma}$ of

$$\tilde{w}(0) = \frac{\tilde{u}_o}{|\tilde{u}_o|}, \qquad \tilde{u}_o = ((d\exp_{\tilde{p}})_{\ell\tilde{\gamma}'(0)})^{-1}(\tilde{v}).$$

It follows that, from the expression for $I_\ell(\tilde{J},\tilde{J})$ given in the proof of the Index Lemma, that

$$\begin{aligned} I_\ell(\tilde{J},\tilde{J}) &= \langle \tilde{J}, \tilde{J}' \rangle(\ell) = \frac{1}{2}\frac{d}{dt}\left|\tilde{J}(t)\right|^2\Bigg|_{t=\ell} = \frac{1}{2}\frac{d}{dt}\frac{\sinh^2(t\sqrt{-b})}{\sinh^2(\ell\sqrt{-b})}\Bigg|_{t=\ell} \\ &= \frac{\sinh(t\sqrt{-b})\cosh(t\sqrt{-b})}{\sinh^2(\ell\sqrt{-b})}\sqrt{-b}\Bigg|_{t=\ell} \\ &= (\coth \ell\sqrt{-b})\sqrt{-b} > \sqrt{-b}. \end{aligned}$$

In addition, if $b = 0$, $I_\ell(\tilde{J}, \tilde{J}) > 0$. Therefore, from the Index Lemma, we obtain the following estimate for $I_\ell(V, V)$:

$$I_\ell(V,V) \geq I_\ell(\tilde{V},\tilde{V}) \geq I_\ell(\tilde{J},\tilde{J}) > \sqrt{-b}.$$

Because the geodesics γ_s in the variation f are shorter than γ, we have

$$\ell E(\gamma) = (L(\gamma))^2 \geq (L(\gamma_s))^2 = \ell E(\gamma_s),$$

hence

$$\begin{aligned} 0 \geq \frac{1}{2}E''(0) &= I_\ell(V,V) + \langle S_{\gamma'(\ell)}V(\ell), V(\ell)\rangle \\ &> \sqrt{-b} + \langle S_{\gamma'(\ell)}V(\ell), V(\ell)\rangle. \end{aligned}$$

Since $V(\ell) = v$, we have

$$\langle B(v,v), \gamma'(\ell)\rangle = \langle S_{\gamma'(\ell)}v, v\rangle < -\sqrt{-b}.$$

Therefore, for all $v \in T_qM$, $|v| = 1$, we have that

$$\|B(v,v)\| > \sqrt{-b}. \tag{6}$$

On the other hand, from the Gauss formula, if v and w are orthonormal vectors in T_qM, we have, by the hypothesis of the Theorem,

$$\langle B(v,v), B(w,w)\rangle - \|B(v,w)\|^2 = K(v,w) - \overline{K}(v,w) \leq -b. \tag{7}$$

The fact that the conditions (6) and (7) are incompatible with the condition $\dim \overline{M} < 2\dim M$ follows from the next algebraic lemma, essentially due to Otsuki. The proof of the lemma will complete, therefore, the proof of the theorem.

3.3 Lemma. *(Otsuki). Let $B\colon \mathbf{R}^n \times \mathbf{R}^n \to R^k$ be a symmetric bilinear form such that, for some $b \leq 0$,*

$$\langle B(v,v), B(w,w)\rangle - \|B(v,w)\|^2 \leq -b,$$

$$\|B(v,v)\| > \sqrt{-b},$$

for all orthonormal pairs $v, w \in \mathbf{R}^n$. Then $k \geq n$.

Proof of the Lemma. We suppose that $k < n$ and obtain a contradiction. Let $S \subset \mathbf{R}^n$ be the unit sphere and consider the map $f: S \to \mathbf{R}$, given by

$$f(v) = \langle B(v,v), B(v,v)\rangle.$$

Let $v \in S$ be a point where f assumes a minimum. Then taking the curve $v(t) = v\cos t + w \sin t$ in S, $t \in (-\varepsilon, \varepsilon)$, we have

$$\begin{aligned}(8) \qquad 0 = df_v(w) &= 2\langle \frac{d}{dt} B(v(t), v(t))\Big|_{t=0}, B(v,v)\rangle \\ &= 2\langle 2B(v'(0), v), B(v,v)\rangle = 4\langle B(w,v), B(v,v)\rangle,\end{aligned}$$

and, since $v''(0) = -v$,

$$\begin{aligned}(9) \qquad 0 \leq d^2 f_v(w,w) &= 4\langle B(v'(0), v), 2B(v'(0), v)\rangle \\ &+ 4\langle B(v''(0), v), B(v,v)\rangle + 4\langle B(w, v'(0)), B(v,v)\rangle \\ &= 8\, \|B(w,v)\|^2 - 4\langle B(v,v), B(v,v)\rangle \\ &+ 4\langle B(w,w), B(v,v)\rangle.\end{aligned}$$

Consider now the map $L: T_vS \to \mathbf{R}^k$, given by

$$L(w) = B(v,w).$$

The equation (8) above yields that $\langle L(w), B(v,v)\rangle = 0$, hence $\dim L(T_vS) \leq k-1$. Because $k < n$, the kernel of L has dimension at least 1, hence there exists $w_o \neq 0$, $w_o \perp v$, with $L(w_o) = 0 = B(v, w_o)$.

Introducing w_o in equation (9), we obtain

$$0 \leq \langle B(w_o, w_o), B(v,v)\rangle - \langle B(v,v), B(v,v)\rangle.$$

By the hypothesis of the Lemma, we conclude that

$$\begin{aligned}0 &\leq \langle B(w_o, w_o), B(v,v)\rangle - \|B(v,v)\|^2 \\ &< \langle B(w_o, w_o), B(v,v)\rangle - (\sqrt{-b})^2 \leq -b - (\sqrt{-b})^2 = 0,\end{aligned}$$

which produces the contradiction. □

3.4 COROLLARY. (C. Tompkins, "Isometric embeddings of flat manifolds in euclidean space", Duke Math. J. 5 (1939), 58–61). *Suppose that M is compact with zero curvature. If $k < n$, there does not exist an isometric immersion $f: M^n \to \mathbf{R}^{n+k}$; in particular, the flat torus T^n (see Example 2.7, Chap. 1) cannot be immersed isometrically in $\mathbf{R}^{2n-1}$.*

3.5 COROLLARY. (B. O'Neill, "Immersions of manifolds of non-positive curvature", Proc. Amer. Math. Soc., 11 (1960),132–134). *Suppose that $\overline{M}$ is complete, simply connected and $K_{\overline{M}} \leq 0$. Let M be compact, $\dim \overline{M} < 2 \dim M$ and $K_M \leq K_{\overline{M}}$. Then there does not exist an isometric immersion $f: M \to \overline{M}$.*

3.6 REMARK. The hypothesis that M be compact is essential in Theorem 3.1, as is shown by the example of complete surfaces with $K \leq 0$ in $\mathbf{R}^3$. On the other hand, it is known that the hyperbolic plane H^2 (a complete surface with $K \equiv -1$) cannot be isometrically immersed in $\mathbf{R}^3$ (Theorem of Hilbert, see M. do Carmo [dC 2], p. 446). It is an open problem to know if the hyperbolic space H^n can be immersed isometrically in $\mathbf{R}^{2n-1}$. This would be a natural generalization of the Theorem of Hilbert mentioned above.

3.7 REMARK. Essentially the same proof that we used in Theorem 3.1 can serve as a proof of the following theorem of S.S. Chern and N. Kuiper, "Some theorems on the isometric embedding of compact Riemannian manifolds in euclidean spaces", Ann. of Math. 56 (1952), 422-430: *Suppose that M^n is compact and that, for each point $p \in M$ there exists a subspace of dimension m, $V^m \subset T_pM$, such that the sectional curvature with respect to the planes contained in V are non-positive. If $k < m$, there does not exist an isometric immersion $f: M^n \to \mathbf{R}^{n+k}$.* The historical importance of the article of Chern-Kuiper is that they stressed the fundamental fact that the existence of an isometric immersion $f: M^n \to \overline{M}^{n+k}$ implies a relation between the nullity of the second fundamental form of f and the nullity of an intrinsic operator on M defined from the curvature of M. For more details see M. Dajczer, [Da].

3.8 REMARK. The torus T^{n+1} with the flat metric (Cf. Example 2.7, Chap. 1) contains the torus T^n as a totally geodesic submanifold. This shows that the hypothesis that $\overline{M}$ be simply connected in Theorem 3.1 is necessary.

4. Focal points and an extension of Rauch's Theorem

The notion of a conjugate point to a given point $p \in M$ extends to the idea of a focal point to a submanifold $N \subset M$ of a Riemannian manifold M. The idea is to consider variations

$$f: (-\varepsilon, \varepsilon) \times [0, \ell] \to M$$

of a geodesic $\gamma: [0, \ell] \to M$ with $\gamma(0) = p \in N$ and $\gamma'(0) \in (T_pN)^{\perp}$, satisfying the following conditions:

1) The curve $t \to f_s(t)$, $t \in [0, \ell]$ is a geodesic.
2) For all $s \in (-\varepsilon, \varepsilon)$, $f_s(0) = \alpha(s) \in N$ and

$$A(s) = \frac{\partial f}{\partial t}(s, 0) \in (T_{\alpha(s)}N)^{\perp}.$$

It is clear that $J(t) = \frac{\partial f}{\partial s}(0, t)$ is a Jacobi field along γ.

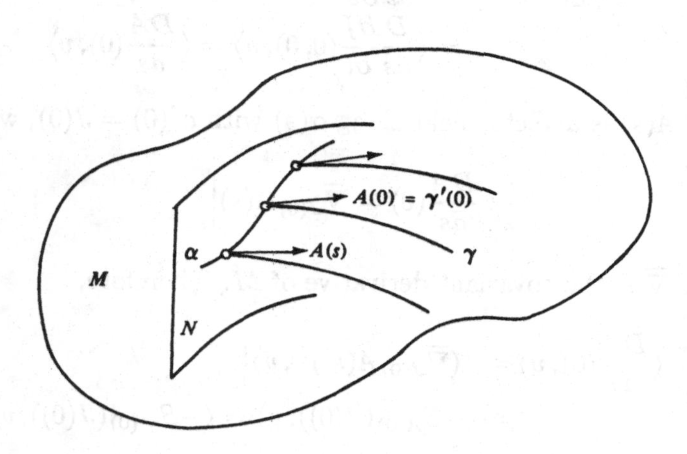

Figure 4

4.1 LEMMA. *The Jacobi field J constructed above satisfies the following properties:*

i) $J(0) \in T_pN$,
ii) $J'(0) + S_{\gamma'(0)}(J(0)) \in (T_pN)^{\perp}$,

where $S_{\gamma'(0)}$ is the linear operator on T_pN given by the second fundamental form of $N \subset M$.

Conversely, if J is a Jacobi field along the geodesic γ with $\gamma(0) = p \in N$, $\gamma'(0) \in (T_pN)^{\perp}$, satisfying (i) and (ii), then there exists a variation f of γ satisfying (1) and (2), whose variational field is J.

Proof. It is clear that

$$J(0) = \frac{\partial f}{\partial s}(0,0) = \frac{d}{ds}(\alpha(s))\Big|_{s=0} \in T_pN,$$

which verifies (i). To prove (ii), let $v \in T_pN$ be an arbitrary vector. We shall show that

$$\langle J'(0) + S_{\gamma'(0)}(J(0)), v\rangle = 0. \tag{10}$$

Indeed,

$$\begin{aligned}\langle \frac{DJ}{dt}(0), v\rangle &= \langle \frac{D}{dt}\frac{\partial f}{\partial s}(0,0), v\rangle \\ &= \langle \frac{D}{ds}\frac{\partial f}{\partial t}(0,0), v\rangle = \langle \frac{DA}{ds}(0), v\rangle.\end{aligned}$$

Since $A(s)$ is a vector field along $\alpha(s)$ with $\alpha'(0) = J(0)$, we have

$$\frac{DA}{ds}(0) = \overline{\nabla}_{J(0)} A(s)\Big|_{s=0}$$

where $\overline{\nabla}$ is the covariant derivative of M. Therefore,

$$\begin{aligned}\langle \frac{DA}{ds}(0), v\rangle &= \langle (\overline{\nabla}_{J(0)} A(s))^T, v\rangle\Big|_{s=0} \\ &= \langle -S_{A(0)}(J(0)), v\rangle = \langle -S_{\gamma'(0)}(J(0)), v\rangle,\end{aligned}$$

which implies (10) and verifies (ii).

To prove the converse, let $s \to \alpha(s)$ be a curve in N with $\alpha(0) = p$, $\alpha'(0) = J(0)$. We can choose a vector field W along α such that $W(0) = \gamma'(0)$ and $\frac{DW}{ds}(0) = \frac{DJ}{dt}(0)$. Write $W(s) = V(s) + U(s)$, where

$$V(s) \in (T_{\alpha(s)}(N))^{\perp}, \quad U(s) \in T_{\alpha(s)}(N),$$

and define $f(s,t) = \exp_{\alpha(s)}(tV(s))$. We show that f is the variation of γ which satisfies (1) and (2) and whose variational field is J.

In fact, for all s, the curve

$$t \mapsto f_s(t) = \exp_{\alpha(s)} tV(s)$$

is a geodesic. In addition,

$$f_s(0) = \alpha(s) \in N,$$

and

$$A(s) = \frac{\partial f}{\partial t}(s,0) = (d\exp_{\alpha(s)})_o(V(s)) = V(s) \in (T_{\alpha(s)}(N))^{\perp}.$$

It remains to show that $\frac{\partial f}{\partial s}(0,t) = J(t)$. It is enough to verify that $\frac{\partial f}{\partial s}(0,0) = J(0)$ and $\frac{D}{dt}\frac{\partial f}{\partial s}(0,0) = \frac{DJ}{dt}(0)$. Because

$$\frac{\partial f}{\partial s}(0,0) = \alpha'(0) = J(0)$$

and

$$\begin{aligned}\frac{D}{dt}\frac{\partial f}{\partial s}(0,0) &= \frac{D}{ds}\frac{\partial f}{\partial t}(0,0) = \frac{DV}{ds}(0) = \frac{DW}{ds}(0) - \frac{DU}{ds}(0)\\ &= \frac{DJ}{dt}(0) - \frac{DU}{ds}(0),\end{aligned}$$

it suffices to show that $\frac{DU}{ds}(0) = 0$. But $J(t)$ and $\frac{\partial f}{\partial s}(0,t)$ satisfy:

$$\frac{DJ}{dt}(0) + S_{\gamma'(0)}(J(0)) \in (T_pN)^{\perp},$$

$$\frac{D}{dt}\frac{\partial f}{\partial s}(0,0) + S_{\gamma'(0)}\left(\frac{\partial f}{\partial s}(0,0)\right) \in (T_pN)^{\perp},$$

where the second assertion comes from the first part of the Lemma. Since $J(0) = \frac{\partial f}{\partial s}(0,0)$, we can conclude that

$$\frac{DU}{ds}(0) = \frac{DJ}{dt}(0) - \frac{D}{dt}\frac{\partial f}{\partial s}(0,0) \in (T_pN)^{\perp}.$$

On the other hand, $U(0) = 0$ and $U(s) \in T_{\alpha(s)}(N)$. Therefore, given $v \in (T_pN)^{\perp}$ and a vector field $v(s)$ along $\alpha(s)$ with $v(0) = v$ and $v(s) \in (T_{\alpha(s)}(N))^{\perp}$, we have

$$\begin{aligned} 0 = \frac{d}{ds}\langle U(s), v(s)\rangle\Big|_{s=0} &= \langle \frac{DU}{ds}(0), v\rangle + \langle U(0), \frac{Dv}{ds}(0)\rangle \\ &= \langle \frac{DU}{ds}(0), v\rangle. \end{aligned}$$

From the arbitrariness of v, this implies that $\frac{DU}{ds}(0) = 0$, which proves the assertion, and concludes the proof. □

4.2 DEFINITION. Let $N \subset M$ be a submanifold of a Riemannian manifold M. The point $q \in M$ is called a *focal point* of N if there exists a geodesic $\gamma\colon [0,\ell] \to M$, with $\gamma(0) = p \in N$, $\gamma'(0) \in (T_pN)^{\perp}$, $\gamma(\ell) = q$, and a non-zero Jacobi field J along γ, satisfying (i), (ii) and with $J(\ell) = 0$.

4.3 EXAMPLE. If $S^{n-1} \subset S^n$ is the equator of S^n, that is, $S^{n-1} = \{x \in S^n; x = (x_1, \dots, x_n, 0)\}$, then the north pole $(0, 0, \dots, 0, 1)$ and the south pole $(0, \dots, 0, -1)$ are focal points of S^{n-1} in S^n.

To obtain a characterization of the focal points in terms of the exponential map, we introduce the following notation. Let $T(M)$ be the tangent bundle of M, $\pi : T(M) \to M$ its projection onto M. Denote this statement by $T(M) \to M$. Similarly, $T(N) \to N$ denotes the tangent bundle $T(N)$ and its projection onto N, $T(N)^{\perp} \to N$ denotes the normal bundle of the immersion $N \subset M$ and its projection onto N (a point of $T(N)^{\perp}$ is a pair (p, n), where $p \in N$, $n \in (T_pN)^{\perp}$). Remember that the exponential mapping can be thought of as a map $\exp\colon T(M) \to M$, defined by

$$\exp(p, v) = \exp_p(v).$$

Since $T(N)^{\perp} \to N$ is contained in the restriction of $T(M) \to M$ to $N \subset M$, the exponential map can be restricted to $T(N)^{\perp}$; this restriction will be denoted by

$$\exp^{\perp}\colon T(N)^{\perp} \to M.$$

Observe that $\dim T(N)^{\perp} = \dim M$.

4.4 PROPOSITION. *The point $q \in M$ is a focal point of $N \subset M$ if and only if it is a critical value of* $\exp^{\perp}$.

Proof. Suppose that q is a focal point of N. From Lemma 4.1, there exists a geodesic $\gamma\colon [0,\ell] \to M$, with $\gamma(0) = p \in N$, $\gamma(\ell) = q$, $\gamma'(0) \in (T_pN)^{\perp}$ and a variation $f\colon (-\varepsilon,\varepsilon)\times[0,\ell] \to M$ of γ satisfying (1) and (2) and with $\frac{\partial f}{\partial s}(0,\ell) = 0$. It follows that

$$w(s) = (\alpha(s), \ell A(s)), \quad A(s) = \frac{\partial f}{\partial t}(s,0), \quad \alpha(s) = f_s(0),$$

is a curve in $T(N)^{\perp}$ such that:

$$\exp^{\perp}(w(s)) = \exp_{\alpha(s)} \ell A(s) = f(s,\ell),$$

$$\exp^{\perp}(w(0)) = q$$

and

$$(d\exp^{\perp})_{w(0)}(w'(0)) = \frac{\partial f}{\partial s}(0,\ell) = 0.$$

Then $w(0)$ is a critical point of $\exp^{\perp}$, with $\exp^{\perp}(w(0)) = q$.

Conversely, suppose that q is a critical value of $\exp^{\perp}$. Then there exist w_o, w_o' such that $\exp^{\perp}(w_o) = q$ and $(d\exp^{\perp})_{w_o}(w_o') = 0$. Let $w(s) = (\sigma(s), \ell V(s))$ be a curve in $T(N)^{\perp}$, with $w(0) = w_o$ and $w'(0) = w_o'$. Since $q = \exp_{\sigma(0)}(\ell V(0))$, there exists a geodesic $\gamma\colon [0,\ell] \to M$ such that $\gamma(0) = \sigma(0)$, $\gamma(\ell) = q$, $\gamma'(0) = V(0)$. Consider the variation of γ given by

$$f(s,t) = \exp_{\sigma(s)} tV(s).$$

f satisfies the conditions (1) and (2). In fact,

$$f(s,0) = \exp_{\sigma(s)}(0) = \sigma(s) \in N,$$

$$\frac{\partial f}{\partial t}(s,0) = (d\exp_{\sigma(s)})_o(V(s)) = V(s) \in (T_{\sigma(s)}(N))^{\perp}.$$

From the lemma, $J(t) = \frac{\partial f}{\partial s}(0,t)$ is a Jacobi field satisfying (i) and (ii), with

$$J(\ell) = \frac{\partial f}{\partial s}(0,\ell) = \frac{\partial}{\partial s}\exp_{\sigma(s)} \ell V(s)\Big|_{s=0}$$

$$= \frac{d}{ds}\exp^{\perp} w(s)\Big|_{s=0} = (d\exp^{\perp})_{w_o}(w_o') = 0,$$

and therefore, q is a focal point of N. □

4.5 DEFINITION. Let $N \subset M$ be a submanifold of a Riemannian manifold M. The set of focal points of N will be called the *focal set* $F(N) \subset M$ of N.

4.6 EXAMPLE. Let $N^2 \subset \mathbf{R}^3$ be a regular surface in $\mathbf{R}^3$. The connected components of the focal set $F(N) \subset \mathbf{R}^3$ are called the "focal surfaces" (or surfaces of centers) in classical differential geometry. In general such subsets are not regular surfaces and can degenerate to points or curves. For example, if N^2 is a sphere S^2, then $F(S^2)$ is the center of S^2; if N^2 is the right circular cylinder C, $F(C)$ is the axis of C. A way of constructing, geometrically, the focal points of $N^2 \subset \mathbf{R}^3$ is the following.

Let $\mathbf{x}: U \subset \mathbf{R}^2 \to N \subset \mathbf{R}^3$ be a parametrization of N at $p \in N$. If we denote by (u_1, u_2) the coordinates of U and by $n = n(u_1, u_2)$ a unit normal field on $\mathbf{x}(U) \subset N$, we can write the map $\exp^\perp: T(U)^\perp \to \mathbf{R}^3$ as

$$\exp^\perp((u_1, u_2), t) = \mathbf{x}(u_1, u_2) + tn, \quad t \in \mathbf{R}.$$

The basis of the tangent space $T_{(p,tn)}(T(N)^\perp) \approx \mathbf{R}^3$ associated to this parametrization is

$$\frac{\partial \mathbf{x}}{\partial u_1}, \frac{\partial \mathbf{x}}{\partial u_2}, n.$$

In this basis, the linear map $d\exp^\perp$ is given by the matrix

$$d\exp^\perp = \begin{pmatrix} \langle \frac{\partial \mathbf{x}}{\partial u_1} + t\frac{\partial n}{\partial u_1}, \frac{\partial \mathbf{x}}{\partial u_1} \rangle & \langle \frac{\partial \mathbf{x}}{\partial u_2} + t\frac{\partial n}{\partial u_2}, \frac{\partial \mathbf{x}}{\partial u_1} \rangle & \langle n, \frac{\partial \mathbf{x}}{\partial u_1} \rangle \\ \langle \frac{\partial \mathbf{x}}{\partial u_1} + t\frac{\partial n}{\partial u_1}, \frac{\partial \mathbf{x}}{\partial u_2} \rangle & \langle \frac{\partial \mathbf{x}}{\partial u_2} + t\frac{\partial n}{\partial u_2}, \frac{\partial \mathbf{x}}{\partial u_2} \rangle & \langle n, \frac{\partial \mathbf{x}}{\partial u_2} \rangle \\ \langle \frac{\partial \mathbf{x}}{\partial u_1} + t\frac{\partial n}{\partial u_1}, n \rangle & \langle \frac{\partial \mathbf{x}}{\partial u_2} + t\frac{\partial n}{\partial u_2}, n \rangle & \langle n, n \rangle \end{pmatrix}$$

$$= \left(\begin{array}{c|c} g_{ij} - t\langle \frac{\partial^2 \mathbf{x}}{\partial u_i \partial u_j}, n \rangle & 0 \\ \hline \mathrm{B} & 1 \end{array} \right)$$

where the matrix B is not important in what follows. By Proposition 4.4, $\mathbf{x} + tn$ is a focal point of N if and only if the matrix above is singular. Because the coefficients of the second fundamental form H_n in the basis $\frac{\partial \mathbf{x}}{\partial u_1}, \frac{\partial \mathbf{x}}{\partial u_2}$ are given by

$$h_{ij} = H_n\left(\frac{\partial \mathbf{x}}{\partial u_i}, \frac{\partial \mathbf{x}}{\partial u_j}\right) = \langle B\left(\frac{\partial \mathbf{x}}{\partial u_i}, \frac{\partial \mathbf{x}}{\partial u_j}\right), n \rangle = \langle \frac{\partial^2 \mathbf{x}}{\partial u_i \partial u_j}, n \rangle,$$

we conclude that the matrix above is singular if and only if the matrix

$$(g_{ij} - th_{ij})$$

is singular. It is possible to choose the parametrization $\mathbf{x}$ in such a way that at p, (g_{ij}) is the identity matrix. Therefore, $\mathbf{x} + tn$ is a focal point if and only if $\frac{1}{t}$ is an eigenvalue of (h_{ij}), that is, $\frac{1}{t}$ is one of the principal curvatures of N^2 at p.

In summary, to construct a connected component of $F(N)$, fix a principal curvature k_i and mark on the normal at each $p \in N$, in the direction given by $n(p)$, a length equal to $1/k_i(p)$ (this point is the center of the osculating circle in the normal section of N in the principal direction e_i, which justifies the name "surface of centers").

We now pass to an extension of Rauch's Theorem in which the notion of focal point replaces that of conjugate point. We need a definition.

4.7 DEFINITION. Let $\gamma\colon [0,a] \to M$ be a geodesic in M. Let $B_\varepsilon(0) \subset \gamma'(0)^\perp$ be a ball of radius ε and center at the origin 0, contained in the orthogonal complement of $\gamma'(0)$. We say that γ is *focal point free* at $(0,a]$ if there exists some $\varepsilon > 0$ such that γ has no focal points relative to the submanifold $\sum_\varepsilon = \exp_{\gamma(0)}(B_\varepsilon(0))$ (Fig. 5).

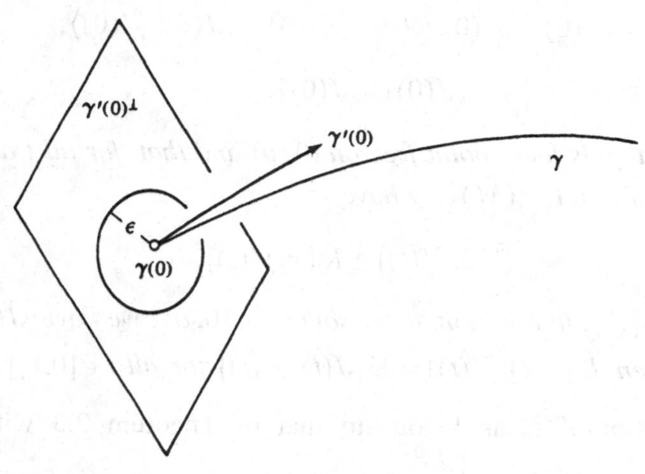

Figure 5

Observe that since Σ_ε is geodesic at p, any Jacobi field J along γ, with $J(0) \neq 0$ and $J'(0) = 0$, automatically satisfies the condition

$$S_{\gamma'(0)}\big(J(0)\big) = 0.$$

The Index Lemma for focal points is as follows.

4.8 LEMMA. *Let γ: $[0, a] \to M^n$ be a geodesic which is focal point free on $(0, a]$. Let J be a Jacobi field along γ, with $\langle J, \gamma' \rangle = 0$, and let V be a piecewise differentiable vector field along γ, with $\langle V, \gamma' \rangle = 0$, Suppose that $J'(0) = 0$ and $J(t_o) = V(t_o)$, $t_o \in (0, a]$. Then*

$$I_{t_o}(J,J) \leq I_{t_o}(V,V)$$

and the equality occurs if and only if $V = J$ on $[0, t_o]$.

Proof. Let $\{J_1, \ldots, J_{n-1}\}$ be a basis of the vector space of Jacobi fields J such that $J'(0) = 0$, $\langle J, \gamma' \rangle = 0$. The fact that γ is focal point free on $(0, a]$ implies that, for each each $t \in (0, a]$, $\{J_1(t), \ldots, J_{n-1}(t)\}$ is a basis for $(\gamma'(t))^{\perp}$.

Starting from there, the proof follows in a manner entirely analogous to the Index Lemma. □

4.9 THEOREM. *Let γ: $[0, a] \to M^n$ and $\tilde{\gamma}: [0,a] \to \tilde{M}^{n+k}$ be geodesics, with the same velocity and let J and $\tilde{J}$ be Jacobi fields along γ and $\tilde{\gamma}$, such that*

$$0 = J'(0) = \tilde{J}'(0), \langle J(0), \gamma'(0) \rangle = \langle J(0), \tilde{\gamma}'(0) \rangle,$$

$$|J(0)| = |\tilde{J}(0)|.$$

Assume that $\tilde{\gamma}$ is focal point free on $(0, a]$ and that, for all t and all $x \in T_{\gamma(t)}(M)$, $\tilde{x} \in T_{\tilde{\gamma}(t)}(\tilde{M})$, we have

$$\tilde{K}\big(\tilde{x}, \tilde{\gamma}'(t)\big) \geq K\big(x, \gamma'(t)\big).$$

Then $|\tilde{J}| \leq |J|$. In addition, if for some $t_o \in (0, a]$, we have $|\tilde{J}(t_o)| = |J(t_o)|$, then $\tilde{K}\big(\tilde{J}(t), \tilde{\gamma}'(t)\big) = K\big(J(t), \gamma'(t)\big)$ for all $t \in [0, t_o]$.

Proof. The proof is analogous to that of Theorem 2.3 with the following modifications: $\dfrac{|J|^2}{|\tilde{J}|^2}$ is well-defined, since $\tilde{\gamma}$ is focal point free and $I_{t_o}\big(\phi(U), \phi(U)\big) \geq I_{t_o}(\tilde{U}, \tilde{U})$ by Lemma 4.8. □

Other useful extensions of the Rauch comparison theorem can be found in F. Warner, *Extensions of the Rauch comparison theorem to submanifolds*, Trans. A.M.S. 122 (1966), 341–356 and in E. Heintze and H. Karcher, *A general comparison theorem with applications to volume estimates for submanifolds*, Ann. Sci. École Norm. Sup., 11 (1978), 451–470.

Rauch's theorem admits an extremely important global generalization, which is called the Theorem of Toponogov. One of its versions can be stated in the following manner.

Theorem. (Toponogov). *Let M be a Riemannian manifold which is complete with sectional curvature $K \geq H$. Let γ_1 and γ_2 be normalized geodesic segments in M with $\gamma_1(0) = \gamma_2(0)$. Denote by $M^2(H)$ a manifold of dimension two with constant curvature H. Assume that the geodesic γ_1 is minimizing and that, if $H > 0$, $\ell(\gamma_2) \leq \frac{\pi}{\sqrt{H}}$. Consider on $M^2(H)$ two normalized geodesics $\bar{\gamma}_1$, $\bar{\gamma}_2$, such that $\bar{\gamma}_1(0) = \bar{\gamma}_2(0)$, $\ell(\gamma_i) = \ell(\bar{\gamma}_i) = \ell_i$, $i = 1, 2$, and $\measuredangle(\bar{\gamma}_1'(0), \bar{\gamma}_2'(0)) = \measuredangle(\gamma_1'(0), \gamma_2'(0))$. Then*

$$d(\gamma_1(\ell_1), \gamma_2(\ell_2)) \leq d(\bar{\gamma}_1(\ell_1), \bar{\gamma}_2(\ell_2)).$$

A proof of this theorem can be found in Cheeger and Ebin [CE].

The Theorem of Toponogov is an essential tool for the study of the relation between topology and curvature mentioned in the Introdution to Chapter 9. One of the culmination points of this study, the sphere theorem, will be presented in Chapter 13 of this book. The proof presented does not use the Theorem of Toponogov. A proof using Toponogov's Theorem, along with various applications of this theorem to the study of the relationship between topology and curvature, can be found in Cheeger and Ebin [CE].

EXERCISES

1. (*Klingenberg's Lemma*). Let M be a complete Riemannian manifold with sectional curvature $K \leq K_o$, where K_o is a pos-

itive constant. Let $p, q \in M$ and let γ_o and γ_1 be two distinct geodesics joining p to q with $\ell(\gamma_o) \leq \ell(\gamma_1)$. Assume that γ_o is homotopic to γ_1, that is, there exists a continuous family of curves α_t, $t \in [0,1]$ such that $\alpha_o = \gamma_o$ and $\alpha_1 = \gamma_1$. Prove that there exists $t_o \in [0,1]$ such that

$$\ell(\gamma_o) + \ell(\alpha_{t_o}) \geq \frac{2\pi}{\sqrt{K_o}}.$$

(Thus, the given homotopy has to pass through a "long" curve. Fig. 6).

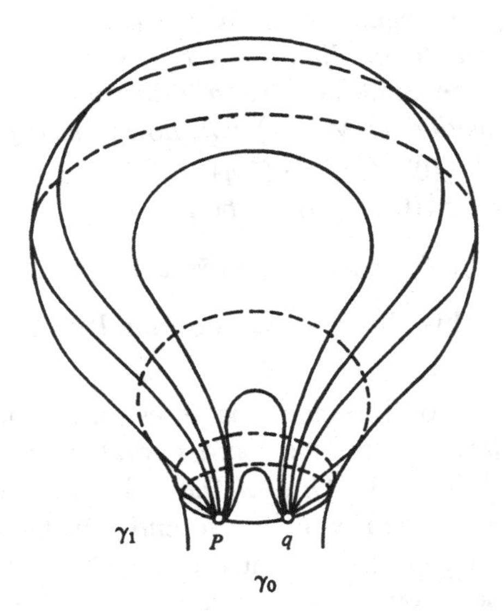

Figure 6

Hint: Assume $\ell(\gamma_o) < \frac{\pi}{\sqrt{K_o}}$ (otherwise, we have nothing to prove). From Rauch's Theorem, $\exp_p: T_pM \to M$ has no critical point in the open ball B of radius $\pi/\sqrt{K_o}$ centered at p. For t small, it is possible to lift the curve α_t to the tangent space T_pM, i.e., there exists a curve $\tilde{\alpha}_t$ in T_pM, joining $\exp_p^{-1}(0) = 0$ to $\exp_p^{-1}(q) = \tilde{q}$, such that $\exp_p \circ \tilde{\alpha}_t = \alpha_t$. It is clear that it is not possible to do the same for every $t \in [0,1]$, since α_1 cannot be lifted keeping the endpoints fixed.

We conclude that for all $\varepsilon > 0$ there exists a $t(\varepsilon)$ such that $\alpha_{t(\varepsilon)}$ can be lifted to $\tilde{\alpha}_{t(\varepsilon)}$ and $\tilde{\alpha}_{t(\varepsilon)}$ contains points with distance $< \varepsilon$ from the boundary ∂B of B. In the contrary case, for some $\varepsilon > 0$, all lifts $\tilde{\alpha}_t$ are at the distance $\geq \varepsilon$ from ∂B; the set of t's for which it is possible to lift α_t will then be open and closed and α_1 could be lifted, which is a contradiction. Therefore, for all $\varepsilon > 0$, we have

$$\ell(\gamma_o) + \ell(\alpha_{t(\varepsilon)}) \geq \frac{2\pi}{\sqrt{K_o}} - 2\varepsilon.$$

Now choose a sequence $\{\varepsilon_n\} \to 0$, and consider a convergent subsequence of $\{t(\varepsilon_n)\} \to t_o$. Then there exists a curve α_{t_o} with

$$\ell(\gamma_o) + \ell(\alpha_{t_o}) \geq \frac{2\pi}{\sqrt{K_o}}.$$

2. Use Klingenberg's Lemma from the last exercise for the proof of Hadamard's Theorem (see Theorem 3.1 of Chap. 7).
Hint: Take $K_o = 1/n$, n an integer, in Klingenberg's lemma and show that if M is simply connected, there exists a unique geodesic joining the two points $p, q \in M$. Because were there to exist two such geodesics, they would be homotopic and, by Klingenberg's lemma, there would exist a sequence of curves of lengths $\geq \pi\sqrt{n}$ in this homotopy.

3. Let M be a complete Riemannian manifold with non-positive sectional curvature. Prove that

$$\left|(d\exp_p)_v(w)\right| \geq |w|,$$

for all $p \in M$, all $v \in T_pM$ and all $w \in T_v(T_pM)$.

4. (*Focal sets of plane curves*).
   a) Let $C \subset \mathbf{R}^2$ be a regular curve. Show that the focal set $F(C) \subset \mathbf{R}^2$ of C is obtained by taking, on the positive normal n at $p \in C$ a length equal to $1/k$, where k is the curvature of C at p.

   Hint: Use the same argument as in Example 4.6.

   b) Show that the focal set of the ellipse $\frac{x^2}{a^2} + \frac{y^2}{b^2} = 1$ is given by

$$\left\{(x, y) \in \mathbf{R}^2; (ax)^{2/3} + (by)^{2/3} = (a^2 - b^2)^{2/3}\right\}$$

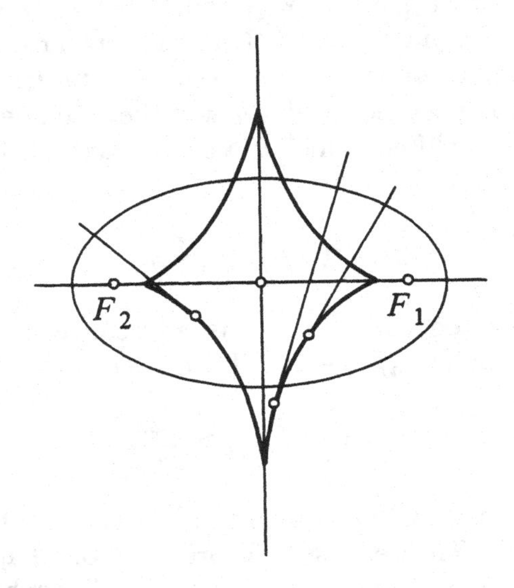

Figure 7

c) Show that the focal set of the curve

$$t \to (\cos t + t \sin t, -\sin t + t cos t)$$

is the circle $t \to (\cos t, -\sin t)$ (Fig. 8).

5. (*The Sturm Comparison Theorem*). In this exercise we present a direct proof of Rauch's Theorem in dimension two, without using material from the present chapter. We will indicate a proof of the Theorem of Sturm mentioned in the Introduction to the chapter.
Let

$$f''(t) + K(t)f(t) = 0, \quad f(0) = 0, \quad t \in [0, \ell],$$

$$\tilde{f}''(t) + \tilde{K}(t)\tilde{f}(t) = 0, \quad \tilde{f}(0) = 0, \quad t \in [0, \ell],$$

be two ordinary differential equations. Suppose that $\tilde{K}(t) \geq K(t)$ for $t \in [0, \ell]$, and that $f'(0) = \tilde{f}'(0) = 1$.

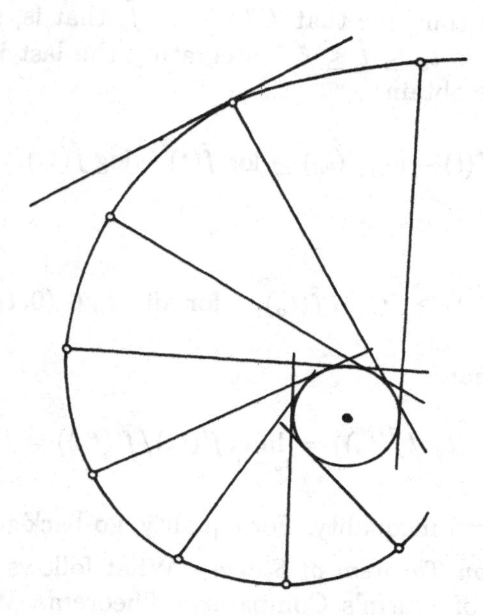

Figure 8

a) Show that for all $t \in [0, \ell]$,

$$(1) \quad 0 = \int_0^t \left\{ \tilde{f}(f'' + Kf) - f(\tilde{f}'' + \tilde{K}\tilde{f}) \right\} dt$$
$$= [\tilde{f}f' - f\tilde{f}']_o^t + \int_0^t (K - \tilde{K}) f\tilde{f} dt.$$

Conclude from this that the first zero of f does not occur before the first zero of $\tilde{f}$ (that is, if $\tilde{f}(t) > 0$ on $(0, t_o)$ and $\tilde{f}(t_o) = 0$, then $f(t) > 0$ on $(0, t_o)$).

Hint: From the initial condition, $f(t)$ is positive in a neighborhood of zero. Assume that $f(t_1) = 0$, $t_1 < t_o$. Then $f'(t_1) < 0$, $\tilde{f}(t_1) > 0$, and this contradicts (1).

b) Suppose that $\tilde{f}(t) > 0$ on $(0, \ell]$. Use (1) and the fact that $f(t) > 0$ on $(0, \ell]$ to show that $f(t) \geq \tilde{f}(t)$, $t \in [0, \ell]$, and that the equality is verified for $t = t_1 \in (0, \ell]$ if and only if $K(t) = \tilde{K}(t)$, $t \in [0, t_1]$.

Verify that this is the Theorem of Rauch in dimension two.

Hint: From (1) conclude that $f'/f \geq \tilde{f}'/\tilde{f}$, that is, $(\log f)' \geq (\log \tilde{f})'$. Let $0 < t_o \leq t \leq \ell$. Integrating the last inequality from t_o to t, we obtain

$$\log f(t) - \log f(t_o) \geq \log \tilde{f}(t) - \log \tilde{f}(t_o),$$

that is,

$$f(t)/\tilde{f}(t) \geq f(t_o)/\tilde{f}(t_o), \quad \text{for all} \quad t_o \in (0, \ell].$$

Now observe that

$$\lim_{t_o \to 0} (f(t_o)/\tilde{f}(t_o)) = \lim_{t_o \to 0} (f'(t_o)/\tilde{f}'(t_o)) = 1,$$

hence the desired inequality. For equality, go back to (1).

6. (*The Oscillation Theorem of Sturm*). What follows is a slight generalization of Sturm's Comparison Theorem. We present the theorem in geometric form.
   Let M^2 be a complete Riemannian manifold of dimension 2, and let $\gamma\colon [0, \infty) \to M^2$ be a geodesic. Let $J(t)$ be a Jacobi field along γ with $J(0) = J(t_o) = 0$, $t_o \in (0, \infty)$, and $J(t) \neq 0$, $t \in (0, t_o)$. Then J is a field normal to γ and can be written $J(t) = f(t)e_2(t)$, where $e_2(t)$ is the parallel transport of a unit vector $e_2 \in T_{\gamma(0)}(M)$ with $e_2 \perp \gamma'(0)$. Because J is a Jacobi field,

$$f''(t) + K(t)f(t) = 0,$$

where K is the Gaussian curvature of M^2. Assume that

$$K(t) \leq L(t),$$

where L is a differentiable function on $[0, \infty)$. Prove that any solution of the equation

$$\tilde{f}''(t) + L(t)\tilde{f}(t) = 0$$

has a zero on $[0, t_o]$, that is, there exists $t_1 \in [0, t_o]$ with $\tilde{f}(t_1) = 0$.

Hint: Suppose that $\tilde{f}(t) \neq 0$ for all $t \in [0, t_o]$. Using expression (1) of the last exercise, we obtain

$$\text{(2)} \qquad \int_0^{t_o} (K - L) f \tilde{f} dt + \tilde{f}(t_o) f'(t_o) - \tilde{f}(0) f'(0) = 0.$$

Suppose, for example, that $\tilde{f}(t) > 0$ and $f(t) < 0$ in $(0, t_o)$. Then $f'(0) < 0$ and $f'(t_o) > 0$. This contradicts (2). The other cases are treated in the same way.

7. (*Kneser's criterion for points conjugate in surfaces*). Let M^2 be a complete Riemannian manifold of dimension two and let $\gamma: [0, \infty) \to M^2$ be a geodesic with $\gamma(0) = p$. Let $K(s)$ be the Gaussian curvature of M^2 along γ. Assume that:

$$\text{(3)} \qquad \int_t^{\infty} K(s) ds \leq \frac{1}{4(t+1)}, \quad \text{for all} \quad t \geq 0,$$

in the sense that the integral converges and has the bound indicated.

a) Define

$$\omega(t) = \int_t^{\infty} K(s) ds + \frac{1}{4(t+1)},$$

and show that $\omega'(t) + (\omega(t))^2 \leq -K(t)$.

b) For $t \geq 0$, put $\omega'(t) + (\omega(t))^2 = -L(t)$ (hence $L(t) \geq K(t)$) and define

$$\tilde{f}(t) = \exp\left(\int_0^t \omega(s) ds\right), \quad t \geq 0.$$

Show that

$$\tilde{f}''(t) + L(t)\tilde{f}(t) = 0, \quad \tilde{f}(0) = 1.$$

c) Observe that $\tilde{f}(t) > 0$ and use the oscillation theorem of Sturm (Exercise 6) to show that there does not exist a Jacobi field $J(s)$ on $\gamma(s)$ with $J(0) = 0$ and $J(s_o) = 0$, for some $s_o \in (0, \infty)$. Therefore *the condition (3) implies that there do not exist conjugate points to p along* γ.

CHAPTER 11

THE MORSE INDEX THEOREM

1. Introduction

In this chapter we wish to prove the Morse Index Theorem. This theorem relates the number of conjugate points on a geodesic segment, counted with their multiplicities, to the index of a certain quadratic form defined in terms of the formula for the second variation (essentially, the expression $I_a(V, V)$).

The Morse Index Theorem is a generalization of a classical theorem of Jacobi (see Cor. 2.9) which states that a geodesic segment minimizes the arc length relative to "neighboring" curves with the same endpoints if and only if such a segment has no conjugate points. The proof of the Index Theorem which we present here uses the Index Lemma of Chapter 10.

2. The Index Theorem

Let γ: $[0, a] \to M^n$ be a geodesic. Denote by $\mathcal{V}(0,a) = \mathcal{V}$ the vector space formed by normal vector fields V along γ, which are piecewise differentiable and vanish at the endpoints of γ, that is, $V(0) = V(a) = 0$.

2.1 DEFINITION. The *index form* of γ is the quadratic form associated to the symmetric bilinear form I_a defined on $\mathcal{V}$ by

$$(1) \qquad I_a(V,W) = \int_0^a \{\langle V', W'\rangle - \langle R(\gamma', V)\gamma', W\rangle\}\, dt,$$

where $V, W \in \mathcal{V}$. (cf. Rem. 2.10 of Chap. 9).

Observe that the symmetry of I_a follows from part (d) of Proposition 2.5 of Chapter 4.

In general, given a symmetric bilinear form B over a vector space $\mathcal{V}$, we define the *index* of B as the the maximal dimension of all subspaces of $\mathcal{V}$ on which the quadratic form associated to B is negative definite. The *nullity* of B is defined to be the dimension of the subspace of $\mathcal{V}$ formed by the elements $V \in \mathcal{V}$ such that $B(V, W) = 0$, for all $W \in \mathcal{V}$; such a subspace is called the *null space* of B. We say that B is *degenerate* if its nullity is strictly positive.

We can now state the main theorem of this chapter.

2.2 Index Theorem. *(Morse). The index of the form I_a is finite and equals the number of points $\gamma(t)$, $0 < t < a$, conjugate to $\gamma(0)$, each counted with its multiplicity.*

Before we start the proof of the theorem, we need a few preliminary propositions.

2.3 PROPOSITION. *An element $V \in \mathcal{V}$ belongs to the null space of I_a if and only if V is a Jacobi field along γ.*

Proof. First observe that we have the following expression for I_a (Compare Rem. 2.10 of Chap. 9):

$$(2) \qquad I_a(V, W) = -\int_0^a \langle V'' + R(\gamma', V)\gamma', W\rangle\, dt - \sum_{j=1}^{k-1} \left\langle \frac{DV}{dt}(t_j^+) - \frac{DV}{dt}(t_j^-), W(t_j)\right\rangle.$$

If V is a Jacobi field, then by (2), V is in the null space of I_a.

Conversely, suppose that $I_a(V, W) = 0$ for all $W \in \mathcal{V}$. Let $0 = t_o < t_1 < \cdots < t_{k-1} < t_k = a$ be a subdivision of $[0, a]$ such that the restriction $V \mid [t_{j-1}, t_j]$ is differentiable, $j = 1, \ldots, k$. Let $f\colon [0, a] \to \mathbf{R}$ be a differentiable function with $f(t) > 0$, for $t \neq t_j$ and $f(t_j) = 0$, $j = 0, \ldots, k$. Define W by

$$W(t) = f(t)(V'' + R(\gamma', V)\gamma').$$

Then

$$0 = I_a(V, W) = -\int_0^a f(t)\, \|V'' + R(\gamma', V)\gamma'\|^2\, dt.$$

It follows that the integrand is zero, and therefore the restriction $V|(t_{j-1}, t_j)$ is a Jacobi field. To see what happens at each t_j, choose $T \in \mathcal{V}$ in such a way that

$$T(t_j) = \frac{DV}{dt}(t_j^+) - \frac{DV}{dt}(t_j^-), \quad j = 1, \ldots, k-1.$$

Since

$$0 = I_a(V, T) = -\sum_{j=1}^{k-1} \left\| \frac{DV}{dt}(t_j^+) - \frac{DV}{dt}(t_j^-) \right\|^2,$$

we conclude that V is of class C^1 at each t_j. By the uniqueness of the solution to an ordinary differential equation, V is C^∞. Therefore V is a Jacobi field. □

2.4 COROLLARY. *I_a is degenerate if and only if the points $\gamma(0)$ and $\gamma(a)$ are conjugate along γ. In this case, the nullity of I_a is equal to the multiplicity of $\gamma(a)$ as a conjugate point.*

For the next proposition, as well as for the proof of the Index Theorem, we need some preliminary considerations.

Since each point of M is contained in a totally normal neighborhood and $\gamma([0, a])$ is compact, we can choose a subdivision

$$0 = t_o < t_1 < \cdots < t_{k-1} < t_k = a$$

of $[0, a]$ such that each $\gamma|[t_{j-1}, t_j]$, $j = 1, \ldots, k$, is contained in a totally normal neighborhood. Thus each $\gamma \mid [t_{j-1}, t_j]$ is a minimizing geodesic and doesn't contain any conjugate points. In what follows, such a subdivision will be called *normal* and will be fixed until mention to the contrary.

Let $\mathcal{V}^-(0, a) = \mathcal{V}^-$ be the vector subspace of $\mathcal{V}$ formed from the fields V such that $V|(t_{i-1}, t_i)$, $i = 1, \ldots, k$, is a Jacobi field; $\mathcal{V}^-$ has finite dimension. Let $\mathcal{V}^+$ be the subspace of $\mathcal{V}$ consisting of vector fields W such that $W(t_1) = W(t_2) = \cdots = W(t_{k-1}) = 0$.

2.5 PROPOSITION. *$\mathcal{V}$ is a direct sum $\mathcal{V} = \mathcal{V}^+ \oplus \mathcal{V}^-$, and the subspaces $\mathcal{V}^+$ and $\mathcal{V}^-$ are orthogonal with respect to I_a. In addition, I_a restricted to $\mathcal{V}^+$ is positive definite.*

Proof. Given $V \in \mathcal{V}$, let W be a vector field in $\mathcal{V}^-$ given by $W(t_j) = V(t_j)$; because $\gamma \mid [t_{j-1}, t_j]$ does not have any conjugate points, such a W exists and is unique. Hence $V - W \in \mathcal{V}^+$ and, therefore $\mathcal{V} = \mathcal{V}^+ \oplus \mathcal{V}^-$. In addition, if $X \in \mathcal{V}^-$ and $Y \in \mathcal{V}^+$, we have

$$I_a(X,Y) = -\sum_{j=1}^{k-1} \left\langle 0, \frac{DX}{dt}(t_j^+) - \frac{DX}{dt}(t_j^-) \right\rangle = 0,$$

that is, $\mathcal{V}^+$ and $\mathcal{V}^-$ are orthogonal, relative to I_a. This proves the first part of Proposition 2.5.

Since $\gamma|[t_{j-1}, t_j]$, $j = 1, \ldots, k$, are minimizing geodesics, they have less energy than any other paths between their endpoints. Therefore, if $V \in \mathcal{V}^+$, then $I_a(V,V) \geq 0$.

It remains to show that $I_a(V,V) > 0$ if $V \in \mathcal{V}^+ - \{0\}$. Suppose, to the contrary, that $I_a(V,V) = 0$ with $V \in \mathcal{V}^+$, $V \neq 0$. We are going to show that this implies that V belongs to the null space of I_a. Indeed, if $W \in \mathcal{V}^-$, then $I_a(V,W) = 0$, from the orthogonality above. If $W \in \mathcal{V}^+$, consider the inequality

$$0 \leq I_a(V + cW, V + cW) = 2cI_a(V,W) + c^2 I_a(W,W),$$

valid for all real c. This says that there exist real numbers $A \geq 0$ and B such that $Ac^2 + 2Bc \geq 0$ for all $c \in \mathbf{R}$, which is only possible when $B = 0$, that is, $I_a(V,W) = 0$. Therefore V belongs to the null space of I_a. Since the null space consists of Jacobi fields and V vanishes at t_j, we conclude that $V = 0$, which is a contradiction and ends the proof. □

2.6 COROLLARY. *The index of I_a is equal to the index of I_a restricted to $\mathcal{V}^-$; in particular, the index of I_a is finite. The same is true for the nullity of I_a.*

Proof of the Index Theorem. It is convenient to introduce the following notation. If $t \in [0,a]$, denote by γ_t the restriction of γ to the interval $[0,t]$; the corresponding index form will be denoted by I_t, and the index of I_t will be denoted by $i(t)$. In this manner, we define a function $i\colon [0,a] \to \mathbf{N}$, whose behavior we wish to study.

Recalling that the subdivision (which is supposed fixed) of $[0,a]$ by the points t_j, $j = 0, \ldots, k$, was chosen in a way that $\gamma|[t_{j-1}, t_j]$ is a minimizing geodesic, we conclude that $i(t)$ is zero in a neighborhood of 0. In addition, $i(t)$ is non-decreasing, that is, if $\bar{t} > t$, then $i(\bar{t}) \geq i(t)$. In fact, by the definition of $i(t)$, there exists a subspace $\mathcal{U} \subset \mathcal{V}(0,t)$ such that I_t is negative definite on $\mathcal{U}$ and $\dim\mathcal{U} = i(t)$. Every element $V \in \mathcal{U}$ can be extended to an element $\overline{V} \in \mathcal{V}(0,\overline{t})$ by defining $\overline{V} = 0$ on $[t,\overline{t}]$. It is clear that $I_t(V,V) = I_{\overline{t}}(\overline{V},\overline{V})$. From the definition of the index-, $i(\overline{t}) \geq i(t)$, as has been claimed.

To obtain other properties of $i(t)$ we proceed in the following manner. First, observe that, from the definition, $i(t)$ does not depend on the choice of normal subdivision of $[0,a]$; we can, therefore, choose such subdivisions in a way that $t \in (t_{j-1}, t_j)$. Next, we observe that the index of I_t is the index of the restriction of I_t to the subspace $\mathcal{V}^-(0,t)$; such a restriction will be again denoted by I_t. Then, as each element of $\mathcal{V}^-(0,t)$ is determined by its value at the points $\gamma(t_1),\ldots,\gamma(t_{j-1})$, we have that $\mathcal{V}^-(0,t)$ is isomorphic to a direct sum

$$\mathcal{V}^-(0,t) = T_{\gamma(t_1)}M \oplus \cdots \oplus T_{\gamma(t_{j-1})}M = S_j.$$

It follows, by letting t vary on (t_{j-1}, t_j), that the spaces $\mathcal{V}^-(0,t)$ are isomorphic to each other and isomorphic to S_j. We can, therefore, consider the quadratic forms I_t as a family of quadratic forms on a fixed space S_j. In addition, since the elements of $\mathcal{V}^-(0,t)$ are "broken" Jacobi fields, it follows from (2) that I_t depends continuously on $t \in (t_{j-1}, t_j)$.

We can now obtain more information about $i(t)$ which will be gathered together in the lemmas below.

2.7 LEMMA. *If $\varepsilon > 0$ is sufficiently small, $i(t-\varepsilon) = i(t)$.*

Proof of Lemma 2.7. Since $i(t)$ is non-decreasing, $i(t) \geq i(t-\varepsilon)$, for all ε. On the other hand, if I_t is negative definite on a subspace $\overline{S} \subset S_j$, with $\dim(\overline{S}) = i(t)$, then, by continuity of I_t, there exists $\varepsilon > 0$ such that $I_{t-\varepsilon}$ is still negative definite on $\overline{S}$, hence $i(t-\varepsilon) \geq i(t)$. Therefore $i(t) = i(t-\varepsilon)$. □

Now let d be the nullity of I_t; observe that $d = 0$ if $\gamma(t)$ is not conjugate to $\gamma(0)$.

2.8 LEMMA. *If $\varepsilon > 0$ is sufficiently small, $i(t+\varepsilon) = i(t) + d$.*

Proof of Lemma 2.8. We show first that $i(t+\varepsilon) \leq i(t) + d$. Indeed, because $\dim(S_j) = n(j-1)$, I_t is positive definite on a subspace of dimension $n(j-1) - i(t) - d$. By continuity, $I_{t+\varepsilon}$ is still positive definite on this subspace, for $\varepsilon > 0$ sufficiently small. Therefore

$$i(t+\varepsilon) \leq n(j-1) - \{n(j-1) - i(t) - d\} = i(t) + d.$$

To prove the inequality in the other direction, it is necessary to use the Index Lemma (Lemma 2.2 of Chap. 10). Let $V \in S_j$, with $V(t_{j-1}) \neq 0$, and denote by V_{t_o} the "broken" Jacobi field which coincides with $V(t_i)$ at t_i, $i = 1, \ldots, j-1$, and which vanishes at the point $t_o \in (t_{j-1}, t_j)$. We claim that

$$I_{t_o}(V_{t_o}, V_{t_o}) > I_{t_o+\varepsilon}(V_{t_o+\varepsilon}, V_{t_o+\varepsilon}).$$

In fact, if we denote by W_{t_o} (see Fig. 1) the vector field defined along $\gamma([0, t_o+\varepsilon])$ by:

$$W_{t_o}(t) = V_{t_o}(t), \qquad t \in [0, t_o],$$

$$W_{t_o}(t) = 0, \qquad t \in [t_o, t_o+\varepsilon],$$

we have, from the Index Lemma,

$$I_{t_o}(V_{t_o}, V_{t_o}) = I_{t_o+\varepsilon}(W_{t_o}, W_{t_o}) > I_{t_o+\varepsilon}(V_{t_o+\varepsilon}, V_{t_o+\varepsilon}),$$

where the last inequality is strict, since W_{t_o} is not a Jacobi field. This proves the assertion made.

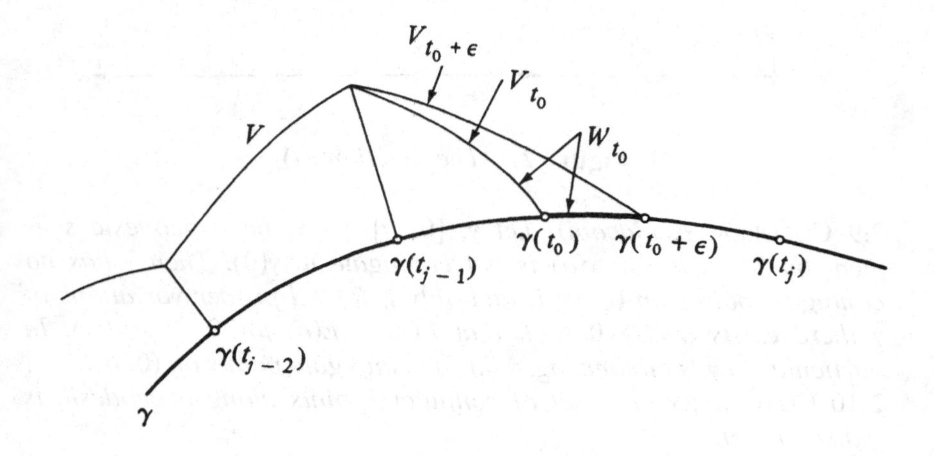

Figure 1

Notice that if $V(t_{j-1}) = 0$, V is either $\equiv 0$ or a "broken" Jacobi field; in both cases, it does not contribute to the nullity d.

Therefore, if $V \in S_j$ and $I_t(V, V) \leq 0$, then $I_{t+\varepsilon}(V, V) < 0$. Hence, if I_t is negative definite on a subspace $\overline{S} \subset S_j$, $I_{t+\varepsilon}$ will still be negative definite on the direct sum of $\overline{S}$ with the null space of I_t. Therefore

$$i(t+\varepsilon) \geq i(t) + d,$$

which, together with the previous inequality implies that $i(t + \varepsilon) = i(t) + d$. □

The information which we obtained about $i(t)$ allows us to describe $i(t)$ as a function which is zero in a neighborhood of the origin, is continuous on the left and has "step" type discontinuities at the conjugate points to $\gamma(0)$, the step being exactly equal to the multiplicity of the conjugate point (see Fig. 2). But this is precisely the statement of the Index Theorem. □

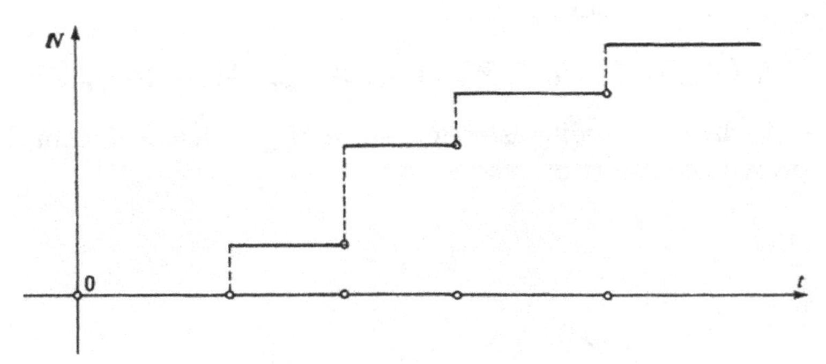

Figure 2. *The function $i(t)$*

2.9 COROLLARY. *(Jacobi). Let γ: $[0, a] \to M$ be a geodesic segment on M such that $\gamma(a)$ is not conjugate to $\gamma(0)$. Then γ has no conjugate points on $(0, a)$ if and only if for all proper variations of γ there exists a $\delta > 0$ such that $E(s) < E(\delta)$ for $0 < |s| < \delta$. In particular, if γ is minimizing, γ has no conjugate points on $(0, a)$.*

2.10 COROLLARY. *The set of conjugate points along a geodesic is a discrete set.*

2.11 REMARK. The index theorem can be generalized to the case in which the points $\gamma(0)$ and $\gamma(a)$ are replaced by submanifolds. For this see W. Ambrose [Am 2]. For an extremely general version of the index theorem, see S. Smale [Sm].

2.12 REMARK. The fact that the index and the nullity of I are finite reflect the fundamental fact that the "infinite dimensional manifold" $\Omega_{p,q} = \Omega$ mentioned in Remark 2.7 of Chapter 9 can be replaced, under certain conditions, by a finite dimensional manifold. This was the original point of view of Morse to prove his theorem. For some applications (cf. the Sphere Theorem, Prop. 3.1), it is convenient to have this construction in mind which can be summarized in the following way. (For more details, see Milnor [Mi], pp. 88–92).

Let Ω^c (resp. $\mathring{\Omega}^c$) be the subset of Ω formed by curves in Ω of energy $\leq c$ (resp. $< c$). It is clear that the curves in Ω^c are contained in a compact subset $S \subset M$. Let $\delta > 0$ be such that given two points of S at a distance less than δ, there exists a unique minimizing geodesic joining these two points. Suppose that the curves in Ω^c are defined on $[0, 1]$ and subdivide $[0, 1]$ by the points t_i, $i = 0, \ldots, k$, in such a way that $|t_i - t_{i-1}| < \frac{\delta^2}{c}$. Let $B \subset \Omega^c$ (resp. $\mathring{B} \subset \mathring{\Omega}^c$) be the set of geodesics broken at t_i, that is, $w \in B$ if $w(0) = p$, $w(1) = q$, and the restriction $w \mid (t_{i-1}, t_i) = w_i$ is a geodesic joining $w(t_{i-1})$ to $w(t_i)$. Such a geodesic is entirely determined by the points $w(t_i)$ since

$$L^2(w_i) = (t_i - t_{i-1})E(w_i) < \delta^2.$$

The correspondence

$$w \to (w(t_1), \ldots, w(t_{k-1})) \in M \times \cdots \times M, \quad w \in \mathring{B},$$

is bijective and takes $\mathring{B}$ into an open subset of $M \times \cdots \times M$. This allows us to introduce a differentiable structure (of finite dimension) on $\mathring{B}$. Observe that the tangent space $T_w(\mathring{B})$ to a broken geodesic w corresponds to the set of Jacobi fields along w, broken at t_i. It is possible to show that there exists a homotopy $h_s\colon \mathring{\Omega}^c \to \mathring{\Omega}^c$, $s \in [0, 1]$ with h_o = ident. and $h_1\colon \mathring{\Omega}^c \to \mathring{B}$, that is, $\mathring{B}$ is a "deformation retract" of $\mathring{\Omega}^c$.

Restricting the energy E to a function $\bar{E}$ on $\mathring{B}$, one verifies that the geodesics γ of $\mathring{\Omega}^c$ are in the manifold $\mathring{B}$ and are precisely the critical points of $\bar{E}$. In addition, the index and the nullity of I at

γ coincide with the index and the nullity, respectively, of the hessian of $\bar{E}$ at γ, since from Corollary 2.6 they coincide with the index and the nullity, respectively, of I restricted to the broken Jacobi fields along γ.

In this way, for many purposes, we can substitute the set $\overset{\circ}{\Omega}{}^c$ by its finite dimensional approximation $\overset{\circ}{B}$.

EXERCISES

1. Prove the following version of the Theorem of Bonnet-Myers: *If M is complete and the sectional curvature K satisfies $K \geq \delta > 0$, then M is compact and* $\operatorname{diam} M \leq \pi/\sqrt{\delta}$, using the Comparison Theorem of Rauch and the Jacobi theorem.
Hint: Comparing M with a sphere of curvature δ, conclude, from Rauch's Theorem, that the first conjugate point to $\gamma(0)$ along a normalized geodesic $\gamma\colon [0,\infty) \to M$ does not occur after $\gamma(\pi/\sqrt{\delta})$. Therefore a geodesic of length larger than $\pi/\sqrt{\delta}$ contains conjugate points. By Jacobi's theorem, such a geodesic does not minimize.

2. Prove the following inequality on real functions (Wirtinger's inequality). Let $f\colon [0,\pi] \to \mathbf{R}$ be a real function of class C^2 such that $f(0) = f(\pi) = 0$. Then
$$\int_0^\pi f^2 dt \leq \int_0^\pi (f')^2 dt,$$
and equality occurs if and only if $f(t) = c\sin t$, where c is a constant. (In the next exercise, we use this fact to prove an interesting geometric fact).
Hint: A geometric solution is the following. Consider a normalized geodesic γ joining the antipodal points p and $-p$ of a unit sphere S^2. Let $v(t)$ be a parallel field along γ, with $\langle v, \gamma'\rangle = 0$, $|v| = 1$. Set $V = fv$ and calculate the second variation for the variational vector V, obtaining
$$I_\pi(V,V) = \int_0^\pi (f')^2 dt - \int_0^\pi f^2 dt.$$

By the Morse index theorem, $I_\pi(V, V) \geq 0$, which establishes the inequality. For the equality, use the fact that $I_\pi(V, V) = 0$ implies that V is a Jacobi field. (For another proof, using Fourier series, see W. Blaschke, Kreis and Kugel, Chelsea, New York, 1949, p. 105).

3. Let M^2 be a complete simply connected 2-dimensional Riemannian manifold. Suppose that for each point $p \in M$, the locus $C(p)$ of (first) conjugate points of p reduces to a unique point $q \neq p$ and that $d(p, C(p)) = \pi$. Prove that, if the sectional curvature K of M satisfies $K \leq 1$, then M is isometric to the sphere S^n with constant curvature 1.
Hint: Let J be a Jacobi field along a normalized geodesic $\gamma: [0, \pi] \to M$ joining p to q with $J(0) = J(\pi) = 0$, $\langle J, \gamma' \rangle = 0$. Choosing fields $e_1, e_2, \ldots, e_{n-1}, \gamma'$ which are parallel and orthonormal along γ, we can write $J = \sum_{i=1}^{n-1} a_i e_i$. Then, letting $K(t) = K(\gamma', J)$, using integration by parts and Exercise 2, we have

$$\begin{aligned} 0 = I_\pi(J,J) &= -\int_0^\pi \langle J'' + R(\gamma', J)\gamma', J\rangle dt \\ &= -\int_0^\pi \left(\sum_i a_i'' a_i\right) dt - \int_0^\pi K(t)\left(\sum_i a_i^2\right) dt \\ &= \int_0^\pi \sum_i (a_i')^2 dt - \int_0^\pi K(t)\left(\sum_i a_i^2\right) dt \\ &\geq \sum_t \int_0^\pi a_i^2 (1 - K(t)) dt \geq 0. \end{aligned}$$

It follows that $K(t) \equiv 1$.

4. Let a: $\mathbf{R} \to \mathbf{R}$ be a differentiable function with $a(t) \geq 0, t \in \mathbf{R}$, and $a(0) > 0$. Prove that the solution to the differential equation

$$\frac{d^2\varphi}{dt^2} + a\varphi = 0$$

with initial conditions $\varphi(0) = 1$, $\varphi'(0) = 0$, has at least one positive zero and one negative zero.

5. Suppose that M^n is a complete Riemannian manifold with sectional curvature strictly positive and let γ: $(-\infty, \infty) \to M$ be

a normalized geodesic in M. Show that there exists $t_o \in \mathbf{R}$ such that the segment $\gamma([-t_o, t_o])$ has index greater or equal to $n-1$.
Hint: Let Y be a parallel field along γ with $\langle \gamma', Y \rangle = 0$, $|Y| = 1$. Set

$$\varphi_Y = \langle R(\gamma', Y)\gamma', Y \rangle,$$
$$K(t) = \inf_Y \varphi_Y(t)$$

and let $a\colon \mathbf{R} \to \mathbf{R}$ be a differentiable function such that

$$0 \leq a(t) \leq K(t), \quad 0 < a(0) < K(0), \quad t \in \mathbf{R}.$$

Let φ be the solution of $\varphi'' + a\varphi = 0$ with $\varphi'(0) = 0$, $\varphi(0) = 1$, and let $-t_1, t_2$ be the two zeros given by Exercise 4. Show that for the field $X = \varphi Y$, the index form satisfies

$$I_{[-t_1,t_2]}(X, X) < -\int_{-t_1}^{t_2} (\varphi'' + a\varphi)\varphi dt = 0.$$

6. A *line* in a complete Riemannian manifold is a geodesic

$$\gamma\colon (-\infty, \infty) \to M$$

which minimizes the arc length between any two of its points. Show that if the sectional curvature K of M is strictly positive, M does not have any lines. By an example show that the theorem is false if $K \geq 0$.

CHAPTER 12

THE FUNDAMENTAL GROUP OF MANIFOLDS OF NEGATIVE CURVATURE

1. Introduction

Let M^n be a complete Riemannian manifold with sectional curvature $K < 0$. A fundamental fact about the topology of M is that the universal covering of M is diffeomorphic to $\mathbf{R}^n$ (Cf. Hadamard's Theorem, Chap. 7). In this chapter, we obtain information about the fundamental group $\pi_1(M)$ of M.

The importance of studying $\pi_1(M)$, when M has negative curvature, comes from the fact that, in a certain sense, all of the topology of M is contained in $\pi_1(M)$. More precisely, in Algebraic Topology one studies certain topological invariants, called the homotopy groups of dimension $k \geq 1$, which generalize the fundamental group (= homotopy group of dimension one). Such groups can be defined, roughly, as homotopy classes of maps, $f: S^k \to M$, of spheres S^k of dimension k into M. It is possible to prove (see M. Greenberg [Gb], p. 32) that if M is covered by $\mathbf{R}^n$, then every such f is homotopic to a constant if $k \geq 2$. This means that the homotopy groups of higher dimension are trivial and that, therefore, at the level of homotopy, the information about the topology of M is contained in $\pi_1(M)$.

The object of this chapter is to prove the Theorem of Preissman which states the following: If M is compact and $K < 0$ then every non-trivial abelian subgroup of $\pi_1(M)$ is infinite cyclic. This shows, for example, that the torus $S^1 \times S^1 \times S^1$ whose fundamental group is $\mathbf{Z} \oplus \mathbf{Z} \oplus \mathbf{Z}$ cannot support a metric of strictly negative curvature.

In Section 2 we introduce some important notions and prove a theorem of E. Cartan on the existence of closed geodesics which does not use any hypothesis on the curvature. In Section 3 we prove the Theorem of Preissman. We prove also that if M is compact and $K < 0$, then $\pi_1(M)$ is not abelian. Finally, we show that, in the statement of Preissman's Theorem, it is possible to replace the condition "abelian subgroup" by the weaker condition "solvable subgroup".

2. Existence of closed geodesics

We denote by $\pi: \tilde{M} \to M$ the universal covering of a complete Riemannian manifold M with the covering metric and denote by $A(\tilde{M})$ the group of covering transformations of $\tilde{M}$ (covering automorphisms). Observe that the elements in $A(\tilde{M})$, different from the identity, are isometries of $\tilde{M}$, which have no fixed points, and that $A(\tilde{M})$ is isomorphic to $\pi_1(M;p)$, $p \in M$ (cf. Massey [Ma]). We should say that this isomorphism depends on the choice of the point $\tilde{p} \in \tilde{M}$, with $\pi(\tilde{p}) = p$, and associates to each $g \in \pi_1(M;p)$ an isometry $\alpha_{\tilde{p}} \in A(\tilde{M})$ defined in the following way: If $\tilde{q} \in \tilde{M}$, join $\tilde{q}$ to $\tilde{p}$ by a path $\tilde{\sigma}$, put $\pi(\tilde{\sigma}) = \sigma$ and $\beta = \sigma g \sigma^{-1}$, where, by abuse of notation, we also denote by g a path in the class g. In what follows, if $\beta: [0,1] \to M$ is a path in M, $\tilde{\beta}_{\tilde{q}}(1)$ will denote the endpoint of the lifting to $\tilde{M}$ of β starting from $\tilde{q}$. Taking $\beta = \sigma g \sigma^{-1}$, we define

$$\alpha_{\tilde{p}}(\tilde{q}) = \tilde{\beta}_{\tilde{q}}(1).$$

2.1 DEFINITION. A set $\mathcal{L}$ of closed paths in M is called a *free homotopy class* if given $f \in \mathcal{L}$ and $g: I \to M$ such that there exists a homotopy

$$F: I \times I \to M, F(0,t) = f(t), F(1,t) = g(t), F(s,0) = F(s,1),$$

then $g \in \mathcal{L}$. The set of such classes will be denoted by $C_1(M)$.

The difference between the definition above and the definition of the fundamental group is that in the free class we allow the origins of the paths to vary in M.

The theorem below shows that in a compact Riemannian manifold M with $\pi_1(M) \neq \{e\}$ there always exists a *closed geodesic*, that is, a closed curve that is geodesic at all of its points. We should distinguish a closed geodesic from a *geodesic lasso* which is

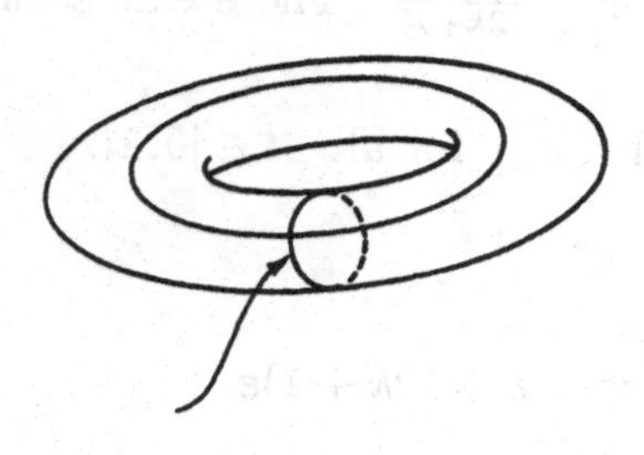
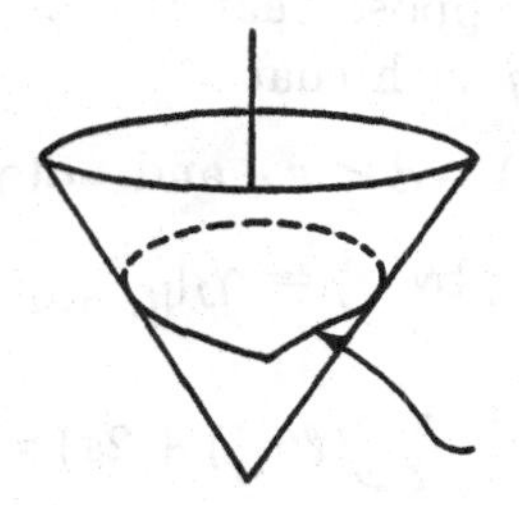

Figure 1

a closed curve that is geodesic at all except one of its points, where it fails to be regular (see Fig 1).

2.2 THEOREM. *(Cartan). If M is compact and $\mathcal{L} \in C_1(M)$ is not the constant class, then there exists a closed geodesic of M in the class $\mathcal{L}$.*

Proof. Let d be the infimum of the lengths of piecewise differentiable curves belonging to $\mathcal{L}$. Since $\mathcal{L}$ is not trivial, $d > 0$. Let γ_j be a sequence of piecewise differentiable curves belonging to $\mathcal{L}$ such that $\ell(\gamma_j) \to d$. We can suppose that γ_j is a broken geodesic defined on the interval $[0,1]$ parametrized proportionally to arc length. Let $L = \sup \ell(\gamma_j)$. Then

$$d(\gamma_j(t_1), \gamma_j(t_2)) \leq \int_{t_1}^{t_2} |\gamma_j'(t)|\, dt \leq L(t_2 - t_1),$$

for all $t_1 \leq t_2 \in [0,1]$. Therefore the set $\{\gamma_j\}$ is equicontinuous. Since M is compact, there exists a subsequence of γ_j, which we denote again by γ_j, which converges uniformly to a continuous closed curve $\gamma_o\colon [0,1] \to M$.

Now let $0 = t_o < t_1 < \cdots < t_k = 1$ be a partition of the interval $[0,1]$ such that $\gamma_o\,|_{[t_{i-1},t_i]}$, $i = 1,\ldots,k$, is contained in a totally normal neighborhood. Let $\gamma\colon [0,1] \to M$ be a piecewise differentiable curve such that $\gamma^i = \gamma|_{[t_{i-1},t_i]}$ is the unique geodesic segment which joins the points $\gamma_o(t_{i-1})$ and $\gamma_o(t_i)$. It is clear that $\gamma \in \mathcal{L}$, hence $\ell(\gamma) \geq d$. We are going to show that $\ell(\gamma) = d$.

Suppose that $\ell(\gamma) > d$ and let $\varepsilon = \frac{\ell(\gamma)-d}{2k+1}$. There exists an integer j such that

$$\ell(\gamma_j) - d < \varepsilon \quad \text{and} \quad d(\gamma_j(t), \gamma_o(t)) < \varepsilon, \quad \text{for all} \quad t \in [0,1].$$

Denoting by $\gamma_j^i = \gamma_j|_{[t_{i-1},t_i]}$, we have

$$\sum_{i=1}^{k}(\ell(\gamma_j^i) + 2\varepsilon) = \ell(\gamma_j) + 2k\varepsilon < d + (2k+1)\varepsilon$$

$$= \ell(\gamma) = \sum_{i=1}^{k} \ell(\gamma^i).$$

Therefore, there exists an integer i, $1 \leq i \leq k$, such that

$$\ell(\gamma_j^i) + 2\varepsilon < \ell(\gamma^i),$$

which contradicts the fact that γ^i is minimizing and proves that $\ell(\gamma) = d$.

We parametrize γ by arc length. Then $\gamma\colon [0,d] \to M$ is a broken geodesic which has minimum length in the class $\mathcal{L}$. We are going to show that γ is regular at the point $p_i = \gamma(t_i)$, for all $i = 0, \ldots, k$.

Suppose to the contrary and let B be a convex ball centered at p_i. Choose points q_1 and q_2 in $\gamma \cap B$ in a way that the geodesic triangle $p_i q_1 q_2$ is homotopic to a point (see Fig. 2). Then the closed curve constituted by the minimizing geodesic $q_1 q_2$ and by the arc of γ between q_1 and q_2 that does not contain p_i is in the class $\mathcal{L}$ and has length smaller than γ, which is a contradiction. □

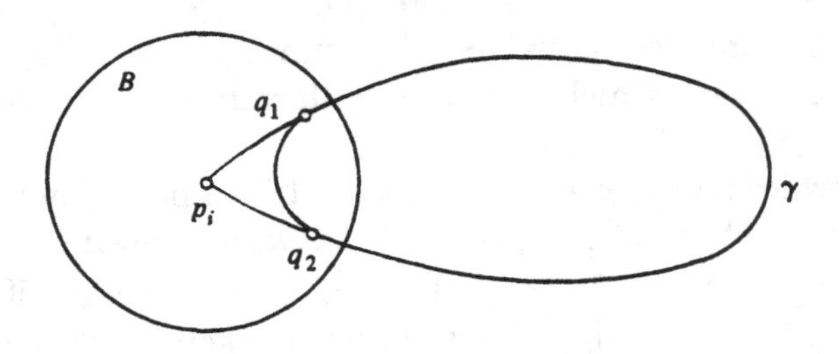

Figure 2

2.3 REMARK. If M is not compact, the theorem is false, as shown by the example of Fig 3. The surface in Fig. 3 represents a surface of revolution generated by a curve which is asymptotic to the axis of revolution. Since there exist curves of arbitrarily small length in the free homotopy classes which are non-trivial, such classes do not contain curves of minimum length.

2.4 REMARK. If M is simply connected (and compact), the existence of a closed geodesic in M, although true, is a more difficult problem. Indeed, the problem of determining the number and the nature of the closed geodesics on a Riemannian manifold is one of the most celebrated chapters in Geometry. A good reference on this topic is Klingenberg [Kℓ 4].

For our purposes, it is important to know when an isometry of $\tilde{M}$ without fixed points leaves invariant a geodesic.

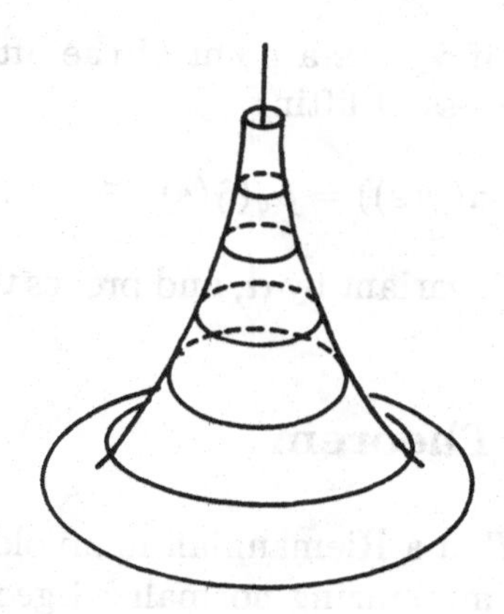

Figure 3. *A surface which has no closed geodesics*

2.5 DEFINITION. An isometry $f: \tilde{M} \to \tilde{M}$ without fixed points is said to be a *translation* of $\tilde{M}$ if it leaves invariant some geodesic $\tilde{\gamma}$ of $\tilde{M}$, that is, if $f(c) = c$, where $c = \tilde{\gamma}((-\infty, \infty))$. In this case, we say that f is a *translation along* $\tilde{\gamma}$.

2.6 PROPOSITION. *Let M be a compact Riemannian manifold and α a covering transformation of $\tilde{M}$, considered with the covering metric. Then α is a translation of $\tilde{M}$.*

Proof. Let $\tilde{p} \in \tilde{M}$ and let $g \in \pi_1(M;p)$, $p = \pi(\tilde{p})$, the element corresponding to α under the isomorphism mentioned in the introduction of this Section. We can assume $\alpha \neq$ ident. By Cartan's Theorem, there exists a closed geodesic γ of M in the free homotopy class determined by g. Choose a point $q \in \gamma$. Then γ is homotopic to the closed path $\sigma g \sigma^{-1}$, where σ is a path joining p to q. Let $\tilde{q} = \tilde{\sigma}_{\tilde{p}}(1)$, that is, $\tilde{q}$ is the endpoint of the lifting of σ starting from $\tilde{p}$. Let $\tilde{\gamma}$ be the lift of γ starting from $\tilde{q}$; we are going to show that α leaves $\tilde{\gamma}$ invariant.

For this, let $\alpha_{\tilde{q}} \in A(\tilde{M})$ be the isometry corresponding to the class $[\gamma] \in \pi_1(M; q)$ and the point $\tilde{q}$, in the isomorphism indicated above. We claim that $\alpha_{\tilde{q}} = \alpha$. In fact, since γ is homotopic to $\sigma g \sigma^{-1}$, its lift starting from $\tilde{q}$ has the same endpoint, that is,

$$\alpha(\tilde{q}) = \alpha_{\tilde{q}}(\tilde{q}).$$

Therefore, $\tilde{q}$ is a fixed point of $\alpha \circ \alpha_{\tilde{q}}^{-1}$, and the claim made follows easily.

It follows that if $\tilde{\gamma}(s)$ is a point of the lift of γ starting from $\tilde{q}$, we have, by uniqueness of lifting,

$$\alpha(\tilde{\gamma}(s)) = \alpha_{\tilde{q}}(\tilde{\gamma}(s)) \in \tilde{\gamma},$$

which shows that $\tilde{\gamma}$ is invariant by α, and proves the Proposition. □

3. Preissman's Theorem

A *geodesic triangle* T in a Riemannian manifold M is a set formed by three segments of minimizing normalized geodesics (called *sides* of the triangle)

$$\gamma_1\colon [0, \ell_1] \to M, \quad \gamma_2\colon [0, \ell_2] \to M, \quad \gamma_3\colon [0, \ell_3] \to M,$$

in such a way that $\gamma_i(\ell_i) = \gamma_{i+1}(0)$, $i = 1, 2$ and $\gamma_3(\ell_3) = \gamma_1(0)$. The endpoints of the geodesic segments are called *vertices* of T. The angle

$$\measuredangle(-\gamma_i'(\ell_i), \gamma_{i+1}'(0)), i = 1, 2,$$

or

$$\measuredangle(-\gamma_3'(\ell_3), \gamma_1'(0)),$$

is called the (interior) *angle* of the corresponding vertex.

3.1 LEMMA. *Let $\tilde{M}$ be a complete simply connected Riemannian manifold, with curvature $K \leq 0$. Let a, b and c be three points of $\tilde{M}$. Such points determine a unique geodesic triangle T in $\tilde{M}$ with vertices a, b, c. Let α, β and γ be the angles of the vertices a, b, c, respectively, and let A, B, C be the lengths of the sides opposite the vertices a, b, c, respectively. Then*

(i) $A^2 + B^2 - 2AB\cos\gamma \leq C^2 \quad (< C^2, \text{ if } K < 0)$

(ii) $\alpha + \beta + \gamma \leq \pi \quad (< \pi, \text{ if } K < 0)$.

Proof. Let γ_A, γ_B, γ_C be the geodesics of lengths $\ell(\gamma_A) = A$, $\ell(\gamma_B) = B$, $\ell(\gamma_C) = C$ which form the sides of T. Let $\Gamma_A = \exp_c^{-1}(\gamma_A)$, $\Gamma_B = \exp_c^{-1}(\gamma_B)$ and $\Gamma_C = \exp_c^{-1}(\gamma_C)$ be curves in $T_c(\tilde{M})$. Since γ_A and γ_B are radial geodesics issuing from the origin c, we have

$$A = \ell(\gamma_A) = \ell(\Gamma_A), \qquad B = \ell(\gamma_B) = \ell(\Gamma_B).$$

In addition, denoting by Γ_o the segment of the straight line in $T_c(\tilde{M})$ that joins the endpoints of Γ_C, we have that $\ell(\Gamma_o) \leq \ell(\Gamma_C)$ (see Fig 4) and

$$\ell(\Gamma_o)^2 = A^2 + B^2 - 2AB\cos\gamma.$$

Since $K \leq 0$ and $T_c(\tilde{M})$ has zero curvature, we can apply Proposition 2.5 of Chapter 10 (application of Rauch's Theorem) and obtain that

$$\ell(\Gamma_C) \leq \ell(\gamma_C) \qquad (<, \text{if } K < 0).$$

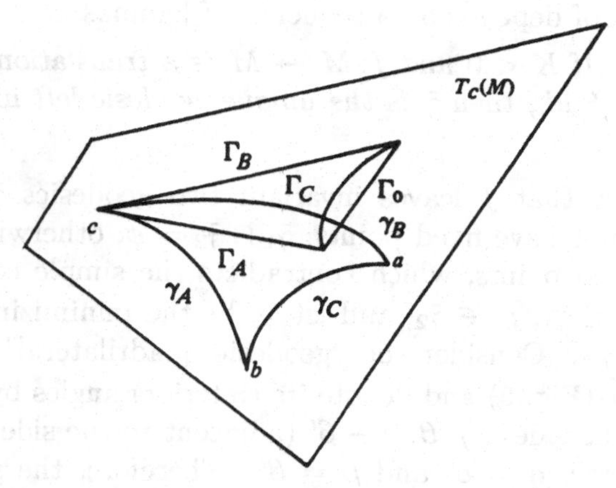

Figure 4

It follows that

$$A^2 + B^2 - 2AB\cos\gamma \le \ell(\Gamma_C)^2 \le \ell(\gamma_C)^2 = C^2 \quad (<, \text{if } K < 0),$$

which proves (i).

To prove (ii), let us observe that

$$C = d(a,b), \qquad B = d(a,c), \quad A = d(b,c)$$

and, therefore, each length A, B or C is bounded by the sum of the other two. We can then consider in the Euclidean space $T_c(\tilde{M})$, a triangle whose sides have lengths A, B and C. Denoting the opposite angles of this triangle by α', β' and γ', respectively, we obtain from (i),

$$\alpha \le \alpha', \qquad \beta \le \beta', \qquad \gamma \le \gamma' \qquad (<, \quad \text{if } K < 0).$$

Since $\alpha' + \beta' + \gamma' = \pi$, (ii) follows. □

From now on, M will denote a complete Riemannian manifold with sectional curvature $K < 0$. As always, $\pi\colon \tilde{M} \to M$ denotes the universal covering of M with the covering metric. Our goal is to prove the following theorem.

3.2 THEOREM. *(Preissman [Pr]) If M is a compact Riemannian manifold with negative curvature, then any abelian subgroup of the fundamental group $\pi_1(M)$, different from the identity, is infinite cyclic.*

The proof depends on a sequence of lemmas.

3.3 LEMMA. *If $K < 0$ and $f\colon \tilde{M} \to \tilde{M}$ is a translation along the geodesic $\tilde{\gamma}$, $f \neq id.$, then $\tilde{\gamma}$ is the unique geodesic left invariant by f.*

Proof. Suppose that f leaves invariant two geodesics $\tilde{\gamma}_1$ and $\tilde{\gamma}_2$. Since f does not have fixed points, $\tilde{\gamma}_1 \cap \tilde{\gamma}_2 = \phi$; otherwise, $\tilde{\gamma}_1 \cap \tilde{\gamma}_2$ has at least two points, which contradicts the simple connectivity of $\tilde{M}$. Let $\tilde{p}_1 \in \tilde{\gamma}_1$, $\tilde{p}_2 \in \tilde{\gamma}_2$, and let $\tilde{\gamma}_3$ be the minimizing geodesic joining $\tilde{p}_1$ to $\tilde{p}_2$. Consider the "geodesic quadrilateral" $\tilde{p}_1$, $f(\tilde{p}_1)$, $\tilde{p}_2$,$f(\tilde{p}_2)$ in $\tilde{M}$ (Fig. 5) and denote its (interior) angles by α, $\pi - \alpha'$ (adjacent to the side $\tilde{\gamma}_1$), β, $\pi - \beta'$ (adjacent to the side $\tilde{\gamma}_2$). Since f is an isometry, $\alpha = \alpha'$ and $\beta = \beta'$. Therefore, the sum of the interior angles of such a quadrilateral is equal to 2π. Now divide this quadrilateral into two triangles T_1 and T_2 and denote by $\sum_i$ the sum of the interior angles of the triangle T_i, $i = 1, 2$. It is clear that, at each vertex, the sum of the angles of the triangles which meet there is greater than or equal to the angle of the quadrilateral at this vertex. Therefore $\sum_1 + \sum_2 \geq 2\pi$. It follows that one of the two triangles has its sum of interior angles $\geq \pi$, which contradicts Lemma 3.1 (ii). □

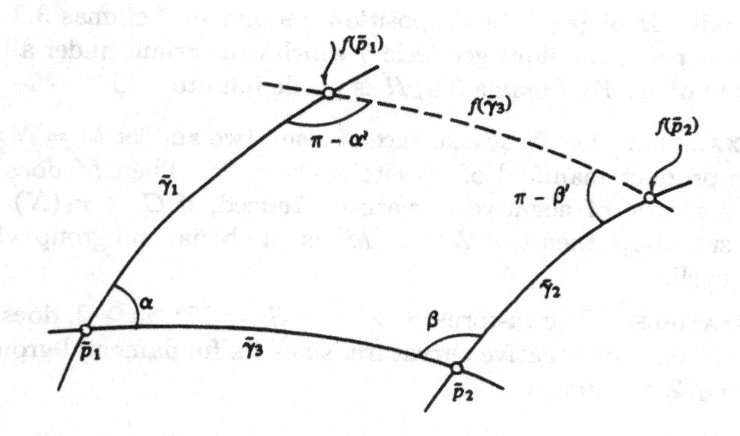

Figure 5

3.4 LEMMA. *If $K < 0$ and $g: \tilde{M} \to \tilde{M}$ is an isometry without fixed points which commutes with a translation f along $\tilde{\gamma}$, $f \neq id.$, then g is a translation along $\tilde{\gamma}$.*

Proof. It suffices to observe that

$$f \circ g(\tilde{\gamma}) = g \circ f(\tilde{\gamma}) = g(\tilde{\gamma}),$$

hence, by uniqueness of the previous lemma $g(\tilde{\gamma}) = \tilde{\gamma}$. □

3.5 LEMMA. *If all the elements of a non-trivial subgroup $H \subset \pi_1(M)$, considered as isometries of $\tilde{M}$, leave invariant a fixed geodesic $\tilde{\gamma}$, then H is infinite cyclic.*

Proof. Fix a point $\tilde{p} \in \tilde{\gamma}$ as origin and consider the map $\theta: H \to \mathbf{R}$ given by $\theta(h) = \pm d(\tilde{p}, h(\tilde{p}))$, where the sign $-$ or $+$ is used acordingly as $h(\tilde{p})$ is "before" or "after" $\tilde{p}$, respectively, in the orientation of $\tilde{\gamma}$. Since the elements of H are isometries which leave $\tilde{\gamma}$ invariant, θ is a homomorphism of H into the additive group of the reals $\mathbf{R}$. θ is an injective homomorphism; otherwise, $h_1(\tilde{p}) = h_2(\tilde{p})$, with $h_1 \neq h_2$, hence $\tilde{p}$ would be a fixed point of $h_1 h_2^{-1}$. Therefore, H is a additive subgroup of $\mathbf{R}$. But any additive subgroup of $\mathbf{R}$ is either dense in $\mathbf{R}$ or infinite cyclic. Since the isometries of H operate in a totally discontinuous way, H is not dense. Hence H is infinite cyclic. □

Proof of Preissman's Theorem. Let $H \subset \pi_1(M)$ be an abelian subgroup with $H \neq \{e\}$. By Proposition 2.9 and by Lemmas 3.3 and 3.4, there exists a unique geodesic $\tilde{\gamma}$ which is invariant under all the elements of H. By Lemma 3.5, H is cyclic infinite. □

3.6 EXAMPLE. Let N be a surface of genus two and let $M = N \times S^1$ be the product manifold of N with a circle S^1. Then M does not carry a metric of negative curvature. Indeed, if $C \subset \pi_1(N)$ is a cyclic subgroup, then $C \oplus \mathbf{Z} \subset \pi_1(M)$ is an abelian subgroup which is not cyclic.

3.7 EXAMPLE. The m-torus $S^1 \times \cdots \times S^1 = T^m$, $m \geq 2$, does not carry a metric of negative curvature, since its fundamental group is $\mathbf{Z} \oplus \cdots \oplus \mathbf{Z}$ (m times).

The same technique used in Preissman's Theorem allows us to obtain a slightly more general theorem. To that end, we need another theorem on the fundamental group, also due to Preissman.

3.8 THEOREM. *If M is compact and $K < 0$, then $\pi_1(M)$ is not abelian.*

Proof. To prove what is needed, we establish the lemma below which is a little more general.

3.9 LEMMA. *If M is complete, $K \leq 0$, and there exists a geodesic invariant under the elements of $A(\tilde{M})$, then M is not compact.*

Proof of the Lemma. Let $\tilde{\gamma}$ be the geodesic invariant under the elements of $A(\tilde{M})$. Fix a point $\tilde{p} \in \tilde{\gamma}$, a real number $t > 0$, and consider the normalized geodesic $\tilde{\beta}\colon [0, t] \to \tilde{M}$, $\tilde{\beta}(0) = \tilde{p}$, perpendicular to $\tilde{\gamma}$ at $\tilde{p}$. Let $\beta = \pi \circ \tilde{\beta}$, $\gamma = \pi \circ \tilde{\gamma}$, $p = \pi(\tilde{p})$, and α_t be a minimizing geodesic in M joining $\beta(t)$ to p. We are going to show that the length $\ell(\alpha_t) = t$.

Let $\tilde{\alpha}_t$ be the lift of α_t starting from $\tilde{\beta}(t)$. Since $\tilde{\gamma}$ is invariant, the endpoint of $\tilde{\alpha}_t$ belongs to $\tilde{\gamma}$. (see Fig. 6). Since $K \leq 0$, by

Lemma 3.1 (i), $\ell(\tilde{\alpha}_t) \geq \ell(\tilde{\beta})$. On the other hand,

$$\ell(\tilde{\alpha}_t) = \ell(\alpha_t) \leq \ell(\beta) = \ell(\tilde{\beta}) = t,$$

hence $\ell(\alpha_t) = t$, as was claimed.

Since t is arbitrary, M is not bounded, which concludes the proof of Lemma 3.9. □

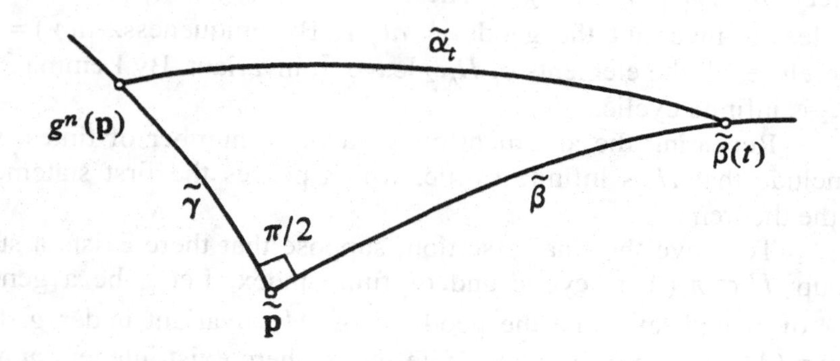

Figure 6

The theorem follows now from the fact that if M is compact, $K < 0$ and $\pi_1(M)$ is abelian, there exists a geodesic invariant under all the elements of $A(\tilde{M})$. □

We can now refine Preissman's Theorem and prove the following fact.

3.10 THEOREM. *(Byers, [By]). If M is compact, $K < 0$, and H is a solvable subgroup of $\pi_1(M)$, $H \neq \{e\}$, then H is infinite cyclic. In addition, $\pi_1(M)$ does not have a cyclic subgroup of finite index.*

Proof. Since H is solvable, there exists a finite sequence of subgroups

$$H = H_o \supset H_1 \supset \cdots \supset H_{k-1} \supset H_k = \{e\}$$

such that H_{i+1} is normal in H_i and H_i/H_{i+1} is abelian. Then H_{k-1} is abelian, hence, by Preissman's Theorem, infinite cyclic.

Let $g \in \pi_1(M)$ be a generator of H_{k-1} and let $\tilde{\gamma}$ be the geodesic in $\tilde{M}$ invariant under g. Let $a \in H_{k-2}$ and $b \in H_{k-1}$. Since $a^{-1}b^{-1}ab \in H_{k-1}$, we have, for some integer n,

$$a^{-1}b^{-1}ab = g^n.$$

It follows that

$$a^{-1}b^{-1}ab(\tilde{g}) = g^n(\tilde{\gamma}) = \tilde{\gamma}.$$

Since $b \in H_{k-1}$, $b(\tilde{\gamma}) = \tilde{\gamma}$. Therefore $b^{-1}a(\tilde{\gamma}) = a(\tilde{\gamma})$, that is, b^{-1} leaves invariant the geodesic $a(\tilde{\gamma})$. By uniqueness, $a(\tilde{\gamma}) = \tilde{\gamma}$. Therefore, all the elements of H_{k-2} leave $\tilde{\gamma}$ invariant. By Lemma 3.5, H_{k-2} is infinite cyclic.

Repeating the argument above a finite number of times, we conclude that H is infinite cyclic, which proves the first statement of the theorem.

To prove the final assertion, suppose that there exists a subgroup $H \subset \pi_1(M)$, cyclic and of finite index. Let g be a generator of H and let $\tilde{\gamma}$ be the geodesic of $\tilde{M}$ invariant under g. Let $a \in \pi_1(M) - H$. Since H has finite index, there exist integers m and n such that $a^n = g^m$. Therefore

$$a^n(\tilde{\gamma}) = g^m(\tilde{\gamma}) = \tilde{\gamma}.$$

By uniqueness, $a(\tilde{\gamma}) = \tilde{\gamma}$, for all $a \in \pi_1(m) - H$. It follows that every element of $\pi_1(M)$ leaves invariant the geodesic $\tilde{\gamma}$. By Lemma 3.5, $\pi_1(M)$ is infinite cyclic, which contradicts Theorem 3.8. □

3.11 REMARK. Riemannian manifolds of non-positive curvature form a substantial topic in Riemannian Geometry, which we have barely touched. To the reader, interested in some of the recent developments, we recommend the excellent survey by P. Eberlein, "Structure of manifolds of nonpositive curvature", in Global Differential Geometry and Global Analysis 1984, Proceedings, Berlin, edited by D. Ferus, R. Gardner, S. Helgason and U. Simon, Lecture Notes in Math. 1156, Springer Verlag, 1985, pp. 86–153. Other relationships between the geometry of a manifold M of non-positive curvature and $\pi_1(M)$ are described in Section 7 of this survey.

CHAPTER 13

THE SPHERE THEOREM

1. Introduction

One of the most beautiful theorems of Global Differential Geometry is the Sphere Theorem which states the following:

1.1 THEOREM. *Let M^n be a compact simply connected, Riemannian manifold, whose sectional curvature K satisfies*

$$0 < hK_{\max} < K \leq K_{\max}. \tag{1}$$

Then if $h = 1/4$, M is homeomorphic to a sphere.

The number h is called the "pinching" of M. Multiplying the metric by a constant, we can suppose that $K_{\max} = 1$, and (1) can be written, without loss of generality, as

$$0 < h < K \leq 1. \tag{1'}$$

The Sphere Theorem was proved for the first time by Rauch [R 1] for $h \sim 3/4$. A fundamental contribution was made by Klingenberg [Kℓ 1] who introduced into the problem the consideration of the "cut locus" (see the definition in Section 2). In the case for which the dimension of M is even, Klingenberg obtained in [Kℓ 1] an estimate for the distance from a point to its "cut locus" and he proved the theorem for $h \sim 0.55$. Using the theorem of Toponogov and the estimate mentioned above, Berger [Br 1] obtained the theorem, still in even dimension, with $h = 1/4$. The use of Toponogov's theorem is unnecessary, as was shown by Tsukamoto [Ts]. Finally, Klingenberg [Kℓ 2] extended his estimate from even dimension to odd dimension, which together with the work of Berger [Br 1] yielded the theorem as stated above.

In the case of even dimension, the theorem is false if we replace (1′) by

$$0 < 1/4 \leq K \leq 1. \tag{2}$$

It is possible to show (see Exercise 12, Chap. 8) that the complex projective space $P^n(\mathbf{C})$, $n > 1$, is simply connected, has a metric whose sectional curvature satisfies (2) and is not homeomorphic to a sphere. As a matter of fact, in [Br 1] Berger proved that if (2) is satisfied then: either diam(M) $> \pi$, and M is homeomorphic to a sphere, or diam(M) $= \pi$, and M is isometric to a symmetric space. (For more details, see Cheeger and Ebin [CE].)

In the odd dimensional case, it is known that the theorem is still true substituting (1′) by (2), but it is not known whether the result can be improved.

For dimensions two and three, the theorem is valid supposing only that $h \geq 0$, that is, if M^n, $n = 2, 3$, is compact simply connected and has positive sectional curvature, then M^n is homeomorphic to S^n. For $n = 2$ this is an immediate consequence of the Gauss-Bonnet Theorem, and for $n = 3$, it follows from a theorem of Hamilton [H].

In this chapter we present an exposition of the Sphere Theorem not using the theorem of Toponogov. In Section 2 we introduce the notion of the "cut locus" and study some of its properties. In Section 3 we obtain an estimate for the distance from a point to its "cut locus". At a crucial point of the proof of this estimate it is necessary to use some basic facts from Morse Theory. We state explicitly the necessary facts, with references. For even dimension, the proof of this estimate can be made without the use of Morse Theory. In order to present the sphere theorem, at least in even dimensions, independently of Morse Theory, we shall give that proof here. In Section 4 we prove the sphere theorem and in Section 5 we mention some recent developments.

Unless otherwise stated, the manifolds considered in this chapter will be complete and the geodesics will be normalized.

2. The cut locus

Let M be a complete Riemannian manifold, let $p \in M$ be a point of M, and let $\gamma\colon [0,\infty) \to M$ be a normalized geodesic with $\gamma(0) = p$. We know that if $t > 0$ is sufficiently small, $d(\gamma(0), \gamma(t)) = t$, that is, $\gamma([0,t])$ is a minimizing geodesic. In addition, if $\gamma([0,t_1])$ is not minimizing, the same is true for all $t > t_1$. By continuity, the set of numbers $t > 0$ for which $d(\gamma(0), \gamma(t)) = t$ is of the form $[0, t_0]$ or $[0, \infty)$. In the first case, $\gamma(t_0)$ is called *the cut point of p along γ*; in the second case, we say that such a cut point does not exist.

We define the *cut locus of p* denoted by $C_m(p)$, as the union of the cut points of p along all of the geodesics that start from p.

Observe that if M is compact, its diameter is finite, hence there exists a cut point for any point $p \in M$ along any geodesic starting from p. The converse of this fact is true and will be proved a little further along (see Cor. 2.11).

2.1 Remark. The cut locus, for surfaces, was introduced under the name of "ligne de partage" by H. Poincaré in 1905 [Po]. For Riemannian manifolds, the cut locus was introduced by J. H. C. Whitehead in 1935, [Wh], in which he proved some of the properties that we present in this section (see pp. 700–704 of [Wh]). The interest in the cut locus was revived in 1959 when Klingenberg showed in [Kℓ 1] that its introduction is useful in improving the "pinching" of the sphere theorem.

Before presenting some examples, it is convenient to prove a fundamental property of the cut locus.

2.2 Proposition. *Suppose that $\gamma(t_0)$ is the cut point of $p = \gamma(0)$ along γ. Then:*

a) either $\gamma(t_0)$ is the first conjugate point of $\gamma(0)$ along γ,

b) or there exists a geodesic $\sigma \neq \gamma$ from p to $\gamma(t_0)$ such that $\ell(\sigma) = \ell(\gamma)$.

Conversely, if (a) or (b) is satisfied, then there exists $\tilde{t}$ in $(0, t_0]$ such that $\gamma(\tilde{t})$ is the cut point of p along γ.

Proof. Let t_o satisfy the condition asserted, and let $\{t_o + \varepsilon_i\}$ be a sequence, in which $\varepsilon_i > 0$ and $\{\varepsilon_i\} \to 0$. Consider a sequence of minimizing geodesics σ_i joining p to $\gamma(t_o + \varepsilon_i)$ and let $\{\sigma_i'(0)\} \in T_pM$ be its corresponding sequence of tangent vectors at p. Since the sphere in T_pM is compact, we can suppose, taking a subsequence, if

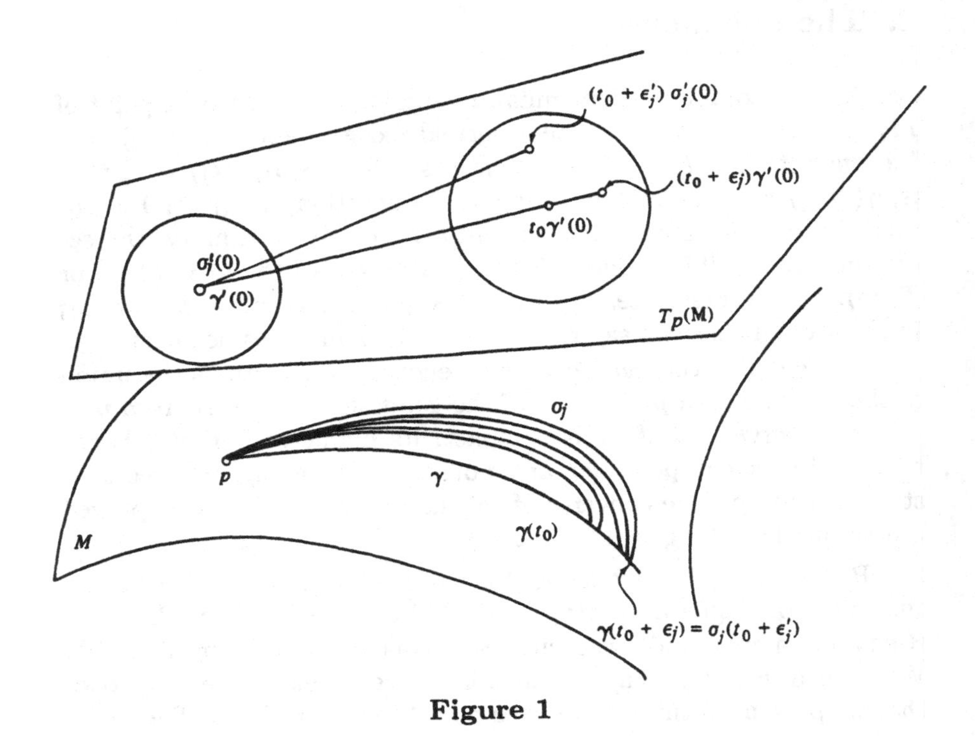

Figure 1

necessary, that $\{\sigma_i'(0)\}$ converges. Therefore there exists a geodesic σ such that $\sigma_i'(0) \to \sigma'(0)$. By continuity, σ is a minimizing geodesic joining p to $\gamma(t_o)$, that is, $\ell(\sigma) = \ell(\gamma)$. If $\sigma \neq \gamma$, the assertion (b) is verified. If $\sigma = \gamma$, we are going to show that (a) holds. Since γ is minimizing up to $\gamma(t_0)$, it suffices to show that $d\exp_p$ is singular at $t_o\gamma'(0)$.

Suppose therefore that $\sigma'(0) = \gamma'(0)$ and that $d\exp_p$ is not singular at $t_o\gamma'(0)$. Then there exists a neighborhood U of $t_o\gamma'(0)$ where $\exp_p$ is a diffeomorphism (Fig. 1). By definition of σ_j, $\gamma(t_o + \varepsilon_j) = \sigma_j(t_o + \varepsilon_j')$, where $\varepsilon_j' \leq \varepsilon_j$, because σ_j is minimizing. Take ε_j sufficiently small so that $(t_o + \varepsilon_j')\sigma_j'(0)$ and $(t_o + \varepsilon_j)\gamma'(0)$ belong to U. Then

$$\begin{aligned}\exp_p(t_o + \varepsilon_j)\gamma'(0) &= \gamma(t_o + \varepsilon_j)\\ &= \sigma_j(t_o + \varepsilon_j') = \exp_p(t_o + \varepsilon_j')\sigma_j'(0),\end{aligned}$$

hence $(t_o + \varepsilon_j)\gamma'(0) = (t_o + \varepsilon_j')\sigma_j'(0)$, that is, $\gamma'(0) = \sigma_j'(0)$, which contradicts the definition of t_o, and concludes the proof of the first part of the proposition.

Conversely, if (a) holds, since a geodesic does not minimize distance after the first conjugate point, (see Corollary 2.9 to the Index Theorem, Chap. 11), the cut point of p along γ occurs at $\gamma(\tilde{t})$, $\tilde{t} \leq t_o$.

Suppose now that (b) holds. Let $\varepsilon > 0$ be sufficiently small so that $\sigma(t_o - \varepsilon)$ and $\gamma(t_o + \varepsilon)$ are both contained in a totally normal neighborhood of $\gamma(t_o)$, and consider the unique minimizing geodesic τ which joins $\sigma(t_o - \varepsilon)$ to $\gamma(t_o + \varepsilon)$ (Fig. 2). The curve $p\sigma(t_o - \varepsilon) \cup \tau$ has arc length strictly less than $t_o + \varepsilon$. Therefore, the cut point of p along γ occurs at $\gamma(\tilde{t})$, $\tilde{t} \leq t_o$. □

Now we are in a position to present some examples.

2.3 EXAMPLE. If M^n is a sphere S^n, and $p \in S^n$, the cut locus of p is its antipodal point. Observe that, in this case, the cut locus of p coincides with the locus of conjugate points to p.

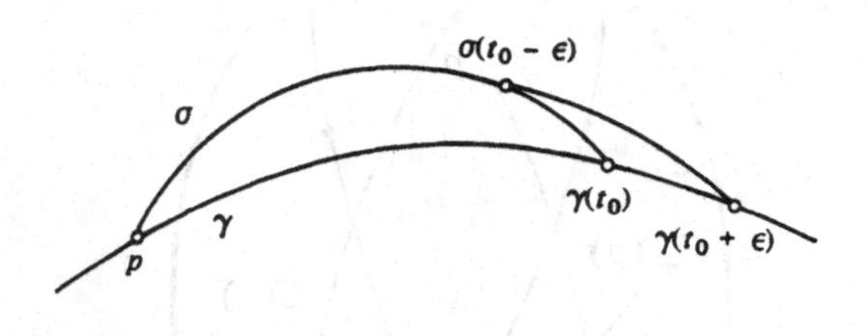

Figure 2

2.4 EXAMPLE. If M^n is the real projective space $P^n(\mathbf{R})$, obtained by identifying the antipodal points of S^n, the cut locus of $[p, -p] \in P^n(\mathbf{R})$ is the subset $P^{n-1}(\mathbf{R}) \subset P^n(\mathbf{R})$ obtained by identifying the antipodal points of the "equator" of p. Observe that the conjugate locus of $[p, -p]$ is $[p, -p]$ itself.

2.5 EXAMPLE. Consider the flat torus T obtained by identifying the oppposite sides of the quadrilateral $ABCD$ of Fig. 3. It is easy to verify that the cut locus of the point A is the subset formed by the medians of the segments BA and BC. Observe that the conjugate locus is empty. Similarly, the cut locus of a point p of a cylinder in R^3 is the "opposite generator" to that which passes through p.

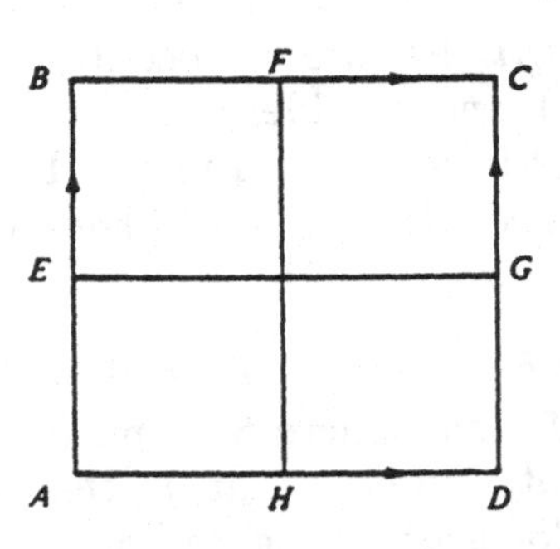

Figure 3

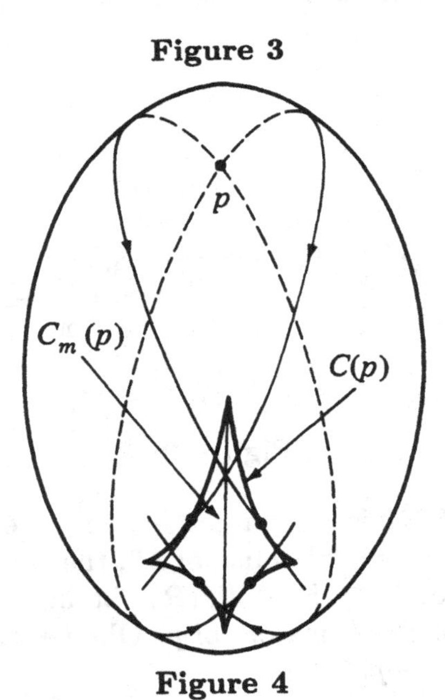

Figure 4

2.6 REMARK. In Example 2.3 the situations (a) and (b) of Proposition 2.2 occur simultaneously. In Example 2.4 the situation (b) occurs before the situation (a). In Example 2.5, only the situation

(b) occurs. In the example of the ellipsoid of Figure 4, except for two directions, the situation (b) occurs before the occurence of (a).

2.6′ REMARK. A compact Riemannian manifold M for which the cut locus $C_m(p)$ of any point $p \in M$ reduces to a unique point is called a *wiedersehen manifold*. There is a theorem due to L. Green [Gn] to the effect that wiedersehen surfaces ($\dim M = 2$) are isometric to the sphere. This theorem was generalized, recently, by M. Berger and J. Kazdan [Be, pp. 236–246] to the case in which $\dim M$ is even, and by C.T. Yang [Yg] to the case in which $\dim M$ is odd. For an exposition of these questions, see Kazdan [Ka].

2.7. Corollary to Proposition 2.2. *If q is a cut point of p along γ, then p is the cut point of q along $-\gamma$; in particular, $q \in C_m(p)$ if and only if $p \in C_m(q)$.*

Proof. If q is the cut point of p along the geodesic γ, then, by Proposition 2.2, either q is conjugate to p, or there exists a geodesic $\sigma \neq \gamma$, joining p to q, such that $\ell(\sigma) = \ell(\gamma) = d(p,q)$. In both cases, the cut point of q along $-\gamma$ does not occur before p. Because $\ell(-\gamma) = d(p,q)$, we conclude that p is the cut point of q along $-\gamma$. □

2.8 Corollary to Proposition 2.2. *If $q \in M - C_m(p)$, there exists a unique minimizing geodesic joining p to q.*

The Corollary 2.8 shows that $\exp_p$ is injective on an open ball $B_r(p)$ centered at p if and only if the radius r is less than or equal to the distance from p to $C_m(p)$. For this reason, we usually call

$$i(M) = \inf_{p \in M} d(p, C_m(p))$$

the *injectivity radius of M*.

Corollary 2.8 shows also that $M - C_m(p)$ is homeomorphic to an open ball of Euclidean space. In a certain sense, this indicates that the topology of M is contained in its cut locus.

The proposition below shows that the distance from a point $p \in M$ to its cut point along γ depends continuously on the initial direction of γ. In fact, the proposition is a little more general. Let T_1M be the unit tangent bundle of M and define a function

$f: T_1 M \to R \cup \{\infty\}$ by:

$$f(\gamma(0), \gamma'(0)) = \begin{cases} t_o, & \text{if } \gamma(t_o) \text{ is the cut point of } \gamma(0) \\ & \text{along } \gamma, \\ \infty, & \text{if the cut point along } \gamma \\ & \text{does not exist.} \end{cases}$$

Introduce on $R \cup \{\infty\}$ the topology whose base of open sets is given by joining the open intervals $(a, b) \subset R$ with the subsets of the form $(a, \infty] = (a, \infty) \cup \{\infty\}$. Observe that the set $[a, \infty]$ is compact in this topology, and that a sequence $\{t_n\} \to \infty$ in this topology when $\lim_{n \to \infty} t_n = \infty$ in the usual sense.

2.9 PROPOSITION. *The function f, defined above, is continuous.*

Proof. Consider sequences $\gamma_i(0) \to \gamma(0)$ and $\gamma_i'(0) \to \gamma'(0)$, and let $\gamma_i(t_o^i)$ and $\gamma(t_o)$ be the cut points of $\gamma_i(0)$ and $\gamma(0)$ along γ_i and γ, respectively, where $t_o^i, t_o \in R \cup \{\infty\}$. We have to prove that $\lim t_o^i = t_o$.

First let us prove that $\limsup t_o^i \leq t_o$. If $t_o = \infty$, there is nothing to prove. Let then $t_o < \infty$ and let $\varepsilon > 0$. Observe that there do not exist infinitely many indices j such that $t_o + \varepsilon < t_o^j$. Otherwise,

$$d(\gamma_j(0), \gamma_j(t_o + \varepsilon)) = t_o + \varepsilon$$

and, by continuity of d, $d(\gamma(0), \gamma(t_o + \varepsilon)) = t_o + \varepsilon$, which contradicts the fact that $\gamma(t_o)$ is a cut point of $\gamma(0)$ along γ. Therefore $\limsup t_o^i \leq t_o + \varepsilon$, for all $\varepsilon > 0$, which proves the claim made.

Now let $\bar{t} = \liminf t_o^i$. Since

$$\bar{t} = \liminf t_o^i \leq \limsup t_o^i \leq t_o,$$

it suffices to show that $\bar{t} \geq t_o$ to complete the proof.

For this, suppose $\bar{t} < \infty$ (if $\bar{t} = \infty$ nothing need be proved), and consider a subsequence of the sequence t_o^j, again denoted by t_o^j, which converges to $\bar{t}$. Because an accumulation point of conjugate points is a conjugate point, if for any such subsequence, the points $\gamma_j(t_o^j)$ are conjugate to $\gamma_j(0)$ along γ_j, then $\gamma(\bar{t})$ is conjugate to $\gamma(0)$ along γ, hence $\bar{t} \geq t_o$, and we have concluded the proof.

Suppose, therefore, that there exists a subsequence $t_o^j \to \bar{t}$ in which, for all j, $\gamma_j(t_o^j)$ is not conjugate to $\gamma_j(0)$ along γ_j. By Proposition 2.2, there exist geodesics $\sigma_j \neq \gamma_j$ with

$$\sigma_j(0) = \gamma_j(0), \quad \sigma_j(t_o^j) = \gamma_j(t_o^j), \quad \ell(\sigma_j) = \ell(\gamma_j).$$

Taking, if necessary, a subsequence, we can suppose that $\sigma_j \to \sigma$, where σ is a geodesic joining $\gamma(0)$ to $\gamma(\bar{t})$. If $\sigma \neq \gamma$, by Proposition 2.2, we have $t_o \leq \bar{t}$, as required. If $\sigma = \gamma$, then, by a similar argument as in Proposition 2.2, we show that $\gamma(\bar{t})$ is conjugate to $\gamma(0)$, hence $\bar{t} \geq t_o$. □

2.10 COROLLARY. *For all $p \in M$, $C_m(p)$ is closed; in particular, if M is compact, $C_m(p)$ is compact.*

Proof. It is easy to verify that

$$C_m(p) = \{\gamma(t); t = f(p, \gamma'(0))\},$$

where $\gamma\colon [0, \infty) \to M$ is a normalized geodesic with $\gamma(0) = p$ and f is the function of Proposition 2.9. Therefore, if q is an accumulation point of $C_m(p)$, there exists a sequence $\gamma_j(t_j)$, with $t_j = f(p, \gamma_j'(0))$, such that $\lim \gamma_j(t_j) = q$. Passing to a subsequence, if necessary, the sequence $\gamma_j'(0)$ converges to a unit vector v. Let γ be the geodesic with $\gamma(0) = p$, $\gamma'(0) = v$. Then, since f is continuous,

$$\begin{aligned} q = \lim \gamma_j(t_j) &= \lim \gamma_j(f(p, \gamma_j'(0))) \\ &= \lim \exp_p(f(p, \gamma_j'(0))\gamma_j'(0)) = \exp_p(f(p, \gamma'(0))\gamma'(0)) \\ &= \gamma(f(p, \gamma'(0))) \in C_m(p), \end{aligned}$$

which shows that $C_m(p)$ is closed. □

2.11 COROLLARY. *Suppose that M is complete and that there exists $p \in M$ which has a cut point for every geodesic starting from p. Then M is compact.*

Proof. Observe that

$$M = \cup \{\gamma(t); t \leq f(p, \gamma'(0))\}$$

for all geodesics γ of M with $\gamma(0) = p$. The hypothesis of the corollary and the continuity of f imply that f is bounded. Therefore M is bounded and the corollary follows from the Hopf-Rinow Theorem.

2.12 PROPOSITION. *Let $p \in M$. Suppose that there exists a point $q \in C_m(p)$ which realizes the distance from p to $C_m(p)$. Then:*

a) *either there exists a minimizing geodesic γ from p to q along which q is conjugate to p,*

b) *or there exist exactly two minimizing geodesics γ and σ from p to q; in addition, $\gamma'(\ell) = -\sigma'(\ell)$, $\ell = d(p,q)$.*

Proof. Let γ be a minimizing geodesic joining p to q. By Proposition 2.2, either q is conjugate to p along γ, and (a) holds, or there exists another minimizing geodesic $\sigma \neq \gamma$, joining p to q with $\ell(\sigma) = \ell(\gamma)$. Suppose then that q is not conjugate to p along γ and σ, and that $\gamma'(\ell) \neq -\sigma'(\ell)$, from which we are going to derive a contradiction. In particular, this shows that, in this situation, there can be only two such geodesics.

Because $\gamma'(\ell) \neq -\sigma'(\ell)$, there exists $V \in T_q(M)$ such that

$$\langle V, \gamma'(\ell)\rangle < 0, \qquad \langle V, \sigma'(\ell)\rangle < 0.$$

Let $\tau\colon (-\varepsilon, \varepsilon) \to M$ be a curve with $\tau(0) = q$, $\tau'(0) = V$. Since q is not conjugate to p along γ, there exists a neighborhood $U \subset T_pM$ of $\ell\gamma'(0)$ where $\exp_p$ is a diffeomorphism. Let $v\colon (-\varepsilon, \varepsilon) \to U$ be a curve such that $\exp_p v(s) = \tau(s)$, $s \in (-\varepsilon, \varepsilon)$, and let $\gamma_s(t) = \exp_p \frac{t}{\ell} v(s)$, $t \in [0, \ell]$, be a variation of γ (see Fig. 5). By the formula for the first variation,

$$\frac{d}{ds}\ell(\gamma_s)\bigg|_{s=0} = \langle V, \gamma'(\ell)\rangle < 0.$$

Because q is not conjugate to p along σ, we obtain in an analogous manner a variation $\sigma_s(t)$ of σ such that:

$$\frac{d}{ds}(\ell(\sigma_s))\bigg|_{s=0} = \langle V, \sigma'(\ell)\rangle < 0,$$

$$\sigma_s(\ell) = \tau(s).$$

Therefore, if $s > 0$ is sufficiently small, $\ell(\gamma_s) < \ell(\gamma)$ and $\ell(\sigma_s) < \ell(\sigma)$.

If $\ell(\gamma_s) = \ell(\sigma_s)$, from Proposition 2.2, $\tau(s) = \gamma_s(\ell)$ is a cut point of p. Since

$$d(p, \gamma_s(\ell)) = \ell(\gamma_s) < d(p, C_m(p)),$$

this contradicts the fact that q realizes the distance from p to $C_m(p)$. If $\ell(\gamma_s) < \ell(\sigma_s)$, σ_s is not minimizing, hence there exists a cut point $\sigma_s(\tilde{t})$, $\tilde{t} < \ell$, of p along σ_s. But this contradicts again the fact that q realizes the distance from p to $C_m(p)$. The case $\ell(\gamma_s) > \ell(\sigma_s)$ is similar. □

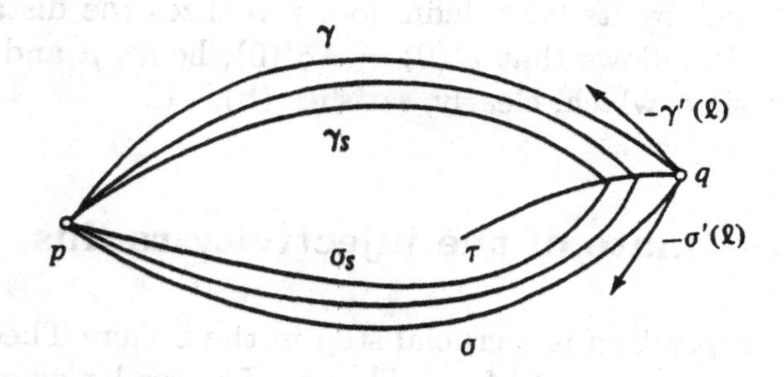

Figure 5

The proposition below shows, if M is compact and has positive sectional curvature, that the determination of a lower bound for the distance from a point to its cut locus depends on the determination of a lower bound for the lengths of closed geodesics in M.

2.13 PROPOSITION. *If the sectional curvature K of a complete Riemannian manifold M satisfies*

$$0 < K_{\min} \leq K \leq K_{\max},$$

then:

a) $i(M) \geq \pi/\sqrt{K_{\max}}$, *or*

b) *there exists a closed geodesic γ in M, whose length is less than that of any other closed geodesic in M, and which is such that*

$$i(M) = \frac{1}{2}\ell(\gamma).$$

Proof. By the Theorem of Bonnet-Myers (Theor. 3.1 of Chap. 9), M is compact. Since T_1M is compact, it follows from Proposition 2.9 that there exists $p \in M$ such that $d(p, C_m(p)) = \inf_{r\in M} d(r, C_m(r))$. Because $C_m(p)$ is compact, there exists $q \in C_m(p)$ such that q assumes the distance from p to $C_m(p)$.

If q is conjugate to p, $d(p,q) \geq \pi/\sqrt{K_{\max}}$ from Proposition 2.4 of Chapter 10. If q is not conjugate to p, there exists, by Proposition 2.12, two minimizing geodesics μ and σ from p to q, with $\mu'(\ell) = -\sigma'(\ell)$, $\ell = d(p,q)$. Since $q \in C_m(p)$, we have that $p \in C_m(q)$ and, by its very definition, p realizes the distance from q to $C_m(q)$. It follows that $\mu'(0) = -\sigma'(0)$, hence μ and σ form a closed geodesic γ which, clearly, satisfies (b). □

3. The estimate of the injectivity radius

The next proposition is a crucial step in the Sphere Theorem, and its proof uses elements of Morse Theory. The reader who wishes to restrict himself to the Sphere Theorem in even dimension can omit its proof and pass directly to Proposition 3.4.

3.1 PROPOSITION. *(Klingenberg [Kℓ 2]). Let M^n, $n \geq 3$, be a simply connected, compact Riemannian manifold, such that $1/4 < K \leq 1$. Then, $i(M) \geq \pi$.*

We need a few lemmas.

3.2 LEMMA. *Let M and $\tilde{M}$ be Riemannian manifolds of the same dimension and such that their sectional curvatures K and $\tilde{K}$, respectively, satisfy $\sup \tilde{K} \leq \inf K$. Consider a geodesic $\sigma\colon [0,\ell] \to M$, $\sigma(0) = p$, and fix a point $\tilde{p} \in \tilde{M}$. Let $i\colon T_pM \to T_{\tilde{p}}\tilde{M}$ be a linear isometry and let $\tilde{\sigma}\colon [0,\ell] \to \tilde{M}$ be given by*

$$\tilde{\sigma}(t) = \exp_{\tilde{p}} t(i\sigma'(0)), \qquad t \in [0,\ell].$$

Then index $\sigma \geq$ index $\tilde{\sigma}$.

Proof. Let $\tilde{W}$ be a piecewise differentiable vector field along $\tilde{\sigma}$. Define, as usual (Cf. Chap. 10), a vector field W along σ by

$$W(t) = \phi^{-1}(\tilde{W})(t) = P_t \circ i^{-1} \circ \tilde{P}_t^{-1}(\tilde{W}(t)),$$

where, for example, P_t is the parallel transport along σ from 0 to t. It is clear that (Cf. Eq. (4) of Chap. 10)

$$\langle W, \sigma' \rangle = \left\langle \tilde{W}, \tilde{\sigma}' \right\rangle, \quad |W(t)| = \left|\tilde{W}(t)\right|, \quad \langle W', W' \rangle = \left\langle \tilde{W}', \tilde{W}' \right\rangle.$$

Because $\sup \tilde{K} \leq \inf K$, we conclude that $I(W, W) \leq I(\tilde{W}, \tilde{W})$. Therefore, if $I(\tilde{W}, \tilde{W}) < 0$, we have that $I(W, W) < 0$, which proves the lemma. □

For the next lemma we need the following facts. Let $f\colon M^n \to \mathbf{R}$ be a differentiable function on a differentiable manifold M. A point $p \in M$ is a *critical point* of f if $df(p) = 0$; $f(p)$ is then called a *critical value* of f. If p is a critical point and $(x_1, \ldots, x_n)$ is a coordinate system on M around p, the *hessian* of f, defined by the matrix $(\frac{\partial^2 f}{\partial x_i \partial x_j})(p)$, represents a symmetric bilinear form on T_pM that does not depend on the chosen system of coordinates. We say that the critical point is *non-degenerate* if the determinant of this matrix is different from zero. Such a critical point is isolated, and it is possible to choose a coordinate system $(x_1, \ldots, x_{n-\lambda}, y_1, \ldots, y_\lambda)$ in such a way that, in the coordinate neighborhood considered, f can be written

$$f(x_1, \ldots, y_\lambda) = f(p) + x_1^2 + \cdots + x_{n-\lambda}^2 - y_1^2 - \cdots - y_\lambda^2.$$

The integer λ is called the *index* of the non-degenerate critical point p. (For more details see Milnor [Mi].)

3.3 Lemma. *Let M^n be a differentiable manifold and let $f\colon M^n \to R$ be a differentiable function whose critical points are all non-degenerate. Given $p, q \in M$ and a differentiable curve $\gamma\colon [0,1] \to M$ with $\gamma(0) = p$, $\gamma(1) = q$, let $a = \max\{f(p), f(q)\}$, and put*

$$M^a = \{x \in M; f(x) \leq a\},$$

$b = \max_{t\in[0,1]}(f \circ \gamma(t))$. Assume that $f^{-1}([a, b])$ is compact and does not contain any critical points of index zero or one. Then, for all $\delta > 0$, γ is homotopic, keeping endpoints fixed, to a curve $\tilde{\gamma}$ such that $\tilde{\gamma}([0,1]) \subset M^{a+\delta}$.

Proof. Choose a Riemannian metric on M. We use, in the course of the proof, the following fact (cf. Milnor [Mi], p. 12). If $f^{-1}([a_1, a_2])$ is compact and does not contain critical points of f, using the vector field grad f (which vanishes exactly at the critical points of f), we obtain a map $h: M^{a_2} \to M^{a_1}$, which is a *deformation retract*, that is, $h \mid M^{a_1} =$ identity, and there exists a homotopy $h_s: M^{a_2} \to M^{a_2}$, $s \in [0,1]$, such that $h_o =$ identity and $h_1 = h$.

Let b be as in the statement. We can suppose $b > a$; otherwise, there is nothing to prove. Let $p_1, \ldots, p_k$ be the critical points of f in $M^b - M^a$, with values $f(p_i) = c_i$, $i = 1, \ldots, k$. Since the critical points are isolated, we can perturb f on $M^b - M^a$ in such a way that all the c_i's are distinct. Suppose therefore that $b \geq c_1 > c_2 > \cdots > c_k > a$.

If b is not a critical value of f, $f^{-1}([c_1+\varepsilon, b])$ does not contain critical points of f for sufficiently small $\varepsilon > 0$. From the remark above, there exists a deformation retract $h_1: M^b \to M^{c_1+\varepsilon}$ and the curve $\gamma_1 = h_1 \circ \gamma$ is in $M^{c_1+\varepsilon}$. If $b = c_1$ is a critical value of f, we proceed in the following way.

By hypothesis, p_1 has index $\lambda \geq 2$. Let H be a neighborhood of p_1 where f can be written in the form

$$f = c_1 + x_1^2 + \cdots + x_{n-\lambda}^2 - y_1^2 - \cdots - y_\lambda^2.$$

Suppose that H and $\varepsilon > 0$ are sufficiently small so that $M^{c_1+\varepsilon} - (M^{c_1-\varepsilon} \cup H)$ does not contain critical points. Then, using the vector field grad f we obtain a deformation retract $\bar{h}_1: M^{c_1+\varepsilon} \to M^{c_1-\varepsilon} \cup H$. Let $\bar{\gamma}_1 = \bar{h}_1 \circ \gamma \subset M^{c_1-\varepsilon} \cup H$.

Now let L be the set of points of H such that $y_1 = y_2 = \cdots = y_\lambda = 0$ (see Fig. 6). Since

$$\dim \bar{\gamma}_1' + \dim L \leq 1 + n - 2 < \dim M,$$

it is possible, by transversality, to perturb $\bar{\gamma}_1$ in such a way that $\bar{\gamma}_1([0,1]) \cap L = \phi$; such a perturbation can be done in H, in a way that fixes the endpoints of $\bar{\gamma}_1$. Since, for points not in L, the trajectories of grad f go away from the critical point p_1, there exists a deformation retract $h_2: M^{c_1-\varepsilon} \cup (H - L) \to M^{c_1-\varepsilon}$. It is clear that $\gamma_2 = h_2 \circ \bar{\gamma}_1 \subset M^{c_1-\varepsilon}$.

Proceeding inductively, we obtain $\gamma_j \subset M^{c_k-\varepsilon}$. Given $\delta > 0$, there do not exist critical points in $M^{c_k-\varepsilon} - M^{a+\delta}$ and we can obtain a deformation retract $h_{j+1}: M^{c_k-\varepsilon} \to M^{a+\delta}$. It is clear that $\tilde{\gamma} = h_{j+1} \circ \gamma_j \subset M^{a+\delta}$, as required. □

With the same argument, we see that if there are critical points of index zero or one in $f^{-1}([a,b]$ and c is the largest value of such critical points, then γ is homotopic to a curve $\tilde{\gamma}$ with $\tilde{\gamma}([0,1]) \subset M^{c+\delta}$.

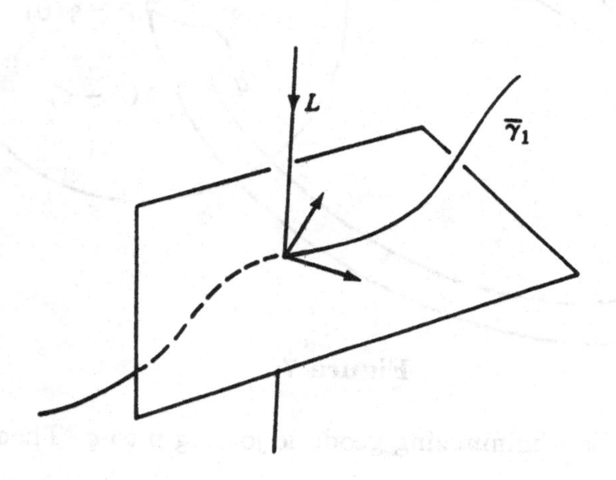

Figure 6

Proof of Proposition 3.1. Suppose $i(M) < \pi$. From Proposition 2.13, there exists a closed geodesic γ in M of length $\ell = \ell(\gamma) < 2\pi$.

From the Corollary 2.10 of Chap. 11, the set of conjugate points to $\gamma(0)$ along γ is discrete. Choose an $\varepsilon > 0$ satisfying the following conditions ($B_a(q)$ denotes the open ball with center q and radius a):

1) $\gamma(\ell - \varepsilon)$ is not conjugate to $p = \gamma(0)$ along γ;
2) $\exp_p$ is a diffeomorphism on $B_{2\varepsilon}(p)$;
3) $3\varepsilon < 2\pi - \pi/\sqrt{\bar{K}}$, where $\bar{K} = \inf_M K$;
4) $3\varepsilon < 2\pi - \ell$;
5) $5\varepsilon < 2\pi$.

From Sard's Theorem, there exists a regular value $q \in B_\varepsilon(\gamma(\ell - \varepsilon))$ of $\exp_p$. By (1), it is possible to choose q in such a way that $q = \gamma_1(t)$, where γ_1 is a geodesic starting from p with $3\varepsilon < \ell(\gamma_1) < \ell$ (v. Fig. 7).

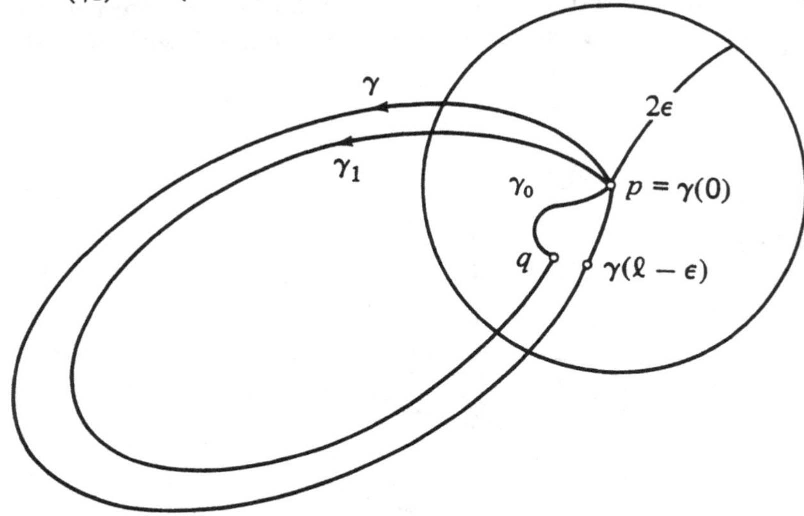

Figure 7

Let γ_o be a minimizing geodesic joining p to q. Then by (2),

$$\ell(\gamma_o) \leq d(p, \gamma(\ell - \varepsilon)) + d(\gamma(\ell - \varepsilon), q) < 2\varepsilon,$$

hence $\gamma_o \neq \gamma_1$.

Consider the space $\Omega^c_{p,q}$ of Remark 2.12 of Chapter 11 and its finite dimensional approximation B. Because q is a regular value of $\exp_p$, all of the critical points in $\Omega^c_{p,q}$ are non-degenerate. From the fact that M is simply connected, there exists a homotopy γ_s between γ_o and γ_1, keeping fixed the points p and q. Since B is a deformation retract of $\Omega^c_{p,q}$, the homotopy γ_s (which is a curve in $\Omega^c_{p,q}$) is deformed to a curve in B. From the remark after Lemma 3.3, for all $\delta > 0$, γ_s can be deformed by a homotopy $\tilde{\gamma}_s$ from γ_o to γ_1 in such a way that the curve of larger length $\tilde{\gamma}$ of $\tilde{\gamma}_s$ satisfies the following condition: $\ell(\tilde{\gamma}) < a + \delta$, where $a = \max\{\ell(\gamma_o), \ell(\gamma_1), \ell(\sigma)\}$ and σ is the geodesic of larger length with index $\sigma < 2$ in $\Omega^c_{p,q}$.

Put $\delta = \varepsilon$. We have already seen that $\ell(\gamma_o) < 2\varepsilon$ and that $\ell(\gamma_1) < \ell < 2\pi - 3\varepsilon$, by (4). To estimate $\ell(\sigma)$, apply Lemma 3.2 to the case in which $\tilde{M}$ is a sphere with curvature $\tilde{K} = \inf_M K > 1/4$ obtaining

$$\text{index } \tilde{\sigma} \leq \text{ index } \sigma < 2.$$

Since $\tilde{\sigma}$ is a geodesic on S^n, $n \geq 3$, it follows from the Morse Index Theorem that if $\ell(\sigma) > \pi/\sqrt{\tilde{K}}$ then index $\tilde{\sigma} \geq 2$. Therefore, by (3), $\ell(\sigma) \leq \pi/\sqrt{\tilde{K}} < 2\pi - 3\varepsilon$. It follows then, by (5), that

$$\ell(\tilde{\gamma}) < a + \varepsilon < 2\pi - 3\varepsilon + \varepsilon = 2\pi - 2\varepsilon.$$

On the other hand, the Lemma of Klingenberg (see Exercise 1 of Chap. 10) yields that in the homotopy $\tilde{\gamma}_s$ there exists a curve $\tilde{\gamma}_{s_o}$ with $\ell(\gamma_o) + \ell(\tilde{\gamma}_{s_o}) \geq 2\pi$. It follows that

$$\ell(\tilde{\gamma}) \geq \ell(\tilde{\gamma}_{s_o}) \geq 2\pi - \ell(\gamma_o) > 2\pi - 2\varepsilon,$$

which contradicts the fact above that $\ell(\tilde{\gamma}) < 2\pi - 2\varepsilon$, and shows that the hypothesis $i(M) < \pi$ untenable. This completes the proof of Proposition 3.1. □

For the particular case of even dimension it is possible to prove a stronger version of Proposition 3.1, and the proof does not use Morse Theory.

3.4 PROPOSITION. *If the sectional curvature K of a compact orientable even dimensional Riemannian manifold M^n satisfies $0 < K \leq 1$, then $i(M) \geq \pi$.*

Proof. By compactness, there exist points $p, q \in M$ such that:

$$q \in C_m(p) \quad \text{and} \quad d(p,q) = i(M).$$

Supposing that $d(p,q) < \pi$, we are going to obtain a contradiction.

If p is conjugate to q, it follows from Proposition 2.4 of Chap. 10 that $d(p,q) \geq \pi$, which contradicts the hypothesis. Therefore p is not conjugate to q and, by Proposition 2.12 and 2.13, there exists a closed geodesic γ passing through $p = \gamma(0)$ and q, and such that $\ell(\gamma) < 2\pi$. The parallel transport along the closed curve γ leaves invariant a vector V orthogonal to γ (cf. Lemma 3.8 of Chap. 9). Calculating the second variation $E''_V(0)$ for such V, we see that it is strictly negative. Therefore there exists a variation through regular closed curves $\gamma_s(t)$ of γ, $s \in [0, \varepsilon]$, such that $\ell(\gamma_s) < \ell(\gamma)$ for all $s \neq 0$(the fact that M has even dimension plays a crucial role here).

Let q_s be a point of γ_s at a maximum distance from $\gamma_s(0)$ (see Fig 8). Since $d(\gamma_s(0), q_s) < d(p, q)$, there exists a unique minimizing geodesic σ_s joining $q_s = \sigma_s(0)$ to $\gamma_s(0)$. Because q is the unique point of γ which is at maximum distance from p, $\lim_{s \to 0} \sigma_s(0) = q$. In addition, by local compactness of the tangent bundle, there exists in $T_q(M)$ an accumulation point w of the vectors $\sigma_s'(0)$. By continuity, the geodesic $\sigma(t) = \exp_q tw$ is a minimizing geodesic joining q to p. On the other hand, let $\sigma_{s,t}$ be a minimizing geodesic joining $\gamma_s(0)$ to a point $\gamma_s(t)$, close to q_s. Then $\sigma_{s,t}$ is a variation of σ_s and, applying the formula for the first variation, we conclude that $\sigma_s'(0)$ is orthogonal to γ_s' at q_s, for all $s \in [0, \varepsilon]$; therefore $\sigma'(0)$ is orthogonal to γ' at q. It follows that there exist three minimizing geodesics joining p to q. Since p is not conjugate to q, this contradicts Proposition 2.12. □

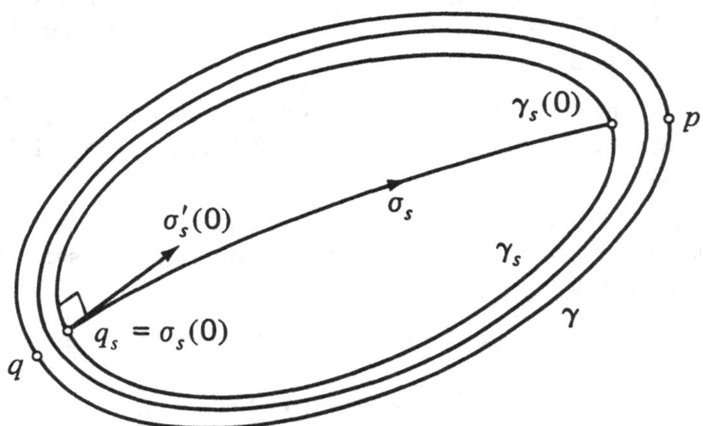

Figure 8

3.5 REMARK. From Synge's theorem (see Chap. 9, Cor. 3.10), the manifolds which satisfy the hypotheses of Proposition 3.4 are simply connected. On the other hand, a simply connected manifold is orientable; otherwise, it has a two-sheeted covering (see Exercise 12 of Chap. 0), which contradicts the fact it is simply connected. Therefore, the hypothesis of orientability in Proposition 4 is equivalent to the assumption that M be simply connected.

4. The sphere theorem

In what follows, the fact that the dimension M is even or odd plays no role. We need a few lemmas.

4.1 LEMMA. *(Berger [Br 1]). Let M be a compact Riemannian manifold, and let $p, q \in M$ be such that $d(p,q) = \operatorname{diam} M$. Then, for all $W \in T_pM$, there exists a minimizing geodesic γ from $p = \gamma(0)$ to q with $\langle \gamma'(0), W\rangle \geq 0$.*

Proof. Let $\lambda(t) = \exp_p tW$ and let $\gamma_t\colon [0, \ell(\gamma_t)] \to M$ be a minimizing geodesic such that $\gamma_t(0) = \lambda(t)$, $\gamma_t(\ell(\gamma_t)) = q$. Suppose, first, that for each integer n there exists a t_n, $0 \leq t_n \leq \frac{1}{n}$, such that $\langle \gamma'_{t_n}(0), \lambda'(t_n)\rangle \geq 0$. Passing to a subsequence, if necessary, the sequence γ_{t_n} converges to a minimizing geodesic γ which satisfies

$$0 \leq \langle \gamma'(0), \lambda'(0)\rangle = \langle \gamma'(0), W\rangle\,,$$

and the proof would be finished.

Suppose now that there exists an n such that for all t, $0 \leq t \leq 1/n$, $\langle \gamma'_t(0), \lambda'(t)\rangle < 0$ holds. We are going to show that this leads to a contradiction.

Consider a totally normal neighborhood U of $\lambda(t)$, and let $q_o \in U$ be a point of γ_t. Let $\varepsilon > 0$ be sufficiently small, $s \in (t-\varepsilon, t+\varepsilon)$ and let σ_s be a minimizing geodesic joining q_o to $\lambda(s)$ (see Fig. 9). Using the formula for the first variation of energy, we obtain

$$\frac{1}{2}\frac{d}{ds}E(\sigma_s)\bigg|_{s=t} = -\langle \gamma'_t(0), \lambda'(t)\rangle > 0.$$

Therefore, for $s < t$, $d(q_o, \lambda(s)) < d(q_o, \lambda(t))$ holds, hence

$$\begin{aligned} d(q, \lambda(s)) \leq d(q, q_o) + d(q_o, \lambda(s)) &< d(q, q_o) \\ &+ d(q_o, \lambda(t)) = d(q, \lambda(t)). \end{aligned}$$

On the other hand, since p is at a maximum distance from q, $d(q, \lambda(t)) \leq d(q, \lambda(0))$, which furnishes the desired contradiction. □

The following lemma was originally proved by Berger [Br 1] using Toponogov's theorem. The proof presented here is due to Tsukamoto and uses Rauch's Theorem.

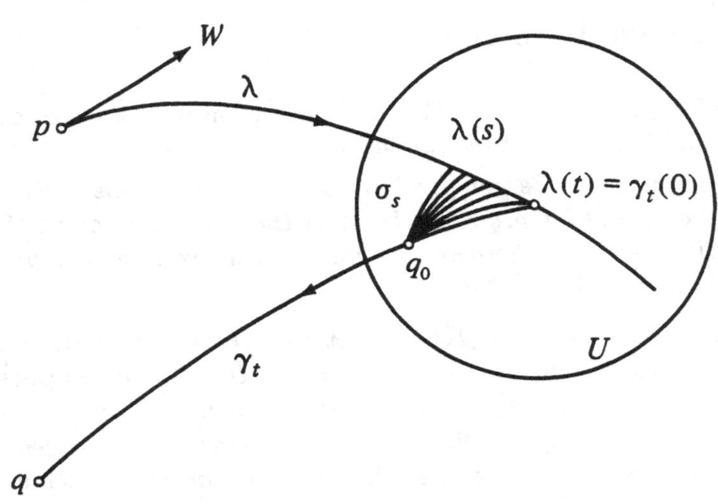

Figure 9

4.2 LEMMA. *Let M^n be a compact, simply connected Riemannian manifold whose sectional curvature K satisfies*

$$\frac{1}{4} < \delta \leq K \leq 1,$$

and let $p, q \in M$ be such that $d(p,q) = \mathrm{diam}(M)$. Then

$$M = B_\rho(p) \cup B_\rho(q),$$

where $B_r(p) \subset M$ denotes the open geodesic ball of radius r and center $p \in M$ and ρ is such that $\pi/2\sqrt{\delta} < \rho < \pi$.

Proof. (Tsukamoto [Ts]). By the estimate of the injectivity radius of Section 3, $B_\rho(p)$ does not contain points of the cut locus of p. Therefore $B_\rho(p)$ is diffeomorphic, through the exponential map, to an Euclidean ball, and the same thing happens for $B_\rho(q)$. It is at this point that the estimate $i(M) \geq \pi$ enters, in a crucial manner, in the Sphere Theorem.

Suppose that there exists $r \in M$ such that $d(p,r) \geq \rho$, $d(q,r) \geq \rho$. Then we are going to obtain a contradiction. We can assume that $d(p,r) \geq d(q,r) \geq \rho$.

A minimizing geodesic from q to r intersects $\partial(B_\rho(q))$ in a point $q' \notin B_\rho(p)$; otherwise,

$$d(r,q') > d(r, B_\rho(p)) \geq d(r, B_\rho(q)) = d(r,q'),$$

which is absurd. On the other hand, from the Theorem of Bonnet-Myers, $\operatorname{diam}(M) \leq \pi/\sqrt{\delta} < 2\rho$. Therefore, if q'' is a point of intersection of the minimizing geodesic from q to p with $\partial B_\rho(q)$, then $q'' \in B_\rho(p)$, since,

$$d(p, q'') = d(p, q) - d(q, q'') < 2\rho - \rho = \rho.$$

Since $\partial B_\rho(q)$ is path connected (because $B_\rho(q)$ is homeomorphic to an Euclidean ball), it follows that $\partial B_\rho(p) \cap \partial B_\rho(q) \neq \phi$, hence there exists $r_o \in M$ such that $d(r_o, p) = d(r_o, q) = \rho$. The desired contradiction obtains from the existence of such an r_o.

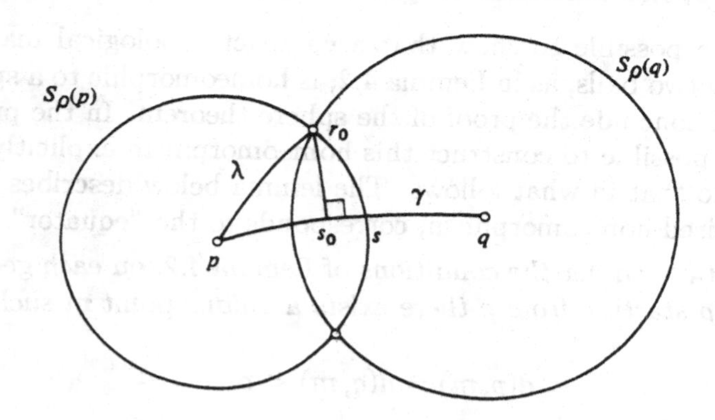

Figure 10

Consider a minimizing geodesic λ joining p to r_o. By Berger's Lemma, there exists a minimizing geodesic γ from p to q with $\langle \gamma'(0), \lambda'(0) \rangle \geq 0$ (Fig. 10). Let s be a point of γ such that $d(p, s) = \rho$. Apply Rauch's Theorem to this situation (see Proposition 2.5 of Chap. 10), comparing M^n with a sphere S^n of constant curvature δ. Since the angle $\sphericalangle r_o p s \leq \pi/2$ and $d(r_o, s)$ is less than or equal to the length of any curve in M joining r_o to s, we conclude that

$$d(r_o, s) \leq \pi/2\sqrt{\delta}.$$

Since $d(r_o, p) = d(r_o, q) = \rho$, and there exists a point s of γ with $d(r_o, s) < \rho$, the distance from r_o to γ is realized by a point

s_o in the interior of γ. The minimizing geodesic from r_o to s_o is orthogonal to γ, and

$$d(r_o, \gamma) = d(r_o, s_o) \leq \pi/2\sqrt{\delta}.$$

Since $d(p,q) \leq \pi/\sqrt{\delta}$, we have that either $d(p, s_o) \leq \pi/2\sqrt{\delta}$ or $d(q, s_o) \leq \pi/2\sqrt{\delta}$. In either of the two cases, we obtain the desired contradiction. Consider the case $d(p, s_o) \leq \pi/2\sqrt{\delta}$, the other case being analogous. In this case, since $d(r_o, s_o) \leq \pi/2\sqrt{\delta}$ and the angle $\measuredangle\ p s_o r_o = \pi/2$, we have, by the comparison the Theorem of Rauch cited above, $d(p, r_o) \leq \pi/2\sqrt{\delta} < \rho$, which contradicts the fact that $d(p, r_o) = \rho$, and concludes the proof. □

It is possible to show that a compact topological manifold covered by two balls, as in Lemma 4.2, is homeomorphic to a sphere, which will conclude the proof of the sphere theorem. In the present case, it is possible to construct this homeomorphism explicitly, and we shall do that in what follows. The lemma below describes what, in the desired homeomorphism, corresponds to the "equator" of M.

4.3 LEMMA. *Under the conditions of Lemma 4.2, on each geodesic of length ρ starting from p there exists a unique point m such that*

$$d(p, m) = d(q, m) < \rho.$$

Similarly, on each geodesic starting from q of length ρ there exists a unique point n equidistant from p and q.

Proof. Let $\gamma(s)$ be a geodesic with $\gamma(0) = p$ and consider the difference

$$d(q, \gamma(s)) - d(p, \gamma(s)) = f(s).$$

The function f is evidently continuous and $f(0) = d(q,p) > 0$. Let s_o be such that $\gamma(s_o)$ is the cut point of p along γ. Then by the estimate of Klingenberg (Prop. 3.1, or Prop. 3.4 for the even dimensional case), $d(p, \gamma(s_o)) \geq \pi > \rho$. From Lemma 4.2, $d(q, \gamma(s_o)) < \rho$. Therefore,

$$f(s_o) = d(q, \gamma(s_o)) - d(p, \gamma(s_o)) < 0.$$

It follows that f vanishes for some $s_1 \in (0, s_o)$; that is, there exists $m = \gamma(s_1)$ satisfying the claim.

To prove uniqueness, suppose that there exists $m_1 \neq m_2$, both equidistant from p and q. We can assume that m_1 is between p and m_2. Then

$$d(q,m_2) = d(p,m_2) = d(p,m_1) + d(m_1,m_2) = d(q,m_1) + d(m_1,m_2).$$

Let σ_1 be the (unique) minimizing geodesic joining q to m_1. By the equality above, σ_1 coincides with γ, hence q belongs to γ. Since $d(p,m_1) = d(q,m_1)$, $d(p,m_2) = d(q,m_2)$ and $m_1 \neq m_2$, we conclude that $p = q$, which contradicts the hypothesis. The situation for the point q is entirely analogous. □

4.4 Remark. Note that, due to its uniqueness, the point m depends continuously on the (initial direction of) geodesic which cotains it.

We can now construct explicitly the homeomorphism mentioned in the sphere theorem.

Proof of Theorem 1.1. Let $p, q \in M$ such that $\operatorname{diam} M = d(p,q)$. Let D_1 and D_2 be the subsets of M formed by all geodesic segments $\overline{pm}$ and $\overline{qn}$, respectively, where the points m and n are given by Lemma 4.3. By the continuity of m (see Rem. 4.4), D_1 and D_2 are closed subsets of M. We claim that $D_1 \cup D_2 = M$ and $\partial D_1 = \partial D_2 = D_1 \cap D_2$.

To see this, let $r \in M$. By Lemma 4.2, either $d(p,r) < \rho$ or $d(q,r) < \rho$. Consider the first case, the second being analogous. Since $d(p, C_m(p)) \geq \pi > \rho$, there exists a unique minimizing geodesic γ passing through p and r. By Lemma 4.3, there exists a unique point m on γ with $d(p,m) = d(q,m) < \rho$. If $d(p,r) < d(q,r)$, then $r \in \overline{pm}$, that is, $r \in D_1$. If $d(p,r) = d(q,r)$, by uniqueness, $r = m$, that is, $r \in \partial D_1$. If $d(p,r) > d(q,r)$, then $d(q,r) < \rho$ and, using the previous argument with q in place of p, we have that $r \in \overline{qn}$ and therefore $r \in D_2$. Therefore $r \in D_1 \cup D_2$. If $r \in D_1 \cap D_2$, it follows from the previous argument that $d(p,r) = d(q,r)$, hence $r = m = n$, which proves the assertion made.

We can now define a map $\varphi \colon S^n \to M^n$ in the following way. To a fixed point $N \in S^n$ associate p, and to its antipodal point $S \in S^n$ associate q. Choose a linear isometry $i \colon T_N S^n \to T_p M$. For each point e of the equator E of S^n relative to the north pole N, consider the geodesic $\gamma(s)$ of S^n, $0 \leq s \leq \pi$, given by $\gamma(0) = N$,

$\gamma(\pi/2) = e$. Let m be the point given by Lemma 4.3 on the geodesic of M which passes through p with the velocity vector $i\gamma'(0)$. Define:

$$\begin{cases} \varphi(\gamma(s)) = \exp_p(s\frac{2}{\pi}d(p,m)(i \circ \gamma'(0))), & 0 < s \leq \frac{\pi}{2}, \\ \varphi(\gamma(s)) = \exp_q((2 - \frac{2s}{\pi})d(q,m)V), & \frac{\pi}{2} \leq s \leq \pi, \end{cases}$$

where V is the unit tangent vector at q of the unique minimizing geodesic from q to m.

It is easy to verify that φ maps the closed northern hemisphere bijectively onto D_1, the closed southern hemisphere bijectively onto D_2 and the equator E onto $\partial D_1 = \partial D_2 = D_1 \cap D_2$. By uniqueness of the points m and n given in Lemma 4.3, φ is continuous. Since $M = D_1 \cup D_2$, φ is surjective. Finally, since $D_1 \cap D_2 = \partial D_1 = \partial D_2 = \varphi(E)$, φ is injective. Since S^n is compact, φ is a homeomorphism. □

5. Some further developments

The Sphere Theorem gives rise to a large number of questions. This has produced an intense activity in research which extends up to the present and constitutes one of the most vigorous branches of Differential Geometry.

The first natural question is if it is possible to replace "homeomorphic" by "diffeomorphic" in the statement of the sphere theorem. Observe that the homeomorphism of the Sphere Theorem (see Section 4) is obtained by "glueing" two discs along their boundaries. Such a construction may lead to a differentiable structure on M distinct from the usual structure of the sphere. Therefore, the proof of the Sphere Theorem presented here is not sufficient to establish a diffeomorphism.

In 1966, D. Gromoll [Gℓ] and E. Calabi (unpublished) proved that, for a "pinching" h_n sufficiently small (h_n depending on the dimension n of M), the Sphere Theorem is valid, replacing "homeomorphic" by "diffeomorphic". The "pinching" h_n has been since improved in various works (for example, [Sh] and [SSK]) but, it is not yet known what the best possible value is; the best estimate at the moment is lim $h_n \sim 0{,}68$ when $n \to \infty$ [IHR].

Another natural question is what happens, in case of even dimension, for $h < 1/4$.

A possible point of view for this problem is, following an idea of Rauch [R2], to compare the curvature of the manifold M with the curvature of a symmetric space (compact, simply connected, of rank one) and try to show that if the curvatures are "close", in a certain sense, the topologies are "similar". After some preliminary results of do Carmo [dC 1], Klingenberg [Kl 3], Kobayashi [Ko] and Cheeger [Cr 1], a satisfactory theorem was obtained by Min-Oo and Ruh in [MR].

Another point of view is to show that if $h = (1/4) - \varepsilon$, for some $\varepsilon > 0$, a compact simply connected Riemannian manifold which is h-pinched is still homeomorphic to a sphere or one of the previously mentioned symmetric spaces. This was recently obtained by Berger [Br 3] but no estimate for the value of ε is presently known.

In the case of odd dimension, one can also ask the question of what happens for $h < 1/4$. Nevertheless, the problem in this case appears rather difficult. One of the difficulties is that the estimate $i(M) \geq \pi$ for the injectivity radius (see Prop. 3.1), valid in even dimension for any value h (see Prop. 3.4), cannot be valid in odd dimension without some restriction on h. By an example of Berger (see Cheeger-Ebin [CE] pp. 70–71 for details), the best one can hope for is that the estimate be valid for $h = 1/9$, but again this has not yet been proved.

In 1977, Grove and Shiohama [GS] proved a sphere theorem, in which the hypothesis of "pinching" (1') is replaced by a hypothesis on the diameter: *If M is compact, $K \geq 1$, and* $\mathrm{diam}(M) > \pi/2$ *then M is homeomorphic to a sphere*. The case $\mathrm{diam}(M) = \pi/2$ (where the theorem is false, as shown by the example of real projective space) was essentially classified by Gromoll and Grove $[G\ell - G]$.

Motivated by the Sphere Theorem, we can ask if a Riemannian manifold M^n, compact with sectional curvature K satisfying $-1 - \varepsilon \leq K \leq -1$, for some $\varepsilon > 0$, admits a Riemannian metric of constant curvature -1. In 1978, Gromov [Gv 0] showed that if the diameter of M is bounded above by D, there exists an $\varepsilon = \varepsilon(n, D)$, depending on the dimension n and the bound D of the diameter, which answers affirmatively the above question.

A surprising theorem on the topology of compact, connected, manifolds of non-negative curvature was obtained by Gromov [Gv 2], and asserts that, in this case, there exists a constant $C(n)$, depending only on the dimension n of the manifold, such that the sum $\sum b_i$, $1 \leq i \leq n$, of the Betti numbers b_i (over any field) is bounded above by $C(n)$. This shows that the connected sum of a sufficiently large number of copies of products $S^n \times S^{n-p}$, $0 < p < n$, (S^k is the sphere of dimension k) does not admit a metric of non-negative curvature.

The experience with the problems suggested by the Sphere Theorem provoked a change of the initial viewpoint, and led to the formulation of the following question: How many distinct topological types can have Riemannian metrics with given restrictions? The initial point of view was to characterize the topology of a manifold by means of the curvature, or, if necessary, by means of the curvature and other simple geometric elements, such as the diameter and the volume. In the point of view mentioned above, we try to look at the set of Riemannian manifolds with certain properties in order to determine how many topologies can arise.

A surprising and fundamental theorem in this direction is due to Cheeger [Cr 2] (see also Weinstein [We 1] and Peters [Pe]) and asserts that, for fixed dimension, there exists a finite number of homeomorphism types in the set of compact Riemannian manifolds with volume bounded below, diameter bounded above and sectional curvature bounded in absolute value. The lower bound on the volume can be replaced by a lower bound on the radius of injectivity; all of the conditions are, however, necessary.

The viewpoint adopted in the theorem above has shown to be extremely fruitful. M. Gromov [Gv 1] proved that the set of Riemannian manifolds (dimension fixed) which satisfies the hypothesis of the theorem of Cheeger has compact closure in a suitable topology. This theorem has a variety of applications among which is the $(\frac{1}{4} - \varepsilon)$-theorem of Berger, cited above. For more details on the subject, we recommend the exposition of Pansu in the Seminar Bourbaki [Pa].

It is impossible, in a few pages, to do justice to the importance of the Sphere Theorem. Because of the lack of space and time, we have to leave without mentioning various important developments that are related to the Sphere Theorem or have been motivated by it (the non-compact complete case; the use of other types of curvature, such as the Ricci curvature or the scalar curvature; other hypotheses on the "pinching", etc.). A portion of this material is covered in an excellent survey of M. Berger [Br 3], where the "leitmotiv" is the role of the Sphere Theorem in the development of Riemannian Geometry. An ample list of references can be found in the survey of Sakai [Sa]. Finally, another proof of the sphere theorem, due to Gromov, following a path completely distinct from the one presented here, can be found in Eschenburg [E].

REFERENCES

[Am 1] AMBROSE, W., Parallel translation of Riemannian curvature, Ann. of Math. 64 (1956), 337–363.

[Am 2] AMBROSE, W., The index theorem in Riemannian geometry, Ann. of Math. 73 (1961), 49–86.

[Ar] ARMSTRONG, M., Basic topology, Springer-Verlag, Berlin, Heidelberg, 1983.

[Br 1] BERGER, M., Les variétés Riemannienes (1/4)-pincées, Ann. Scuola Norm. Sup. Pisa, Ser. III, 14 (1960), 161–170.

[Br 2] BERGER, M., Sur les variétés riemanniennes pincées juste au-dessous de 1/4, Ann. Institut Fourier 33 (1983), 135–150.

[Br 3] BERGER, M., H.E. Rauch Géomètre Différentiel, in Differential Geometry and Complex Analysis, Eds. I. Chavel, H.M. Farkas, Springer-Verlag, Berlin, Heidelberg, 1985.

[Be] BESSE, A., Manifolds all of whose geodesics are closed, Springer-Verlag, Berlin, Heidelberg, 1978.

[Bo] BONNET, O., Sur quelques propriétés des lignes géodésiques, C.R. Ac. Sc. Paris 40(1855) 1311–1313.

[By] BYERS, W., On a theorem of Preissmann, Proc. Am. Math. Soc. 24 (1970), 50–51.

[dC 1] CARMO, M. do, The cohomology ring of certain kählerian manifolds, Ann. of Math. 81 (1965), 1–14.

[dC 2] CARMO, M. do, Differentiable Curves and Surfaces, Prentice-Hall, New Jersey, 1976.

[dC 3] CARMO, M. do, Differential Forms and Applications, Lecture notes from the College of Differential Geometry at Trieste, 1988. Translation of Formas Diferenciais e Aplicações, IMPA, Rio de Janeiro, 1973.

[Ca] CARTAN, E., Leçons sur la Geométrie des Espaces de Riemann, (2nd edition) Gauthier-Villars, Paris, 1951.

[CE] CHEEGER, J. and EBIN, D., Comparison Theorems in Riemannian Geometry, North-Holland, Amsterdam, 1975.

[Cr 1] CHEEGER, J., Pinching theorems for a certain class of Riemannian manifolds, Amer. J. Math. 91 (1969), 807–834.

[Cr 2] CHEEGER, J., Finiteness theorems for Riemannian manifolds, Amer. J. Math. 92 (1970), 61–74.

[Ch 1] CHERN,S.S., Differentiable manifolds, mimeographed notes, University of Chicago. Reproduced in Textos de Matemática do Recife, 4, 1959.

[Ch 2] CHERN, S.S., Minimal Submanifolds of a Riemannian Manifold, Department of Math., Univ. of Kansas, Lawrence, Kansas, 1968.

[Cf] CHRISTOFFEL, E.B., Über die Transformation der homogenen Differentialausdrücke zweiten Grades, J. Reine Angew. Math. 70 (1869), 46–70.

[Da] DAJCZER, M. et al., Submanifolds and isometric immersions, Math. Lect. Ser., 13, Publish or Peris, Inc., Houston, Texas, 1990.

[E] ESCHENBURG, J., Local convexity and nonnegative curvature - Gromov's proof of the sphere theorem, Inventiones Math., 84(1986), 507–522.

[Gb] GREENBERG, M., Lectures on Algebraic Topology, W.A. Benjamin, Inc., New York, Amsterdam, 1967.

[Ga] GAUSS, K.F., General Investigations of Curved Surfaces, Raven Press, New York, 1965.

[Gn] GREEN, L.W., Auf Wiedersehensflächen, Ann. of Math. 78 (1963), 289–299.

[Gℓ] GROMOLL, D., Differenzierbare Strukturen und Metriken positiver Krümmung auf Sphären, Math. Ann. 164 (1966), 353–371.

[GℓG] GROMOLL, D. and GROVE, K., Rigidity of positively curved manifolds with large diameter, Seminar of Diff. Geometry, ed. S.T. Yau, Annals of Math. Studies 102 (1982), 203–207.

[GKM] GROMOLL, D., KLINGENBERG, W., MEYER, W., Riemannche Geometrie im Grossen, Lecture Notes in Mathematics, n.55, Springer-Verlag, 1968.

[GS] GROVE, K. and SHIOHAMA, K., A generalized sphere theorem, Ann. of Math. 106 (1977), 201-211.

[Gv 0] GROMOV, M., Manifolds of negative curvature, J. Diff. Geom. 13 (1978), 223–230.

[Gv 1] GROMOV, M., Structures métriques pour les variétés riemanniennes, rédigé par Lafontaine et Pansu, Textes mathématiques 1 (1981), CEDIC-Nathan.

[Gv 2] GROMOV, M., Curvature, diameter and Betti numbers, Comm. Math. Helvetici 56 (1981), 179–195.

[H] HAMILTON, R., Three-manifolds with positive Ricci curvature, J. Diff. Geometry 17 (1982), 255–306.

[Hi] HIRSCH, M., Differential topology, Springer-Verlag, Berlin, Heidelberg, 1976.

[HR] HOPF, H. and RINOW, W., Über den Begriff der vollständigen differentialgeometrischen Flächen, Comm. Math. Helv. 3 (1931), 209–225.

[IHR] IM-HOFF, H.C., RUH, E., An equivariant pinching theorem, Comment. Math. Helv. 50 (1975), 389–401.

[Ka] KAZDAN, J.L., An isoperimetric inequality and wiedersehen manifolds, Seminar on Differential Geometry, Ed. S.T. Yau, Annals of Math. Studies 102 (1982).

[Kℓ 1] KLINGENBERG, W., Contributions to Riemannian Geometry in the large, Ann. of Math. 69 (1959), 654–666.

[Kℓ 2] KLINGENBERG, W., Über Riemannsche Mannigfaltigkeiten mit positiver Krümmung, Comm. Math. Helv. 35 (1961), 47–54.

[Kℓ 3] KLINGENBERG, W., Manifolds with restricted conjugate locus, Ann. Math. 78 (1963), 527–547.

[Kℓ 4] KLINGENBERG, W., Lectures on Closed Geodesics, Springer-Verlag, Yellow series n.230, 1978.

[Ko] KOBAYASHI, S., Topology of positively pinched Kähler manifolds, Tôhoku Math. J. 15 (1963), 121–139.

[Ku 1] KULKARNI, R., Curvature and metric, Ann. of Math. 91 (1970), 311–331.

[Ku 2] KULKARNI, R., Curvature structures and conformal transformations, J. Diff. Geom. 4 (1970), 425–451.

[La] LAWSON, B., Lectures on Minimal Submanifolds, vol. 1, Monografias de Matemática, 14, IMPA, 1973, 2nd edition, Publish or Perish, Berkeley, 1980.

[LC] LEVI-CIVITA, T., Nozione di parallelismo in una varietà qualunque, Rend. Circ. Mat. Palermo, 42 (1917), 173–205.

[Ma] MASSEY, W., Algebraic Topology, An Introduction, Harcourt, New York, 1967.

[Me] MEEKS III, W.H., A survey of the geometric results in the classical theory of minimal surfaces, Bol. Soc. Bras. Mat. 12 (1981), 29–86.

[MR] MIN-OO., RUH, E., Comparison theorems for compact symmetric spaces, Ann. Scient. école Norm. Sup. 12 (1979), 335–353.

[Mi] MILNOR, J., Morse Theory, Annals of Mathematics Studies, 51, Princeton University Press, Princeton, N.J. 1963.

[Mo] MOORE, J.D., An application of second variation to submanifold theory, Duke Math. Journal 42 (1975), 191–193.

[My] MYERS, S.B., Riemannian manifolds with positive mean curvature, Duke Math. J. 8 (1941), 401–404.

[Os] OSSERMAN, R., A Survey of Minimal Surfaces, Van Nostrand-Reinhold, New York, 1969, 2nd ed., Dover Pub., New York, 1986.

[Pa] PANSU, P., Séminaire Bourbaki, 36e. annèe, 1983/84, n.618.

[Pe] PETERS, S., Cheeger's finiteness theorem for diffeomorphism classes of Riemannian manifolds, J. reine angew. Math. 349 (1984), 77–82.

[Po] POINCARÉ, H., Sur les lignes géodésiques des surfaces convexes, Trans. Amer. Math. Soc. 17 (1905), 237–274.

[Pr] PREISSMAN, A., Quelques propriétés globales des espaces de Riemann, Comm. Math. Helv. 15 (1943), 175–216.

[R 1] RAUCH, H.E., A contribution to differential geometry in the large, Ann. of Math. 54 (1951), 38–55.

[R 2] RAUCH, H.E., Geodesics, symmetric spaces and differential geometry in the large, Comment. Math. Helv. 27 (1953), 294–320.

[R 3] RAUCH, H.E., Geodesics and Curvature in Differential Geometry in the Large, Yeshiva University, N.Y., 1959.

[Ri] RIEMANN, B., On the hypotheses which lie at the foundations of geometry, translated from the German by Henry S. White. In: A Source Book in Mathematics, by David E. Smith, Dover edition, Vol. 2, Dover, New York, 1959, pp. 411-425.

[Sa] SAKAI, T., Comparison and finiteness theorems in Riemannian Geometry, in Advanced Studies in Pure Math. 3 (1984), Geometry of geodesics and related fields, 125–181.

[Sh] SHIKATA, Y., On the differentiable pinching problem, Osaka J. Math. 4 (1967), 279–287.

[Sm] SMALE, S., On the Morse index theorem, J. Math. Mech. 14 (1965), 1049–1056.

[Sy] SYNGE, J.L., On the connectivity of spaces of positive curvature, Quart. J. Math. (Oxford Series) 7 (1936), 316–320.

[SSK] SUGIMOTO, M., SHIOHAMA, K., KARCHER, H., On the differentiable pinching problem, Math. Ann. 195 (1971), 1–16.

[Ts] TSUKAMOTO, Y., On Riemannian manifolds with positive curvature, Memoirs of Fac. Sci. Kyushu Univ. Ser. A, Math. 15 (1962), 90–96.

[We 1] WEINSTEIN, A., On the homotopy type of positively pinched manifolds, Arch. Math. 18 (1967), 523–524.

[We 2] WEINSTEIN, A., A fixed point theorem for positively curved manifolds, J. Math. Mech. 18 (1968), 149–153.

[Wh] WHITEHEAD, J.H., On the covering of a complete space by the geodesics through a point, Annals of Math. 36 (1935), 679–704.

[Yg] YANG, C.T., Odd-dimensional wiedersehen manifolds are spheres, J. Diff. Geometry, 15 (1980), 91–96.

[Ya] YAU, S.T., Curvature preserving diffeomorphisms, Ann. of Math. 100 (1974), 121–130.

INDEX

Printed in the United States
By Bookmasters

Printed in the United States
By Bookmasters